Encyclopaedia of Mathematical Sciences

Volume 70

Editor-in-Chief: R.V. Gamkrelidze

Yu.G. Reshetnyak (Ed.)

Geometry IV

Non-regular Riemannian Geometry

With 58 Figures

Springer-Verlag
Berlin Heidelberg New York
London Paris Tokyo
Hong Kong Barcelona
Budapest

Title of the Russian edition:
Itogi nauki i tekhniki, Sovremennye problemy matematiki,
Fundamental'nye napravleniya, Vol. 70, Geometriya 4,
Publisher VINITI, Moscow 1989

Mathematics Subject Classification (1991):
52BXX, 53AXX, 53BXX, 53C20, 53C21, 53C45, 58G30

ISBN 3-540-54701-0 Springer-Verlag Berlin Heidelberg New York
ISBN 0-387-54701-0 Springer-Verlag New York Berlin Heidelberg

Library of Congress Cataloging-in-Publication Data
Geometriīa IV. English. Geometry IV: non-regular Riemannian geometry/Yu. G. Reshetnyak (ed.). p. cm.—
(Encyclopedia of mathematical sciences; v. 70)
Includes bibliographical references and indexes.
ISBN 3-540-54701-0 (Berlin: acid-free): DM136.00.—
ISBN 0-387-54701-0 (New York: acid-free)
1. Geometry, Riemannian. I. Reshetnīak, IUriĭ Grigor'evich.
II. Title. III. Title: Geometry 4. IV. Series.
QA649.G46713 1993 516.3'73—dc20 93-13858

Printed in the United States of America
Typesetting: Asco Trade Typesetting Ltd., Hong Kong
41/3140-543210 – Printed on acid-free paper

List of Editors, Authors and Translators

Editor-in-Chief

R.V. Gamkrelidze, Russian Academy of Sciences, Steklov Mathematical Institute, ul. Vavilova 42, 117966 Moscow, Institute for Scientific Information (VINITI), ul. Usievicha 20 a, 125219 Moscow, Russia

Consulting Editor

Yu.G. Reshetnyak, Institute of Mathematics, Siberian Branch of the Russian Academy of Sciences, Universitetskij pr. 4, 630090 Novosibirsk, Russia

Authors

V.N. Berestovskij, Omsk State University, ul. Mira, 55 "A", 644077 Omsk, Russia
I.G. Nikolaev, Institute of Mathematics, Siberian Branch of the Russian Academy of Sciences, Universitetskij pr. 4, 630090 Novosibirsk, Russia
Yu.G. Reshetnyak, Institute of Mathematics, Siberian Branch of the Russian Academy of Sciences, Universitetskij pr. 4, 630090 Novosibirsk, Russia

Translator

E. Primrose, 12 Ring Road, Leicester LE2 3RR, England

Contents

Preface

The book contains a survey of research on non-regular Riemannian geometry, carried out mainly by Soviet authors. The beginning of this direction occurred in the works of A.D. Aleksandrov on the intrinsic geometry of convex surfaces. For an arbitrary surface F, as is known, all those concepts that can be defined and facts that can be established by measuring the lengths of curves on the surface relate to intrinsic geometry. In the case considered in differential geometry the intrinsic geometry of a surface is defined by specifying its first fundamental form. If the surface F is non-regular, then instead of this form it is convenient to use the metric ρ_F, defined as follows. For arbitrary points $X, Y \in F$, $\rho_F(X, Y)$ is the greatest lower bound of the lengths of curves on the surface F joining the points X and Y. Specification of the metric ρ_F uniquely determines the lengths of curves on the surface, and hence its intrinsic geometry. According to what we have said, the main object of research then appears as a metric space such that any two points of it can be joined by a curve of finite length, and the distance between them is equal to the greatest lower bound of the lengths of such curves. Spaces satisfying this condition are called spaces with intrinsic metric. Next we introduce metric spaces with intrinsic metric satisfying in one form or another the condition that the curvature is bounded. This condition is introduced by comparing triangles in space with triangles on a surface of constant curvature having the same lengths of sides.

The book contains two articles. The first is devoted to the theory of two-dimensional manifolds of bounded curvature. This theory at present has a complete character. It is a generalization of two-dimensional Riemannian geometry. For a manifold of bounded curvature there are defined the concepts of area and integral curvature of a set, the length and turn (integral curvature) of a curve.

One of the main results of the theory is the closure of the class of two-dimensional manifolds with respect to the passage to the limit under certain natural restrictions. In particular, this enables us to define two-dimensional manifolds of bounded curvature by means of approximation by polyhedra. The proof of the possibility of such an approximation is one of the main results of the theory. In the account given here it is essential to use the analytic representation of two-dimensional manifolds of bounded curvature by means of a line element of the form $ds^2 = \lambda(z)(dx^2 + dy^2)$. The function $\lambda(z)$ is such that its logarithm is the difference of two subharmonic functions. In contrast to the case of Riemannian manifolds the function λ here may vanish and have points of discontinuity. Some results in the theory of manifolds of bounded curvature do not have a complete analogue in two-dimensional Riemannian geometry. Here we should refer to some estimates and solutions of extremal problems, the theorem on pasting, and so on.

The second article is devoted to the theory of metric spaces whose curvature is contained between certain constants K_1 and K_2, where $K_1 < K_2$. The main result of this theory is that these spaces are actually Riemannian. In each such space we can locally introduce a coordinate system in which its metric is defined by a line element $ds^2 = g_{ij}\, dx^i\, dx^j$, where the functions g_{ij} satisfy almost the same regularity conditions as in ordinary Riemannian geometry. (We say "almost the same" because the functions g_{ij} only have second derivatives, generalized in the sense of Sobolev, that are summable in any degree $p > 0$; this implies that the coefficients g_{ij} are continuous.) The theory of curvature in Riemannian geometry can be transferred to the case of such spaces. Some relations here are satisfied only almost everywhere (for example, the formula for representing the sectional curvature of a manifold). In this article the authors also consider some questions of Riemannian geometry. Applications are given of the theorem on the Riemann property of spaces of two-sided bounded curvature to global Riemannian geometry.

In particular, an axiomatic definition of a Riemannian space is obtained here, based on representations in the spirit of synthetic geometry. A priori it is not required that the spaces under consideration should be manifolds. This fact follows from other axioms.

Yu.G. Reshetnyak

I. Two-Dimensional Manifolds of Bounded Curvature

Yu.G. Reshetnyak

Translated from the Russian
by E. Primrose

Contents

Chapter 1
Preliminary Information

§ 1. Introduction

1.1. General Information about the Subject of Research and a Survey of Results. The theory of two-dimensional manifolds of bounded curvature is a generalization of two-dimensional Riemannian geometry. Formally a two-dimensional manifold of bounded curvature is a two-dimensional manifold in which there are defined the concepts of the length of a curve, the angle between curves starting from one point, the area of a set, and also the integral curvature of a curve and the integral curvature of a set. For the case when the given manifold is Riemannian, the integral curvature of a curve is equal to the integral of the geodesic curvature along the length of the curve, and the integral curvature of a set is equal to the integral of the Gaussian curvature of the manifold with respect to the area. The remaining concepts in this case have the meaning that is usual in Riemannian geometry. For an arbitrary two-dimensional manifold of bounded curvature the integral curvature is a completely additive set function, which may not admit representations in the form of an integral with respect to area.

Another particular case of two-dimensional manifolds of bounded curvature consists of surfaces of polyhedra (not necessarily convex) in three-dimensional Euclidean space. For them the integral curvature is an additive set function concentrated on some discrete set, namely the set of vertices of the polyhedron. If the set consists of a unique point, a vertex of the polyhedron, then its integral curvature is equal to $2\pi - \theta$, where θ is the total angle of the polyhedron at this vertex, that is, the sum of the angles of all its faces that meet at this vertex.

Three methods are known for introducing two-dimensional manifolds of bounded curvature. The first of them is *axiomatic*. A two-dimensional manifold of bounded curvature is defined as a metric space satisfying some special axioms. The second method is based on *approximation* by two-dimensional Riemannian manifolds or manifolds with polyhedral metric. It turns out that under certain natural assumptions the limit of the sequence of two-dimensional manifolds of bounded curvature is also a manifold of bounded curvature. Exact formulations are given later; here we just mention that a certain condition of boundedness of the curvature is fundamental in these assumptions. In particular, the limit of a sequence of manifolds with polyhedral metric is a manifold of bounded curvature. This fact can be used for the definition of the class of two-dimensional manifolds of bounded curvature.

A *two-dimensional Riemannian manifold* is a smooth manifold such that for each local coordinate system there is defined in it a differential quadratic form

$$ds^2 = \sum_{i=1}^{2} \sum_{j=1}^{2} g_{ij}(x_1, x_2)\, dx_i\, dx_j.$$

The main concept of two-dimensional Riemannian geometry is the *Gaussian curvature*. In order that it can be defined it is necessary to require that the coefficients g_{ij} $(i, j = 1, 2)$ have partial derivatives of the first and second orders. There naturally arises the idea of considering "generalized" Riemannian geometries obtainable if we weaken the requirements of regularity imposed on the coefficients of the quadratic form ds^2. It turns out that two-dimensional manifolds of bounded curvature can be defined in such a way. We shall show later how to do this. Here we consider the case when the line element of the manifold has a certain special structure, namely such that $g_{11} = g_{22}$, $g_{12} \equiv 0$. For the case of Riemannian manifolds the line element can always be reduced to such a form by a transformation of the coordinates. The system of coordinates for which $g_{11} \equiv g_{22}$, $g_{12} \equiv 0$ is called *isothermal*. Using such a form of the coordinate system, we obtain a third *analytic* method of introducing two-dimensional manifolds of bounded curvature.

The general plan of the theory of two-dimensional manifolds of bounded curvature is due to A.D. Aleksandrov, who developed the geometrical aspects of this theory (see Aleksandrov (1948b), (1948c), (1949b), (1950), (1954), (1957a), (1957b), Aleksandrov and Burago (1965), Aleksandrov and Strel'tsov (1953), (1965), Aleksandrov, Borisov and Rusieshvili (1975). An account of the theory constructed by Aleksandrov is given in a monograph of Aleksandrov and Zalgaller (1962). An analytic approach to the introduction and study of two-dimensional manifolds of bounded curvature is due to Yu.G. Reshetnyak (Reshetnyak (1954), (1959), (1960), (1961b), (1962), (1963a), (1963b)). Other authors also took part in the development of individual aspects of the theory (the corresponding references are given later in the main text).

The concept of a two-dimensional manifold of bounded curvature was introduced as a development of the research of Aleksandrov on the intrinsic geometry of convex surfaces (Aleksandrov (1944), (1945a), (1945b), (1947), (1948a)), and presented completely in his monograph Aleksandrov (1948a).

Chapter I of this article has an auxiliary character. Two-dimensional manifolds of bounded curvature are defined as metric spaces satisfying certain special conditions. One of these conditions is that the metric of the space must be intrinsic. In § 2 we give necessary conditions and a summary of the basic facts relating to the theory of metric spaces with intrinsic metric. In § 3 we consider two-dimensional manifolds with intrinsic metric. Here we go into details on the definition of the operations of cutting up and pasting such manifolds. In addition, the concept of a side of a simple arc in a two-dimensional manifold has important significance for what follows.

In § 4 we give a summary of the basic results of two-dimensional Riemannian geometry. The main information concerning two-dimensional manifolds with polyhedral metric is contained in § 5. In particular, for such manifolds we define the concepts of integral curvature (or the turn) of a curve and the curvature of a set, and we study the structure of a shortest curve on a two-dimensional polyhedron. Polyhedra play a special role in the theory of manifolds of bounded curvature. By approximating an arbitrary manifold by polyhedra, in many cases it turns out to be possible to reduce the solution of this or that problem to the case of polyhedra, for which it becomes a problem with respect to the formulation belonging to elementary geometry. This enables us to use for the solution of such problems arguments based on intuitive geometric representations. For example, some extremal problems are related to a number of problems for which such a way of action leads to success.

The definition of two-dimensional manifolds of bounded curvature is given in § 6. We regard the axiomatic definition as fundamental. The following fact relating to classical Riemannian geometry is well known. Let T be a *triangle* in a two-dimensional Riemannian manifold, that is, a domain homeomorphic to a disc whose boundary is formed by three geodesics. We denote by α, β and γ the angles of this domain at the vertices of the triangle T and let $\omega(T)$ be the integral over T of the Gaussian curvature with respect to area. We put $\delta(T) = \alpha + \beta + \gamma - \pi$. The quantity $\delta(T)$ is called the *excess* of the triangle T. As we know, $\delta(T) = \omega(T)$. (This statement is a special case of the Gauss-Bonnet theorem; see § 4.) If the Gaussian curvature is non-negative, then it follows that $\delta(T) \geqslant 0$ for any triangle.

Let U be an arbitrary domain in a two-dimensional Riemannian manifold. For any system of pairwise non-overlapping geodesic triangles $T_i \subset U$, $i = 1, 2, \ldots, m$, we have the inequality

$$\sum_{i=1}^{m} \omega(T_i) \leqslant \int_U [\mathscr{K}(x)]^+ d\sigma(x),$$

where $\mathscr{K}$ is the Gaussian curvature, $d\sigma$ is the element of area, $a^+ = \max\{a, 0\}$. This property is taken as the basis for constructing the axiomatics of a two-dimensional manifold of bounded curvature. A two-dimensional manifold of bounded curvature is defined as a certain metric space. A geodesic is a curve, any sufficiently small arc of which is a shortest curve, that is, such that its length is equal to the distance between the ends. The concept of a shortest curve is naturally defined for the case of metric spaces. It is also clear what we need to call a triangle. In order to define the concept of the excess for a triangle in an arbitrary metric space we need to know what the angle between two curves starting from one point is, in the given case the angle between the sides of the triangle. The corresponding definition is given in § 6. A manifold of bounded curvature can be defined as a metric space that is a two-dimensional manifold and is such that for any point of it there is a neighbourhood U for which the sum of the excesses of pairwise non-overlapping geodesic triangles contained in U does not exceed some constant $C(U) < \infty$, however these triangles are chosen. The exact formulations are given in § 6. The final version of the axiomatics of two-dimensional manifolds of bounded curvature is defined by the argument that of the different equivalent forms of the axiomatics we must choose the weakest.

One of the main results of the theory of two-dimensional manifolds of bounded curvature is the characterization of such manifolds by means of approximation by two-dimensional polyhedra, or, which reduces to the same thing, by two-dimensional Riemannian manifolds. The difficulties that must be overcome here are connected with the fact that starting from the axioms of a manifold of bounded curvature it is required to establish some very deep properties of it. In § 6 of this article we give an outline of the proof of theorems on the approximation of a two-dimensional manifold of bounded curvature by Riemannian manifolds. The reader can find complete proofs in the monograph Aleksandrov and Zalgaller (1962). The proof of the necessary conditions is based on arguments that are a development of the ideas worked out by Aleksandrov in the study of the intrinsic geometry of convex surfaces. The proof of the sufficient conditions outlined in § 6 is based on arguments different from those given in Aleksandrov and Zalgaller (1962). (For a complete account of this proof, see Reshetnyak (1962).)

The analytic characteristic of two-dimensional manifolds of bounded curvature is given in § 7. We dwell on it in more detail, bearing in mind the fact that for specialists thinking in terms of categories of mathematical analysis it is the shortest path towards determining what is a two-dimensional manifold of bounded curvature.

We first consider Riemannian manifolds. In a neighbourhood of any point of such a manifold we can introduce a coordinate system in which the line element of the manifold is expressed by the formula

$$ds^2 = \lambda(x, y)(dx^2 + dy^2).$$

(As we said above, such a coordinate system is called isothermal.) The Gaussian curvature $\mathscr{K}$ of a given manifold in this coordinate system admits the representation

$$\mathscr{K}(x, y) = -\frac{1}{2\lambda(x, y)} \Delta \ln \lambda(x, y).$$

Using known results of potential theory, we thus obtain

$$\ln \lambda(z) = \frac{1}{\pi} \iint_G \ln \frac{1}{|z - \zeta|} \mathscr{K}(\zeta)\lambda(\zeta)\, d\xi\, d\eta + h(z).$$

Here $z = (x, y)$, $\zeta = (\xi, \eta)$, G is a domain on the plane, and $h(z)$ is a harmonic function. We now observe that for an arbitrary set $E \subset G$

$$\omega(E) = \iint_E \mathscr{K}(\zeta)\lambda(\zeta)\, d\xi\, d\eta$$

is the integral curvature of the corresponding set in the Riemannian manifold. By virtue of this the integral representation for $\ln \lambda(z)$ given above can be written in the form

$$\ln \lambda(z) = \frac{1}{\pi} \iint_G \ln \frac{1}{|z - \zeta|}\, d\omega(\zeta) + h(z). \tag{1}$$

The last relation naturally suggests that if we wish to have generalized Riemannian manifolds in some sense, for which the integral curvature is an arbitrary completely additive set function, then it is sufficient in (1) to substitute such a function for ω, and then to consider the geometry defined by the corresponding line element $ds^2 = \lambda(z)(dx^2 + dy^2)$. Such a path leads to a *two-dimensional manifold of bounded curvature.*

In § 7 we give only drafts of the necessary proofs. A complete account of them can be found in Reshetnyak (1954), (1960), (1961a).

For an arbitrary two-dimensional manifold of bounded curvature there are defined the concepts of integral curvature and area of a set, and the integral curvature (or the turn) of a curve. In § 8 we show how all these concepts can be defined. We rely on the analytic representation of two-dimensional manifolds of bounded curvature described in § 7.

In § 8 we give a survey of the main results of the theory of two-dimensional manifolds of bounded curvature. Here we are concerned first of all with a theorem on pasting of two-dimensional manifolds of bounded curvature and theorems on passage to the limit. The class of two-dimensional manifolds of bounded curvature turns out to be closed with respect to passages to the limit under significantly weaker assumptions than for the class of Riemannian manifolds.

Among the main results of the theory of manifolds of bounded curvature there are, in particular, those that concern extremal problems for manifolds of bounded curvature. One of the main instruments for research is the method of cutting and pasting created by Aleksandrov. This method uses essentially the specific character of two-dimensional manifolds of bounded curvature. The totality of all such manifolds is invariant with respect to operations connected with the method indicated, which we cannot say, for example, about the class of Riemannian manifolds.

In § 9 of this chapter we give a survey of further research into the theory of two-dimensional manifolds of bounded curvature. The author has tried to express everything that is most essential in this topic.

1.2. Some Notation and Terminology. Later we assume that the concepts of topological and metric spaces are known, like all the basic facts of general topology. In particular, we assume that the reader knows what is a neighbourhood of a point in a topological space, an open or closed set, a connected component, and so on.

Let us recall some standard notation, used in what follows.

Let A be a set in a topological space $\mathfrak{R}$. Then $\bar{A}$ denotes the closure of A, A^0 denotes the totality of all interior points of A, and $\partial A = \bar{A} \setminus A^0$ denotes the boundary of A.

The symbol $\mathbb{R}^n$ denotes the n-dimensional arithmetic Euclidean space of points $x = (x_1, x_2, \ldots, x_n)$, where $x_1, x_2, \ldots, x_n$ are arbitrary real numbers. For $x = (x_1, x_2, \ldots, x_n) \in \mathbb{R}^n$ we put

$$|x| = \sqrt{\sum_{i=1}^{n} x_i^2}.$$

For arbitrary points $x, y \in \mathbb{R}^n$ the distance between x and y is assumed to be equal to $|x - y|$. The function $\rho\colon (x, y) \to |x - y|$ is the *metric*. In a well-known way a given metric defines some topology in $\mathbb{R}^n$. Speaking of $\mathbb{R}^n$ as a topological space, we shall always have this topology in mind.

The space $\mathbb{R}^2$ will be called an *arithmetic Euclidean plane*. The symbol $\mathbb{C}$ denotes the set of complex numbers. Later we shall often identify $\mathbb{R}^2$ and $\mathbb{C}$, regarding the point $(x, y) \in \mathbb{R}^2$ and the complex number $z = x + iy$ as one and the same object.

The usual Euclidean plane is denoted later by the symbol $\mathbb{E}^2$. As a metric space $\mathbb{E}^2$ is isometric to $\mathbb{R}^2$.

Let $B(0, 1)$ be the open disc $\{(x, y)|x^2 + y^2 < 1\}$ in the plane $\mathbb{R}^2$, and $\bar{B}(0, 1)$ the closed disc $\{(x, y)|x^2 + y^2 \leqslant 1\}$.

Henceforth the statement that some set in a topological space is homeomorphic to a disc (a closed disc) always means that this set is homeomorphic to the disc $B(0, 1)$ (respectively, the disc $\bar{B}(0, 1)$.

§2. The Concept of a Space with Intrinsic Metric

2.1. The Concept of the Length of a Parametrized Curve. We assume that the concept of a metric space and some of the simplest information relating to it are known.

Let M be a set in which a metric ρ is specified. We shall denote the metric space obtained in this way by the symbol (M, ρ). This notation is appropriate in that later there will often arise the necessity of considering different metrics on the same set. When no misunderstanding is possible we shall simply talk about a metric space M.

Let M be a metric space and ρ its metric. A *parametrized curve* or *path* in the space M is any continuous map $x\colon [a, b] \to M$ of the interval $[a, b]$ of the set of real numbers $\mathbb{R}$ into M. We shall say that the path $x\colon [a, b] \to M$ joins the points $X, Y \in M$ if $x(a) = X$, $x(b) = Y$.

A metric space M with metric ρ is called *linearly connected* if for any two points X, Y of it there is a path joining these points.

A path $x\colon [a, b] \to M$ is called *simple* if it is a one-to-one map of the interval $[a, b]$. A set L in the space M is called a *simple arc* if there is a simple path $x\colon [a, b] \to M$ such that $L = x([a, b])$. Any simple path $x\colon [a, b] \to M$ satisfying this condition is called a *parametrization of the simple arc L*.

A set Γ in a metric space (M, ρ) is called a *simple closed curve* in M if it is a topological image of the circle $S(0, 1)$ on the plane $\mathbb{R}^2$. If Γ is a simple closed curve in the metric space M, then there is a path $x\colon [a, b] \to M$ such that $x(a) = x(b)$, $x([a, b]) = \Gamma$ and for any $t_1, t_2 \in [a, b]$ such that $t_1 \neq t_2$ and at least one of the points t_1 and t_2 is not an end of the interval $[a, b]$ the points $x(t_1)$ and $x(t_2)$ are distinct.

We shall call a set $L \subset M$ a *simple curve* if L is closed and is either a simple closed curve in M or a topological image of an arbitrary interval of the number line $\mathbb{R}$ (which, generally speaking, may not be closed).

Suppose we are given a path $x: [a, b] \to M$ in a metric space M. We specify arbitrarily a finite sequence $\alpha = \{t_0, t_1, \dots, t_m\}$ of points of the interval $[a, b]$ such that $t_0 = a \leqslant t_1 \leqslant \dots \leqslant t_m = b$ and put

$$s(x, \alpha) = \sum_{i=1}^{m} \rho[x(t_{i-1}), x(t_i)].$$

The least upper bound of $s(x, \alpha)$ on the totality of all sequences α satisfying the conditions mentioned above is called the *length of the path* x and denoted by the symbol $s_\rho(x; a, b)$ or simply $s(x; a, b)$ when no misunderstanding is possible. (The notation $s_\rho(x; a, b)$ is necessary for those cases when we consider different metrics in M and compare the lengths of the path $x: [a, b] \to M$ in these metrics.)

We mention the following properties of length that follow immediately from the definition.

I. Any path $x: [a, b] \to M$ in the space (M, ρ) satisfies the inequality

$$\rho[x(a), x(b)] \leqslant s_\rho(x; a, b).$$

II. Suppose we are given a path $x: [a, b] \to M$. Then for any c such that $a < c < b$ we have

$$s_\rho(x; a, b) = s_\rho(x; a, c) + s_\rho(x; c, b).$$

III. Suppose we are given a sequence of paths $(x_\nu: [a, b] \to M)$, $\nu = 1, 2, \dots$ and a path $x_0: [a, b] \to M$. We assume that $x_0(t) = \lim_{t \to 0} x_\nu(t)$ for any $t \in [a, b]$. Then

$$s_\rho(x; a, b) \leqslant \varliminf_{\nu \to \infty} s_\rho(x_\nu; a, b).$$

Let L be a simple arc in the metric space (M, ρ). We specify arbitrarily a parametrization $x: [a, b] \to M$ of the arc L. Then it is easy to establish that $s_\rho(x; a, b)$ does not depend on the choice of parametrization x of the arc L. In this case we shall call $s_\rho(x; a, b)$ the *length of the simple arc* L and denote it by $s_\rho(L)$ or simply $s(L)$.

Similarly, if Γ is a simple closed curve and $x: [a, b] \to M$ is an arbitrary parametrization of it, then $s_\rho(x; a, b)$ does not depend on the choice of this parametrization and is denoted henceforth by $s_\rho(\Gamma)$ or simply $s(\Gamma)$.

Let L be a simple arc in the metric space M, and $x: [a, b] \to L$ a parametrization of L. The points $A = x(a)$ and $B = x(b)$ are called the *end-points* of L. All the remaining points of L are called *interior points* of it. Let $X = x(t_1)$ and $Y = x(t_2)$, $t_1 < t_2$, be two arbitrary points of the simple arc. The set of all points $Z = x(t)$, where $t_1 \leqslant t \leqslant t_2$, is obviously a simple arc. We shall denote it by $[XY]$. From property I of the length of a parametrized curve it follows that for any simple arc L with end-points A and B we have

$$\rho(A, B) \leqslant s(L).$$

From property II it follows that for any interior point C of the simple arc

$$s([AB]) = s([AC]) + s([CB])$$

(A and B are the end-points of L).

Let Γ be a simple closed curve and A an arbitrary point of it. Then there is a parametrization $x: [a, b] \to M$ of Γ such that $x(a) = x(b) = A$. Any two distinct points X, Y of the simple closed curve Γ split it into two simple arcs, which we denote by Γ_1 and Γ_2. From Property II of the length of a parametrized curve it follows that

$$s_\rho(\Gamma) = s_\rho(\Gamma_1) + s_\rho(\Gamma_2).$$

In the given definitions, in principle, there can be an infinite value of the length. If the length of the path $x: [a, b] \to M$ in the metric space (M, ρ) is finite, then the given path is called *rectifiable*. Similarly, a simple arc (simple closed curve) is called rectifiable if its length is finite.

Let L be a simple arc in the space (M, ρ). We assume that L is rectifiable. Then it admits a parametrization $x: [0, l] \to M$ such that s is equal to the length of the arc $[x(0)x(s)]$ for each $s \in [0, l]$.

2.2. A Space with Intrinsic Metric. The Induced Metric. Suppose we are given a metric space M and a set $A \subset M$. We shall say that the path $x: [a, b] \to M$ lies in the set A (or goes into A) if $x(t) \in A$ for all $t \in [a, b]$.

A set A in a metric space (M, ρ) is said to be *metrically connected* if for any two of its points there is a rectifiable path joining these points and lying in the set A. In particular, the space (M, ρ) itself is said to be metrically connected if for any two of its points X, Y there is a rectifiable path joining these points.

A metric space (M, ρ) is called a *space with intrinsic metric* if it is linearly connected and for any two of its points X, Y the quantity $\rho(X, Y)$ is equal to the greatest lower bound of lengths of arcs joining these points.

If (M, ρ) is a space with intrinsic metric, then M is metrically connected.

Suppose, for example, that M is the usual plane $\mathbb{E}^2$. For arbitrary points $X, Y \in \mathbb{E}^2$ suppose that $\rho(X, Y) = 0$ if $X = Y$ and that $\rho(X, Y)$ is equal to the length of the interval with end-points X and Y if $X \neq Y$. The metric defined in this way on the plane $\mathbb{E}^2$ is obviously intrinsic.

Similarly, if M is a sphere Σ_K of radius $r = 1/\sqrt{K}$ in the space $\mathbb{E}^3$, then taking for $\rho(X, Y)$ the length of the shortest arc of the great circle passing through the points X and Y, we obtain an intrinsic metric on the sphere Σ_K. At the same time, the metric $\rho_0(X, Y)$, where $\rho_0(X, Y)$ is the length of the interval in $\mathbb{E}^3$ joining the points X and Y on the sphere Σ_K, is not intrinsic.

The metric spaces known from analysis, namely Hilbert space and, more generally, any normed vector space, are spaces with intrinsic metric.

Let (M, ρ) be a metric space and $a \in M$ an arbitrary point of M. Let us specify arbitrarily a number $r > 0$. We denote the set of all points $x \in M$ such that $\rho(x, a) < r$ by the symbol $B(a, r)$ and call it the *open ball* with centre a and radius r. In certain cases considered later, instead of the word "ball" we shall say "disc".

The totality of all points $x \in M$ for which $\rho(x, a) = r$ is denoted by the symbol $S(a, r)$ and called the *sphere* with centre a and radius r. In those cases when $B(a, r)$ is called a *disc* we shall call the set $S(a, r)$ a *circle*. We put $\bar{B}(a, r) = B(a, r) \cup S(a, r)$. The set $\bar{B}(a, r)$ is called the *closed ball* with centre a and radius r.

We mention the following properties of spaces with intrinsic metric.

Theorem 2.2.1 (Aleksandrov and Zalgaller (1962)). *Let (M, ρ) be a space with intrinsic metric. If M is locally compact (that is, any point $X \in M$ has a neighbourhood whose closure is compact), then for any $r > 0$ the closed ball $\bar{B}(X, r)$ is a compact set.*

The metric space (M, ρ) is called *complete* if any sequence (x_ν), $\nu = 1, 2, \ldots$, of points of this space for which $\lim_{\nu \to \infty, \mu \to \infty} \rho(x_\nu, x_\mu) = 0$ is convergent. According to a well-known theorem of Hausdorff, for any metric space (M, ρ) there is a complete metric space $(\bar{M}, \bar{\rho})$ such that $M \subset \bar{M}$, $\rho(x, y) = \bar{\rho}(x, y)$ for any $x, y \in M$, and the set M is everywhere dense in $\bar{M}$. The space $(\bar{M}, \bar{\rho})$ is unique in the following sense. If (M', ρ') is another metric space connected with (M, ρ) like $(\bar{M}, \bar{\rho})$, then there is a map $j: \bar{M} \to M'$ such that $j(\bar{M}) = M'$, $j(x) = x$ for any $x \in M$, and $\rho'[j(x), j(y)] = \bar{\rho}(x, y)$ for any $x, y \in \bar{M}$. We shall call $(\bar{M}, \bar{\rho})$ the *Hausdorff completion of the space* (M, ρ). Henceforth the metric of the Hausdorff completion will be denoted like the metric of the original space.

Theorem 2.2.2 (Aleksandrov and Zalgaller (1962)). *The Hausdorff completion of a metric space with intrinsic metric is also a space with intrinsic metric.*

We mention here a general scheme for constructing the metric. Suppose we are given a metric space (M, ρ), and let A be a connected set of this space. The set A with metric ρ is itself a metric space – a subspace of (M, ρ). Even if (M, ρ) is a space with intrinsic metric, the metric space (A, ρ) may not be of this kind. Let us define a metric in the set A, which we denote by ρ_A. Namely, for arbitrary points $X, Y \in A$ we denote by $\rho_A(X, Y)$ the greatest lower bound of lengths of paths in the space (M, ρ) joining the points X and Y and lying in the set A.

Theorem 2.2.3. *If $A \subset M$ is a metrically connected set of the space (M, ρ), then the function $(X, Y) \to \rho_A(X, Y)$ of a pair of points of A, defined in the way indicated above, is a metric on the set A. This metric is intrinsic and for any path $x: [a, b] \to M$ lying in the set A we have $s_\rho(x; a, b) = s_{\rho_A}(x; a, b)$.*

The metric ρ_A is called the *induced intrinsic metric* on the set A of the metric space (M, ρ).

Suppose, for example, that M is the three-dimensional Euclidean space $\mathbb{E}^3$, and that the set A is the sphere $S(a, R)$ in this space. It is easy to show that in the given case the quantity $\rho_A(X, Y)$ is equal to the length of the shortest arc of the great circle passing through the points X and Y, that is, the induced intrinsic metric on the sphere $S(a, R)$ coincides with the metric defined above.

Let L be a rectifiable simple arc in the metric space (M, ρ). Then for arbitrary points $X, Y \in L$ the quantity $\rho_L(X, Y)$ is equal to the length of the arc $[XY]$ of the curve L. We assume that Γ is a rectifiable simple closed curve in the metric space (M, ρ). We take arbitrarily the points $X, Y \in \Gamma$. Then $\rho_\Gamma(X, Y) = 0$ if $X = Y$, and if $X \neq Y$, then X and Y split Γ into two arcs Γ_1 and Γ_2 and $\rho_\Gamma(X, Y) = \min\{s_\rho(\Gamma_1), s_\rho(\Gamma_2)\}$.

In the definition of the induced metric we do not exclude the case when $A = M$. If (M, ρ) is a space with intrinsic metric, then the induced metric ρ_M coincides identically with the metric ρ already existing in M.

We note that, generally speaking, the induced intrinsic metric may turn out to be poorly connected with the set A. For example, the topology determined by the induced metric may be different from the natural topology of A as a subset of the metric space. The following example serves to illustrate this statement.

Let H be a plane in the three-dimensional Euclidean space $\mathbb{E}^3$, and Γ a simple closed curve in H such that any arc of Γ has infinite length. We take arbitrarily a point O not lying in H, and let A be the cone formed by the rays OX, where X ranges over the curve Γ. The set A is metrically connected. In fact, if the points Y, Z do not lie on one of the rays OX, $X \in \Gamma$, then the simple arc composed of the intervals $[YO]$ and $[OZ]$ joins these points. It is obviously rectifiable and is contained in A. If Y and Z lie on one ray OX, $X \in \Gamma$, then the interval of this ray with ends Y and Z is also a rectifiable simple arc joining Y and Z in the set A. It is easy to verify that if Y and Z lie on one ray OX, $X \in \Gamma$, then $\rho_A(Y, Z)$ is equal to the length of the interval $[YZ]$. If Y and Z belong to different rays OX_1, OX_2, $X_1, X_2 \in \Gamma$, then $\rho_A(Y, Z)$ is equal to the sum of the lengths of the intervals $[YO]$ and $[OZ]$. The set A of the space $\mathbb{E}^3$ is homeomorphic to the plane $\mathbb{E}^2$. At the same time the metric space (A, ρ_A) does not admit a topological map into a plane. This follows, for example, from the fact that if the sequence (X_ν), $\nu = 1, 2, \ldots$, of points of the curve Γ converges to a point X_0 on the plane H, then $\rho_A(X_\nu, X_0) \to 2|OX_0| > 0$ and so the sequence (X_ν) does not converge to the point X_0 in the space (A, ρ_A).

We mention an important special case when the topology determined by the induced intrinsic metric in the set A coincides with the natural topology of A. Namely, the following proposition is true.

Lemma 2.2.1. *Let (M, ρ) be a metric space with intrinsic metric, and U an open set of the space M. If the set U is connected, then it is metrically connected. Moreover, if the point $X_0 \in U$ and the number $\delta > 0$ are such that the ball $B(X_0, \delta) \subset U$, then for any points $X, Y \in B(X_0, \delta/2)$ we have*

$$\rho_U(X, Y) = \rho(X, Y). \tag{2.1}$$

The proof of (2.1) is based on the argument that if the path $x\colon [a, b] \to M$ joining the points X and Y of the ball $B(X_0, \delta/2)$ contains points lying outside the ball $B(X_0, \delta)$, then its length must be not less than δ. At the same time, $\rho(X, Y) \leqslant \rho(X, X_0) + \rho(X_0, Y) < \delta$, and so by the definition of $\rho(X, Y)$ we can consider only paths that are contained in the ball $B(a, \delta)$, and therefore in the set U. Hence it follows that $\rho_U(X, Y) = \rho(X, Y)$.

Corollary. *In the conditions of Lemma 2.2.1 the metrics ρ_U and ρ in the set U are topologically equivalent (that is, they determine the same topology in U).*

2.3. The Concept of a Shortest Curve. Suppose we are given a metric space (M, ρ). A simple arc L in the space M is called a *shortest curve* if its length is equal to the distance between its end-points A and B,

$$s_\rho(L) = \rho(A, B).$$

The concept introduced in the theory described here is one of the fundamental concepts. Let us state some theorems that establish the conditions that guarantee the existence of a shortest curve joining two arbitrary points of the given metric space (M, ρ).

Theorem 2.3.1. *Let (M, ρ) be a locally compact metric space with intrinsic metric. Then for any point X of the space M we can find a neighbourhood V of X such that for any two points $Y, Z \in V$ there is a shortest curve joining these points.*

Theorem 2.3.2. *Let (M, ρ) be a locally compact metric space with intrinsic metric. If M is complete, then for any two points $X, Y \in M$ there is a shortest curve joining these points.*

It is obvious that a rectilinear interval on the plane $\mathbb{E}^2$ (and more generally in the space $\mathbb{E}^n$ for any $n \geqslant 1$) is a shortest curve. Let M be the sphere $S(a, R)$ in the space $\mathbb{E}^n$ endowed with the metric induced from $\mathbb{E}^n$. Then any arc of a great circle on $S(a, R)$ whose length does not exceed πR is a shortest curve.

2.4. The Operation of Cutting of a Space with Intrinsic Metric. In the investigation of two-dimensional manifolds of bounded curvature a method proposed by A.D. Aleksandrov turns out to be useful; this is based on a transformation of the manifolds under consideration by successive application of the operations of cutting and pasting. The definition of the operation of pasting will be given below. Here we define the operation of cutting. Formally its definition is somewhat simpler and can be stated in a very general form.

Let (M, ρ) be an arbitrary metric space with intrinsic metric, and A a non-empty closed set in M.

Let U be an arbitrary connected component of the set $M \setminus A$, and ρ_U the induced intrinsic metric in U. We thus obtain a collection of metric spaces (U, ρ_U), where U is a connected component of $M \setminus A$. To each of these spaces we adjoin new elements that can be regarded as points lying on the bank of the cut. For this we first construct the completion $(\bar{U}, \rho_U)$ of the space (U, ρ_U). However, $\bar{U}$ may contain points that must be regarded as superfluous in the given case. Suppose that $X \in \bar{U} \setminus U$, and that (X_m), $m = 1, 2, \ldots$, is an arbitrary sequence of points of U that converge to the point X. Since $\rho(X, Y) \leqslant \rho_U(X, Y)$ for any $X, Y \in U$, in the space (M, ρ) this sequence is fundamental. There is no need to assume that the space (M, ρ) is complete, and so the sequence (X_m) in the original space (M, ρ) may not have a limit. If, however, the limit $\lim_{m \to \infty} X_m$ in the space M exists (in this case it belongs to A), then we can call the point $X \in \bar{U} \setminus U$ an *admissible* point of $\bar{U}$. We shall also regard any point $X \in U$ as admissible. Let $\hat{U}$ be the totality of all admissible points of the set $\bar{U}$. As a result we obtain a metric space $(\hat{U}, \rho_U)$. The metric of this space is intrinsic. Carrying out a similar procedure over each connected component of the set $M \setminus A$, we obtain a collection of spaces with intrinsic metric. We shall say that this collection arises as a result of *cutting the space M with respect to the set A*, or cutting out the set A from M. (If A has internal points, then the term "cutting out" seems more natural in the given context.) Let us take arbitrarily a point $X \in \hat{U}$. Let (X_m), $m = 1, 2, \ldots$, be a sequence of points of the set U such that $\rho_U(X, X_m) \to 0$ as $m \to \infty$. This sequence will have a limit in the space M. The value of this limit does not depend on the choice of the sequence (X_m). We denote it by $\pi(X)$. If $X \in U$ it is obvious that $\pi(X) = X$. If $X \notin U$, then $\pi(X) \in A$. We thus obtain a map $\pi\colon \hat{U} \to M$. This map is continuous. A point $X \in A$ is the image of some point $X \in \hat{U}$ with respect to the map π if and only if there is a sequence (X_m), $m = 1, 2, \ldots$, such that $X_m \in U$ for each m and $X_m \to X$ in the space M, and the sequence (X_m) is fundamental with respect to the metric ρ_U in U. If X satisfies this condition, we shall say that X is an accessible point of the set A from $M \setminus A$.

In the general case the construction we have described can be reduced to objects of a rather pathological nature.

Let us give some simple examples. Let M be the plane $\mathbb{E}^2$, endowed with its natural metric, that is, $\rho(x, y)$ is the length of the interval with ends x and y for $x, y \in \mathbb{E}^2$. Let us assume that A consists of a single point P. We put $U = \mathbb{E}^2 \setminus \{P\}$. It is not difficult to see that $\rho_U(X, Y) = |XY|$ for any $X, Y \in U$. In the given case the set $\hat{U} \setminus U = \pi^{-1}(P)$. Let (X_ν), (Y_ν), $\nu = 1, 2, \ldots$, be two arbitrary sequences of points of U converging to points P', P'' of the set $\hat{U} \setminus U$. These sequences on the plane $\mathbb{E}^2$ converge to the point P. Since $\rho_U(X_\nu, Y_\nu) = |X_\nu Y_\nu| \to 0$ as $\nu \to \infty$, we have $\rho_U(P', P'') = 0$ and so we deduce that the set $\pi^{-1}(P)$ consists of the unique point P'. For any point $X \neq P$ we have $\rho_U(X, P') = |XP|$. Consequently we deduce that the metric space $(\hat{U}. \rho_U)$ is isometric to $\mathbb{E}^2$. It is naturally identified with the plane $\mathbb{E}^2$. Thus we deduce that the plane cannot be cut with respect to a one-point set.

We now consider the case when the set A is an interval $[PQ]$, $P \neq Q$ (Fig. 1). The line PQ splits $\mathbb{E}^2$ into two half-planes, which we denote by $\mathbb{E}^2_+$ and $\mathbb{E}^2_-$. In the given case the situation is such that for any internal point X of the interval $[PQ]$ the set $\pi^{-1}(X)$ consists of exactly two elements X' and X''. Here X' is the limit of a sequence (X_ν), $\nu = 1, 2, \ldots$, where $X_n \in \mathbb{E}^2_+$ for all ν, and X'' is the limit of a sequence (Y_ν), $\nu = 1, 2, \ldots$, where $Y_\nu \in \mathbb{E}^2_+$ for each ν. We omit the proof of this simple fact. It is natural to assume that X' belongs to the upper and X'' to the lower bank of the cut of the plane by the interval $[PQ]$. If X is one of the ends of $[PQ]$, then $\pi^{-1}(X)$ consists of a unique point. The points that belong to the upper bank of the cut, if we add to them $P' = \pi^{-1}(P)$ and $Q' = \pi^{-1}(Q)$, form a

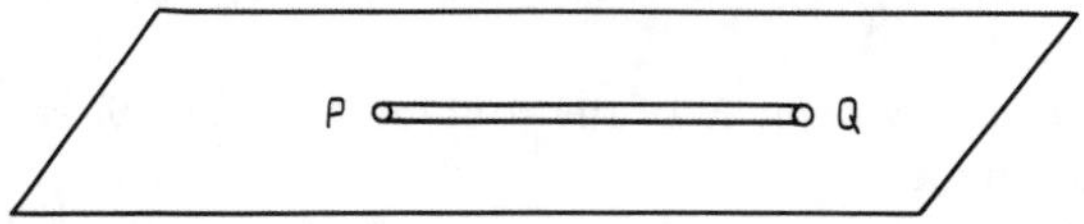

Fig. 1

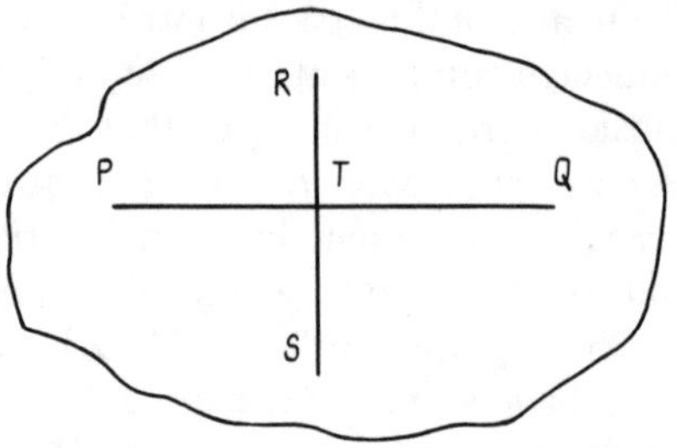

Fig. 2

simple arc, which is mapped one-to-one onto $[PQ]$ by the map π. In the same way the points of the lower bank, if we again add to them the points P and Q, form a simple arc, which is also mapped one-to-one onto $[PQ]$ by the map π.

If A is the union of two intersecting intervals $[PQ]$ and $[RS]$ (Fig. 2) and T is their point of intersection, then in this case for the point T the set $\pi^{-1}(T)$ consists of four elements. In fact, if the sequence (X_ν), $\nu = 1, 2, \ldots$, of points of the set $U = \mathbb{E}^2 \setminus A$, all of whose points lie in one of the four quadrants into which PQ and RS divide the plane, converges to T, then it is convergent in $(\hat{U}, \rho_U)$, as we can easily see, and its limit is a point belonging to $\pi^{-1}(T)$. Let (X_ν), (Y_ν), $\nu = 1, 2, \ldots$, be sequences convergent to T, where all points of each of them lie in one of the four quadrants mentioned. If X_m, $m = 1, 2, \ldots$, and Y_m, $m = 1, 2, \ldots$, lie in the same one of the given angles, then in the space $(\hat{U}, \rho_U)$ they converge to the same element of the set $\pi^{-1}(T)$. If these sequences lie in different angles, then their limits are different points of the set $\pi^{-1}(T)$. It remains to observe that if the sequence (X_m), $m = 1, 2, \ldots$, converges to the point T, and X_m skips from one angle to another for arbitrarily large m, then in the space $(\hat{U}, \rho_U)$ this sequence is not convergent.

§ 3. Two-Dimensional Manifolds with Intrinsic Metric

3.1. Definition. Triangulation of a Manifold. We first give a definition of what such a two-dimensional manifold is.

We recall that a topological space $\mathfrak{R}$ is called a *Hausdorff space* if for any two distinct points X and Y of it we can find a neighbourhood U of X and a neighbourhood V of Y whose intersection is empty.

Let $\mathfrak{R}$ be a topological space and $\mathfrak{B}$ a set of open subsets of $\mathfrak{R}$. We say that $\mathfrak{B}$ is a *base* of $\mathfrak{R}$ if any open set U of $\mathfrak{R}$ can be represented as the union of some set of sets belonging to $\mathfrak{B}$.

A *neighbourhood of a point* X of the topological space $\mathfrak{R}$ is any open set of this space containing X.

The family $\mathfrak{U}$ of neighbourhoods of a point P of the space $\mathfrak{R}$ is called a *basis* of the set of neighbourhoods of P if for any neighbourhood U of P there is a

$V \in \mathfrak{U}$ such that $U \supset V$. The topology of $\mathfrak{R}$ is completely determined if for any point of it we find a basis of the set of neighbourhoods of this point.

Let us introduce the following auxiliary concept. We shall call a subset G of the plane $\mathbb{R}^2$ a *two-dimensional standard domain* if G is either the disc

$$B(0, 1) = \{(x, y) \in \mathbb{R}^2 | x^2 + y^2 < 1\},$$

or the half-disc

$$B^+(0, 1) = \{(x, y) \in \mathbb{R}^2 | x^2 + y^2 < 1, y \geqslant 0\}.$$

The point 0 is called the *centre* of the two-dimensional standard domain. We note that in the case when the two-dimensional standard domain G is a half-disc, all the points of the diameter $\{(x, y) | y = 0, -1 < x < 1\}$ belong to G.

The topological space $\mathfrak{R}$ is called a *two-dimensional manifold with boundary* if it is a Hausdorff space, it has a denumerable basis, and for any point X of it there is a neighbourhood U of X homeomorphic to the standard two-dimensional domain G and a topological map φ of U onto G under which the centre of G corresponds to X. We shall call this neighbourhood U of X a *special neighbourhood* of X, and the homeomorphism φ a *special coordinate system* corresponding to the neighbourhood U and the point X.

Let M be a two-dimensional manifold with boundary. We shall call a point $X \in M$ an *interior point* if it has a special neighbourhood homeomorphic to a disc. Otherwise X is called a *boundary point* of M. By Brouwer's theorem on the invariance of open sets with respect to topological maps, the same point $X \in M$ cannot be simultaneously an interior point and a boundary point of M. The totality of all boundary points of M is called its *boundary* and denoted by ∂M.

We assume that X is an interior point of a manifold M with boundary, and U an arbitrary special neighbourhood of X. Then, obviously, any point $Y \in U$ is also an interior point of the manifold. Hence it follows that the totality of all interior points of M is an open subset of M, and so ∂M is a closed subset of M.

Let X be a boundary point of M, let U be any special neighbourhood of it, and let $\varphi: U \to B^+(0, 1)$ be the corresponding special coordinate system. Obviously, each point $Y \in U$ whose image $\varphi(Y)$ belongs to the bounding diameter of the half-disc $B^+(0, 1)$ is a boundary point.

The boundary of an arbitrary two-dimensional manifold, if it is not empty, consists of at most a denumerable set of connected components, each of which is homeomorphic to either a circle or a straight line. In particular, if the manifold M is compact, then its boundary consists of finitely many (possibly zero) connected components, each of which is homeomorphic to a circle.

Numerous examples of two-dimensional manifolds are known (Fig. 3). A sphere in three-dimensional space, the surface of an ellipsoid, hyperboloids of two sheets and one sheet, a paraboloid, the surface of an infinite circular cylinder in three-dimensional space – all these are two-dimensional manifolds. At the same time a right circular cone, that is, a surface described by a straight line under rotation around an axis intersecting it, is not a manifold. The vertex of the cone – the point of intersection of its generators with the axis of

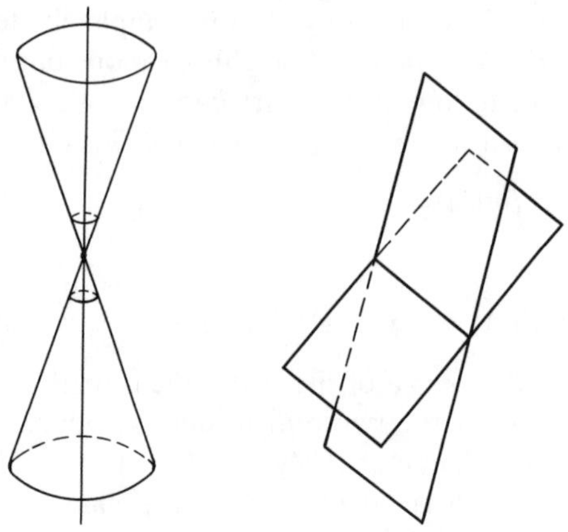

Fig. 3

rotation – does not have a neighbourhood that is homeomorphic to a standard two-dimensional domain. In exactly the same way, a set that is the union of two intersecting planes is not a two-dimensional manifold. In this case the singular points, which do not have a neighbourhood homeomorphic to a two-dimensional standard domain, lie on the line of intersection of the given planes.

A hemisphere and the lateral surface of a finite right circular cylinder give examples of two-dimensional manifolds with boundary.

Let us consider the following example. Let H be the half-plane $\{(x_1, x_2) \in \mathbb{R}^2 | x_2 > 0\}$, and I_0 the interval $(-1, 1)$ lying on the line $x_2 = 0$. The set $H \cup I_0$, endowed with the topology induced from $\mathbb{R}^2$, is a two-dimensional manifold with boundary. Its boundary is the interval I_0. At the same time, if we adjoin to H the closed interval $[-1, 1] = I$, then the resulting set $H \cup I$ is not a two-dimensional manifold, since the points $(-1, 0)$ and $(1, 0)$ do not have neighbourhoods in it that are homeomorphic to a two-dimensional standard domain.

One of the fundamental operations by means of which from some two-dimensional manifolds it is possible to obtain others is the pasting operation. Applications of it in topology are well known. In the theory of two-dimensional manifolds of bounded curvature this operation forms the basis of the method of cutting and pasting developed by A.D. Aleksandrov, which we shall discuss later.

If a manifold M with boundary is connected, then it is linearly connected, that is, for any two points X and Y of it there is a simple arc $L \subset M$ joining X and Y. All the manifolds considered later are assumed to be connected unless we explicitly say otherwise. In the questions studied here this restriction turns out to be immaterial.

Henceforth M denotes an arbitrary two-dimensional manifold with boundary. A set G in M will be called a *simple domain* if G is homeomorphic to a closed disc on the plane $\mathbb{R}^2$.

A *topological triangle* in M is any simple domain $T \subset M$ for which there are three distinct points X_1, X_2, X_3 lying on the boundary of T. The points X_1, X_2, X_3 are called the vertices of the triangle T. They split the boundary of T into simple arcs $L_1 = [X_2X_3]$, $L_2 = [X_3X_1]$, $L_3 = [X_1X_2]$, the pairs of which do not have any points in common except the ends. The arcs L_1, L_2, L_3 are called the sides of the topological triangle T.

A *triangulation* of a two-dimensional manifold M with boundary is any finite or denumerable set $\mathbf{K}$ of topological triangles that satisfies the following conditions.

I. Any point $X \in M$ lies in at least one of the triangles T belonging to $\mathbf{K}$.

II. Each point $X \in M$ has a neighbourhood that intersects only finitely many triangles $T \in \mathbf{K}$.

III. The intersection of any two distinct topological triangles $T', T'' \in \mathbf{K}$ is either empty, or consists of a unique point that is the vertex of each of them, or it is a simple arc that is a side of each of the given triangles.

The topological triangles that form the triangulation $\mathbf{K}$ are called the *faces* of $\mathbf{K}$, their sides are called the *edges* of $\mathbf{K}$, and their vertices are called the *vertices* of $\mathbf{K}$.

Theorem 3.1.1. (Rado's theorem, Rado (1925)). *Any two-dimensional manifold with boundary admits a triangulation.*

The proof of Theorem 3.1.1 is given in Kuratowski (1968) and Rado (1925), for example. It is rather cumbersome and relies on the following strengthened form of Jordan's theorem.

Theorem 3.1.2 (Schoenflies's theorem). *Let Γ be a simple closed curve on the plane $\mathbb{R}^2$, and let D be the bounded planar domain whose boundary is Γ. Then the set $D \cup \Gamma$ is homeomorphic to the closed disc $\bar{B}(0, 1) = \{(x, y) | x^2 + y^2 \leqslant 1\}$ on the plane $\mathbb{R}^2$.*

Later we shall study simple arcs in two-dimensional manifolds. In particular, it is necessary to make precise the concept of a *side of a simple arc* (see 3.4 later). For this we require the following assertion.

Theorem 3.1.3. *Let M be a two-dimensional manifold with boundary, and L a simple arc in M not containing boundary points of M. Then in M there is an open set U containing L and homeomorphic to the disc $B(0, 1)$ on the plane $\mathbb{R}^2$.*

We assume that M is a compact two-dimensional manifold with boundary (the latter, in particular, may be an empty set). From condition II in the definition of triangulation, by a theorem of Borel it follows that any triangulation of M is based on finitely many triangles. Let n_0 be the number of vertices, n_1 the number of edges, and n_2 the number of faces of the triangulation $\mathbf{K}$ of the manifold M. The quantity

$$n_0 - n_1 + n_2 = \chi(M)$$

does not depend on the choice of the triangulation of M and is called its *Euler characteristic*. It is a topological invariant of M. If M is a manifold homeomorphic to a two-dimensional sphere S^2 in the space $\mathbb{E}^3$, then $\chi(M) = 2$. If M is homeomorphic to the closed disc $B(0, 1)$ in $\mathbb{R}^2$, then $\chi(M) = 1$.

3.2. Pasting of Two-Dimensional Manifolds with Intrinsic Metric. One of the main methods used in topology in the study of two-dimensional manifolds consists in the systematic application of the operations of cutting and pasting. As Aleksandrov showed, this method can also be applied successfully in the study of two-dimensional manifolds of bounded curvature. Its effectiveness in the given case is connected with the fact that the conditions that must be satisfied, in order that as a result of applying the operations of cutting and pasting to manifolds of bounded curvature we again obtain manifolds of bounded curvature, have sufficiently general character. For Riemannian manifolds, for example, similar conditions are significantly more restrictive, which makes the application of the operations to Riemannian manifolds less useful.

The *operation of pasting* has quite a simple intuitive sense, although its formal definition, which we give later, is rather cumbersome. The awkwardness of the definition is caused by the desire to describe as far as possible a more general situation. We first consider some examples.

It is well known that the surface of a cube can be pasted from some cross-shaped polygon T. Figure 4 shows how this can be done. The perimeter of the polygon splits into segments. Those that are pasted together in the construction of the cube are denoted by the same letters. The process of pasting the cube from the polygon T is represented on the right of Fig. 4. The segments denoted by one letter are combined so that the arrows marked on each of them are combined. As a result of the pasting we obtain some surface Q. On it we define the intrinsic metric induced by the metric of the space. The distance between points in this metric is equal to the greatest lower bound of the lengths of the curves on the

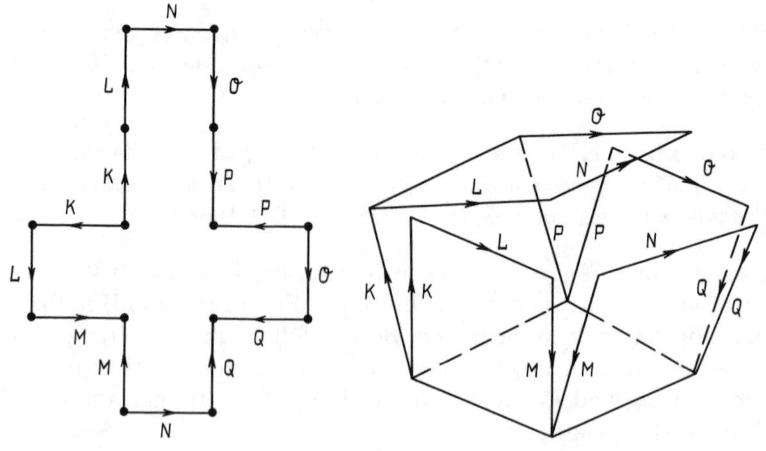

Fig. 4

surface Q that join these points. We shall see how we can define this metric, starting from the polygon T, the development of the surface Q. Any curve on Q under the transition from Q to the polygon T splits into separate pieces and its length is equal to the sum of the lengths of these pieces. The number of such pieces may turn out to be infinite. In the definition of the metric in the given case it is possible, as we can easily show, to restrict ourselves to a consideration of such curves, which on transition from Q to T split into finitely many arcs lying on T. The length of any such curve in space is equal to the sum of the lengths of these arcs.

Figure 5 shows a collection of planar domains, by the pasting of which we can obtain the surface of a right circular cylinder in space. One of these domains is a rectangle of height h and base length $2\pi r$, and the other two are circles of radius r. The boundary of each of the given domains is represented as the union of finitely many simple arcs. Arcs denoted by the same letters have equal lengths and the cylinder is obtained by pasting together such arcs.

Let H be an infinite strip on the plane, bounded by two parallel lines at a distance $2k > 0$ from each other. Rolling this strip in space so as to combine points of the boundary lying on lines perpendicular to the boundary lines (Fig. 6), we obtain a surface isometric to the lateral surface of an infinite right circular cylinder. We note that the boundary of the strip H can be pasted so as to combine points of the boundary obtained in the intersection of the boundary H by a family of parallel lines not perpendicular to the boundary lines (see Fig. 7).

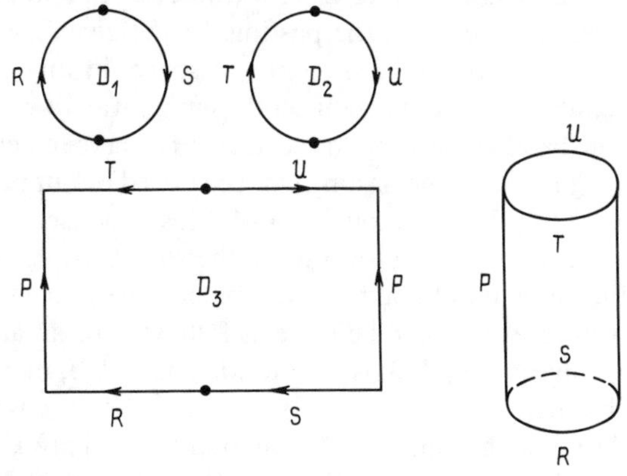

Fig. 5

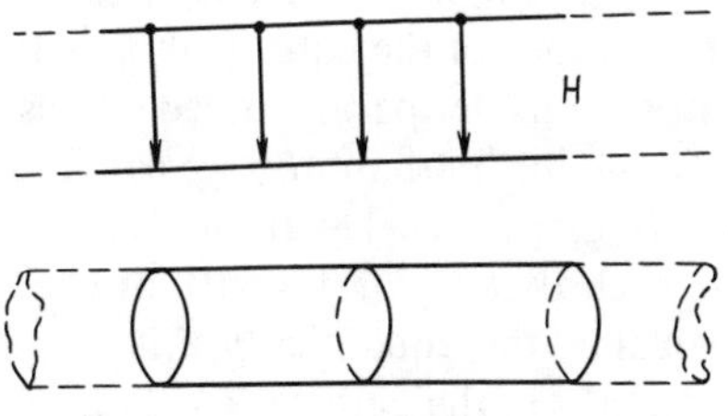

Fig. 6

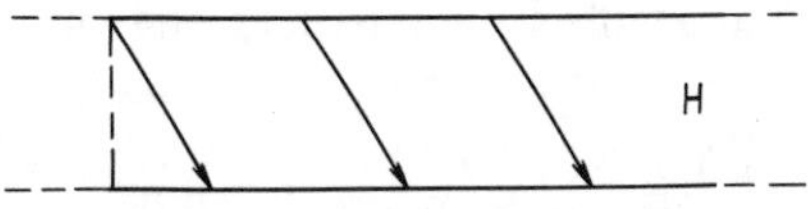

Fig. 7

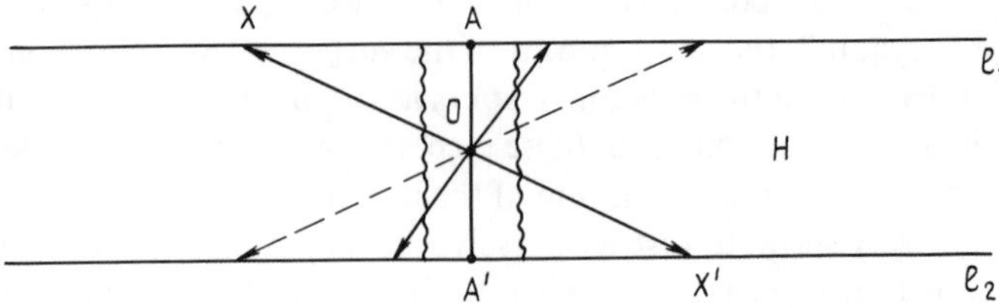

Fig. 8

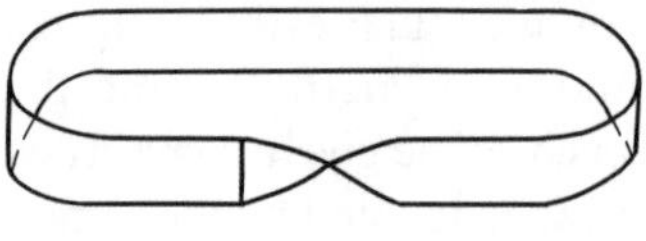

Fig. 9

In this case the pasting can also be brought about in practice by winding H as a band infinite on both sides onto a suitable right circular cylinder. The boundary of the strip H goes over to a spiral curve on the surface of the cylinder.

Figure 8 presents another way of pasting together the edges of a strip H bounded by two parallel lines on the plane. We choose a point O lying in H at the same distance from the boundary lines l_1 and l_2. To any point $X \in l_1$ there corresponds a point X' symmetrical to X with respect to O, and these points X and X' are combined as a result of the pasting. In the given case the pasting cannot be realized by a bending of the strip in space; in any case, it is not obvious how this could be done in a sufficiently simple way (if we cut out from the strip H a narrow band containing the segment AA' perpendicular to l_1 and l_2 and passing through O, then the pasting can be carried out in practice, and as a result we obtain the well-known model of a Möbius band (see Fig. 9), but it is not clear how to realize the pasting together of the whole strip). In this connection there arises the necessity of making more precise the procedure of pasting together in the given case. This can be done as follows. For an arbitrary point $X \in H$ we define an object $p(X)$. If X is an interior point of H, then $p(X) = X$. If X belongs to the boundary of H, then $p(X)$ is the pair (X, X'), where X' is the point symmetrical to X with respect to O. The totality of all $p(X)$, where $X \in H$, is denoted by M. We have a map $X \in H \mapsto p(X) \in M$. In the set M we define a metric ρ. This metric is constructed in the same way as the intrinsic metric of the surface of a cube is defined with respect to its development (that is, a cross-shaped polygon T, by the pasting of which the surface of the cube is obtained). Instead of curves it turns out to be convenient to consider parametrized curves. We shall say that a map $x: [a, b] \to M$ is an admissible path in M if we can find a finite collection of paths going into the strip H, $\xi_i: [a_{i-1}, a_i] \to H$, $i = 1, 2, \ldots, m$, where $a_0 = a < a_1 < \cdots < a_{m-1} < a_m = b$, such that $p[\xi_i(t)] = x(t)$ for each $t \in [a_{i-1}, a_i]$, where $i = 1, 2, \ldots, m$. In particular, the equalities $p[\xi_i(a_i)] = p[\xi_{i+1}(a_i)]$ hold. We shall call the sum of the lengths of the paths ξ_i, $i = 1, 2, \ldots, m$, the length of the path x. For a given admissible path $x: [a, b] \to M$ there can exist different collections of paths ξ_i satisfying the conditions mentioned above, but it is easy to show that the sum of their lengths does not depend on the choice of such a collection. We shall say that an admissible path $x: [a, b] \to M$ in the set M joins the points $P, Q \in M$ if $x(a) = P$, $x(b) = Q$. We denote the greatest lower bound of the lengths of admissible paths in M joining the points $P, Q \in M$ by $\rho(P, Q)$. Thus, a metric is defined on the set M. This metric is intrinsic. Any point $P \in M$ has a neighbourhood that is isometric to a disc on the plane in the metric ρ.

Let us prove the last assertion. Let P be an arbitrary point of M. We first assume that P is an interior point of the strip H. Then there is a $\delta > 0$ such that the disc $B(P, \delta) \subset H$. We show that the disc $U = B(P, \delta/2)$ is the desired neighbourhood of the point P. Any path in H joining an arbitrary point $X \in U$ to a point lying outside the disc $B(P, \delta)$ has length at least $\delta/2$. Hence it follows that any admissible path in M joining two arbitrary points $X, Y \in U$ and not contained in the disc $B(P, \delta)$ has length at least δ. Consequently, in the definition of the distance between the points $X, Y \in U$ it is sufficient to consider only those paths that are contained in the disc $B(P, \delta)$. Hence it is obvious that the distance between the points $X, Y \in U$ is equal to the length of the segment joining the points X and Y, that is, the metric ρ in U coincides with the usual Euclidean metric. In the case when $P = (Q, Q')$, where $Q \in l_1$ and $Q' \in l_2$ are points on the edge of the strip symmetrical with respect to the point O, the desired neighbourhood is obtained as follows. Let $2k > 0$ be the distance between the lines l_1 and l_2. For $t \in (0, k)$ we put $B^+(t) = B(Q, t) \cap H$, $B^-(t) = B(Q', t) \cap H$ and suppose that $U(t) = p[B^+(t) \cup B^-(t)]$. The set $U(t)$ is obtained by pasting together the half-discs $B^+(t)$ and $B^-(t)$ along their diameters. It is easy to establish that in the given case $U(t/2)$ is the required neighbourhood of the point P.

Thus from the strip H we obtain a metric space M. This space is complete, and like the surface of a cylinder it is locally Euclidean. We shall call it a *Möbius surface.*

The operation of cutting is in some sense the opposite of the operation of pasting. Thus, for example, cutting a cube (Fig. 4) along its edges, denoted by the letters K, L, M, N, O, P, Q, and straightening the surface of the cube into a plane, we obtain the cross-shaped polygon from whose pasting the cube was obtained. Cutting the surface of a finite right circular cylinder along the base circles (Fig. 5), we obtain a collection of three surfaces: two discs and the lateral surface of the cylinder. Cutting the latter along a generator and straightening into a plane, we obtain the same collection of surfaces by pasting which (Fig. 8) we obtain the surface of the cylinder, and so on.

We now give the necessary formal definitions.

Suppose we are given a finite or denumerable set (D_k), $k = 1, 2, \ldots$, of two-dimensional manifolds with boundary, each of which is endowed with the intrinsic metric, and for each k the boundary of the manifold D_k is locally rectifiable, that is, any simple arc contained in ∂D_k is rectifiable. Let ρ_k be the metric, specified in D_k. From the given collection of manifolds we can obtain a new manifold with boundary by pasting them together along simple curves, each of which lies on the boundary of one of the given manifolds. Let us describe the pasting rule in detail. We assume that pairs of the sets (D_k), $k = 1, 2, \dots$, do not have elements in common. We introduce the following notation. We put

$$\Delta = \bigcup_{k=1,2,\dots} D_k, \quad \partial\Delta = \bigcup_k \partial D_k.$$

In the set Δ we introduce a topology by the stipulation that a set $U \subset \Delta$ is open in Δ if and only if for each $k = 1, 2, \dots$ the intersection $U \cap D_k$ is an open set in D_k. It is not difficult to see that the set Δ, endowed with such a topology, is a two-dimensional manifold with boundary and $\partial\Delta$ is its boundary. Thus, instead of a collection of manifolds we obtain one manifold; it is true that it consists of separate pieces, not connected with each other. The *pasting rule* is defined in the following way. We first specify a no more than denumerable set R of simple curves, each of which is contained in the set $\partial\Delta$. We assume that any two different curves of R have no points in common except the ends. The curves belonging to R are those along which the pasting is carried out. We then specify the order in which the curves of R are pasted together. For this the set of curves R is split into pairs (L_i, L_i') and for each of them there is specified a topological map φ_i: $L_i \to L_i'$. The following conditions must be satisfied:

1) the union of the curves constituting R is a closed subset of Δ;

2) for each i the curves L_i and L_i' are different and any curve $K \in R$ belongs to one and only one of the pairs (L_i, L_i');

3) for each i, φ_i is a map of L_i onto L_i';

4) any simple arc is transformed by the map φ_i into an arc of the same length.

In the manifold Δ we introduce a relation between its elements, denoted by the symbol $\sim$. Let X and Y be two arbitrary points of Δ. If one of them does not belong to any of the curves of the set R, we suppose that $X \sim Y$ if and only if $X = Y$. If one of the given points X and Y lies on a curve of the set R, we shall suppose that $X \sim Y$ if and only if either $X = Y$ or Y can be obtained from X in finitely many steps by applying the maps φ_i and φ_i^{-1}. It is easy to verify that the relation $\sim$ introduced by the method described is an equivalence relation (that is, it is reflexive, symmetric and transitive). The last condition that the pasting rule must satisfy is:

5) for any point $X \in \Delta$ the set of all $Y \in \Delta$ such that $X \sim Y$ is finite.

Figure 10 (A, B, C) represents some examples that illustrate the general scheme described here.

We denote by D the set obtained from Δ by identifying any two points $X, Y \in \Delta$ such that $X \sim Y$. Formally the elements of the set D are equivalence classes in the set Δ with respect to the relation $\sim$. Let p be a canonical map of Δ onto D. For an arbitrary $X \in \Delta$, $p(X)$ is the element of D generated by this X, that is, $p(X)$ is the totality of all $Y \in \Delta$ such that $X \sim Y$. The topology in D is defined by the stipulation that a set $U \subset D$ is open if and only if $p^{-1}(U)$ is an open set in Δ.

It is easy to verify that the topological space D constructed as described above is a two-dimensional manifold. We make the following remarks about

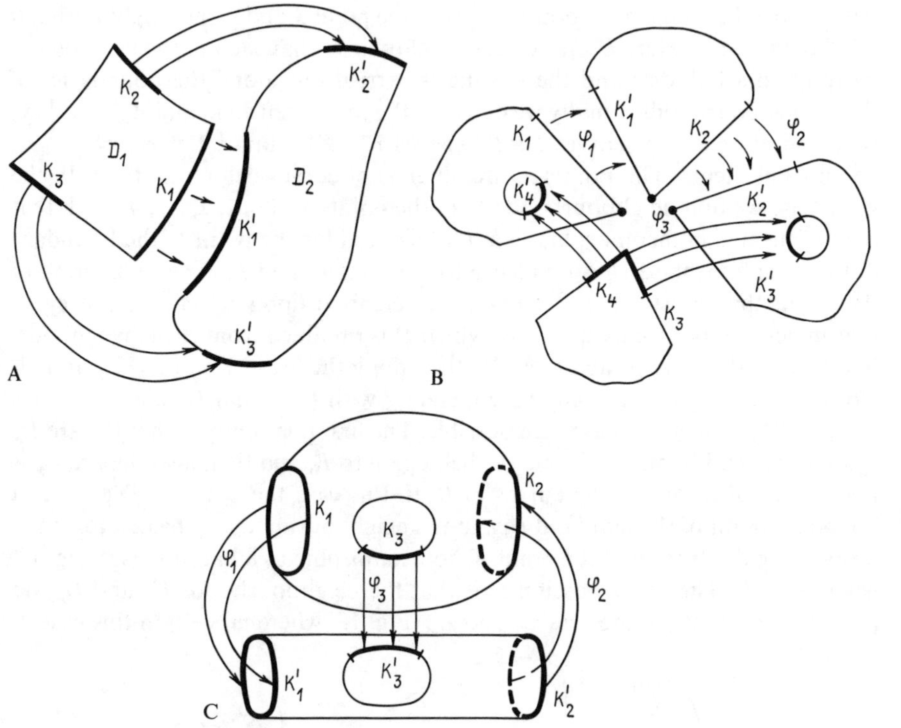

Fig. 10

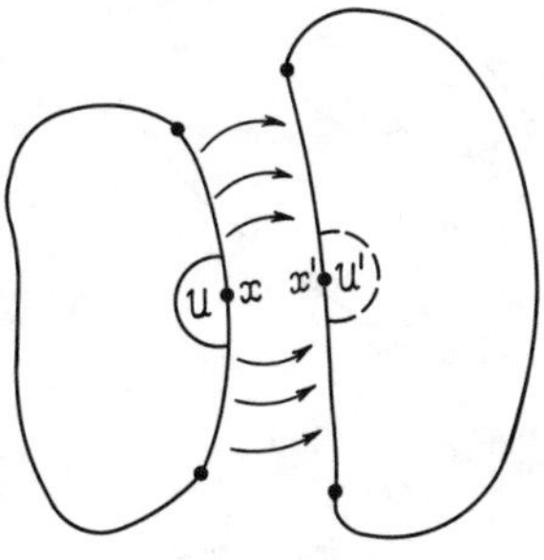

Fig. 11

how D is constructed. Let us take an arbitrary point $X \in D$. Let x be any point of the manifold Δ such that $p(x) = X$. If x does not belong to one of the curves of the set R, then $p^{-1}(x)$ consists of a unique element – the given point x, and the map p maps a neighbourhood U of the point x one-to-one onto a neighbourhood of the point X. We suppose that $x \in p^{-1}(X)$ is an internal point of some curve $L \in R$ (see Fig. 11). Then $p^{-1}(X)$ consists of two distinct elements $x \in L$ and $x' \in L'$, where the curves L and L' are identified according to the given pasting rule. In the given case the point X has a neighbourhood obtained by pasting together some neighbourhood U of the point x and some neighbourhood U' of the point x'. Here U and U' can be chosen so that each of them is homeomorphic to a half-disc and the pasting is carried out along the boundaries of these neighbourhoods. Finally we consider the case when some point $x \in p^{-1}(X)$ is an end of one of the curves $L \in R$. The set $p^{-1}(X)$ is finite. Let $x_1, x_2, \ldots, x_m$ be all its elements. The points x_i are all ends of curves belonging to R. In the given case we find neighbourhoods U_i of the points x_i, $i = 1, 2, \ldots, m$, such that each of them is homeomorphic to a half-disc and for the point x_i the boundary of the neighbourhood U_i is divided into two arcs L_i and L_i'. The numbering of the points that constitute $p^{-1}(X)$ and the neighbourhoods U_i can be chosen so that under the pasting by means of which D is obtained from Δ the neighbourhoods U_1 and U_2 are pasted to each other along the arcs L_1' and L_2, U_2 and U_3 along the arcs L_2' and L_3, and so on, ending with U_{m-1} and U_m along the arcs L_{m-1}' and L_m. Then two cases are possible. The first possibility is that the arc L_m' is not contained in one of the curves belonging to R, and then also the arc L_1 is not contained in one of the curves of R. In this case the point x (Fig. 12) is a boundary point of the manifold D, and the neighbourhoods U_i, pasted together, form a neighbourhood of the point X homeomorphic to a half-disc (see Fig. 12, where $m = 4$). The other possibility is that the neighbourhoods U_1 and U_m are pasted together along the arcs L_1 and L_m' (Fig. 13, where $m = 3$). In this case X

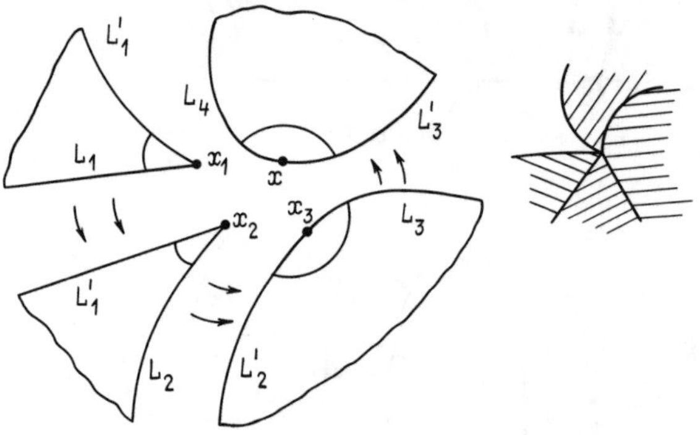

Fig. 12

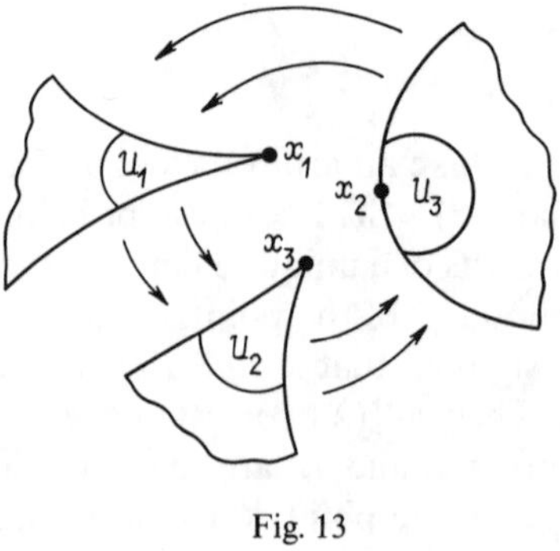

Fig. 13

is an interior point of the manifold D, and by pasting together the neighbourhoods U_i we obtain a neighbourhood of the point X homeomorphic to a disc. The case $m = 1$ is allowed by the construction. If this holds, then the arcs L_1 and L_1' that constitute the boundary of the neighbourhood U_1 are pasted together.

All the constructions described above have a purely topological character. (We observe that condition 4 of the equality of lengths of pasted arcs has not been used.) Questions relating to the metric have not yet been considered. We now show how we can introduce an intrinsic metric in the manifold D. In principle the manifold D, constructed as we have described, may consist of separate pieces in no way joined to each other, and in this case an intrinsic metric in D cannot be defined. We shall assume henceforth that the manifold D is connected.

Let X and Y be two arbitrary points of the manifold D. The sequence $x_1, y_1, x_2, y_2, \ldots, x_n, y_n$ of points of the manifold Δ will be called a *chain* joining the points X and Y if $p(x_i) = X$, $p(y_n) = Y$, $p(y_i) = p(x_{i+1})$ when $i = 1, 2, \ldots, n-1$, and for each $i = 1, 2, \ldots, n$ the points x_i and y_i belong to the same manifold D_{s_i}. We shall denote the greatest lower bound of the sums

$$\sum_{i=1}^{n} \rho_{s_i}(x_i, y_i)$$

on the totality of all chains in the manifold Δ joining the points X and Y by the symbol $\rho(X, Y)$. It is easy to verify that the function $\rho(X, Y)$ of a pair of points of D defined in this way is a metric. This metric is intrinsic and has the following property. Let D_s^0 be the totality of all interior points of the manifold D_s. The map p is one-to-one on the set D_s^0; in this connection we shall identify any point $X \in D_s^0$ with the point $p(X)$ of the manifold D, and in accordance with this we shall further identify D_s^0 with $p(D_s^0)$. The metric induced in the domain D_s^0 of the metric space (D, ρ) coincides with the metric ρ_s. The metric ρ in the manifold D, defined as described above, is uniquely determined by this property and the requirement that it is an intrinsic metric.

Let us make one more remark concerning condition 4 in the description of the pasting rules. From the point of view of topology it is superfluous. In our case the fulfilment of condition 4 is that to each of those arcs along which pasting takes place, in the manifold obtained as a result of the pasting there corresponds an arc of the same length. In principle the fulfilment of condition 4 need not be required in the "metrical case", which is of interest to us. But then the application of the pasting operation will lead to manifolds of a rather pathological character, which are of no interest to us.

As an example we mention a special case of the pasting construction, which is interesting in that in many cases it enables us to reduce the case of a manifold with boundary to the case when the boundary of the manifold is empty. Let D be an arbitrary two-dimensional manifold with intrinsic metric, where the boundary of D is not empty and any simple arc contained in ∂D is rectifiable. Let D_0 and D_1 be two different copies of D. Formally we can represent the pair of manifolds D_0 and D_1 as the direct product of D and a topological space consisting of two elements – the numbers 0 and 1. The elements of D_0 are pairs of the form $(x, 0)$, where $x \in D$, and the elements of D_1 are pairs of the form $(x, 1)$, where $x \in D$. The boundary of D consists of a no more than denumerable set of simple curves (K_m), $m = 1, 2, \ldots$ Let K_m^0 be the set of all points $(x, 0) \in D_0$, where $x \in K_m$, and K_m^1 the set of all points $(x, 1) \in D_1$, where $x \in K_m$. Let R be the totality of all curves $K_m^0, K_m^1, m = 1, 2, \ldots$ In R we naturally select the set of pairs (K_m^0, K_m^1). For any m let φ_m be the map $(x, 0) \in K_m^0 \to (x, 1) \in K_m^1$. For each m the map is topological. It is not difficult to see that conditions 1–5, which the pasting rule must satisfy, are all fulfilled in the given case. Identifying the points corresponding to each other under the map φ_m, $m = 1, 2, \ldots$, we consequently obtain a two-dimensional manifold $\tilde{D}$. We shall call it the *twice covered manifold* D. It is not difficult to see that the boundary of $\tilde{D}$ is empty. As a result of pasting together the curves K_m^0 and K_m^1 we obtain a curve $\tilde{K}_m \subset \tilde{D}$.

The curves $\tilde{K}_m$, $m = 1, 2, \ldots$, split D into two domains D' and D'', each of which is homeomorphic to D. The set $\bigcup \tilde{K}_m$ is the common boundary of these domains. If we introduce in D' and D'' the metrics induced from $\tilde{D}$, then each of the domains D' and D'' will also be isometric to the interior of D.

3.3. Cutting of Manifolds. A formal definition, moreover in a very general situation, was given in 2.4. Here we just make some remarks related to the case of two-dimensional manifolds.

In the case when the space (M, ρ) with intrinsic metric is a two-dimensional manifold, and the set A is constructed rather simply in the topological respect, we can determine a natural condition for the accessibility of a boundary point of the set A. Let us introduce an auxiliary concept.

Let U be a connected open subset of the metric space (M, ρ) with intrinsic metric. We introduce in U the induced intrinsic metric ρ_U. (We recall that by definition $\rho_U(X, Y)$, where $X, Y \in U$, is the greatest lower bound of lengths of curves joining the points X and Y and lying in the domain U.) The greatest lower bound of $\rho_U(X, Y)$, where $X, Y \in U$, is called the *internal diameter* of the set U and denoted henceforth by $\Delta(U)$.

Theorem 3.3.1. *Let (M, ρ) be a two-dimensional manifold with intrinsic metric. We assume that $A \subset M$ is a closed set such that $A \neq M$ and the boundary of A can be represented as the union of a finite or denumerable set of simple arcs L_k, $k = 1, 2, \ldots$, so that any two of them do not have points in common other than the ends and any compact subset $E \subset M$ intersects finitely many arcs L_k. Let X be an arbitrary boundary point of A. If for any $\varepsilon > 0$ we can find a neighbourhood V of the point X such that the internal diameter of each component of the set $V \backslash A$ is less than ε, then X is an accessible point.*

In fact, we assume that the set A satisfies the conditions of the theorem. We take arbitrarily a boundary point X of the set A. We represent the boundary of A as the union of a no more than denumerable set of simple arcs satisfying the conditions of the theorem. Let $L_1, L_2, \ldots, L_m$ be those arcs that contain the point X. Then there is a neighbourhood V_1 of X and a homeomorphism

$\varphi: V_1 \to \mathbb{R}^2$ such that $\varphi(V)$ is the disc $B(0, 1)$, the point X is transformed by φ to the centre of $B(0, 1)$, and the intervals $L_i \cap V_1$ of the arcs L_i contained in V_1 are taken by φ into radii of the disc. The existence of a neighbourhood V_1 and a homeomorphism φ satisfying all these conditions is easily established by means of Schoenflies's theorem. The arcs $V_1 \cap L_i$ split V_1 into sectors that are transformed under the map φ into corresponding sectors of $B(0, 1)$. Each of these sectors is contained either in the set A or in $M \backslash A$. Let $G_1, G_2, \ldots, G_l$ be those of them contained in $M \backslash A$. Let (x_ν), $\nu = 1, 2, \ldots$, be an arbitrary sequence of points lying in the set G_j that converges to a point X. The set G_j is connected, and so it is contained in some connected component of $M \backslash A$. Let U_0 be this component. We specify $\varepsilon > 0$ arbitrarily and find with respect to it a neighbourhood W of the point X such that any connected component of $W \backslash A$ has an interior diameter less than ε (we assume that the point $X \in \partial A$ is such that there is such a neighbourhood for it, whatever $\varepsilon > 0$ is). We put $V_\delta = \varphi^{-1}[B(0, \delta)]$, where $0 < \delta \leqslant 1$. We fix the value $\delta \in (0, 1)$ such that $V_\delta \subset W$. The intersection $V_\delta \cap G_j$ is connected, since it is transferred to some sector of the disc $B(0, \delta)$ by the homeomorphism φ. The set $V_\delta \cap G_j$ is contained in some connected component of $W \backslash A$. Let W_0 be this component. We find ν_0 such that $x_\nu \in V_\delta$ when $\nu \geqslant \nu_0$. For any $\nu \geqslant \nu_0$, $\mu \geqslant \nu_0$ we have $\rho_{V_0}(x_\nu, x_\mu) \leqslant \rho_{W_0}(x_\nu, x_\mu) < \varepsilon$ and since $\varepsilon > 0$ is arbitrary, we have thus proved that the sequence (x_ν) is fundamental in the space (U_0, ρ_{U_0}). Hence it follows that the point X is accessible, which we needed to prove.

We note that, as is easy to see, each of the sectors G_j that figure in the argument advanced above determines exactly one point in the collection of metric spaces obtained by cutting the manifold M along the set A, and all these points are pairwise distinct.

The operation of cutting is in some sense the reverse of the pasting operation. Namely, we assume that the manifold M with intrinsic metric ρ is obtained by pasting together the manifolds D_k. Let A be the closed set in M consisting of all points of M that correspond to points lying on these curves along which the pasting is carried out. Cutting the space (M, ρ) along the given set A, we obtain the original collection of two-dimensional manifolds.

3.4. A Side of a Simple Arc in a Two-Dimensional Manifold. Let M be an arbitrary two-dimensional manifold with boundary, and L a simple arc in M. We introduce here some concepts that enable us to give an exact meaning to words "on a given side of L" or "on the same side of L" (see p. 21).

We require the concept of the orientation of a simple closed curve on a plane. It has the following intuitive meaning. A simple closed curve Γ is oriented if we are given a direction for going round the curve Γ. According to Jordan's theorem the curve Γ splits the plane into two domains G and G', one of which (suppose it is G) is bounded. If on going round Γ in the given direction G turns out to lie on the left of Γ, we shall say that *the curve is positively oriented*. If the domain lies to the right of Γ on going round Γ in the given direction, we shall say that *Γ is negatively oriented*. For example, the curve Γ_1 in Fig. 14 is positively oriented, and Γ_2 is negatively oriented.

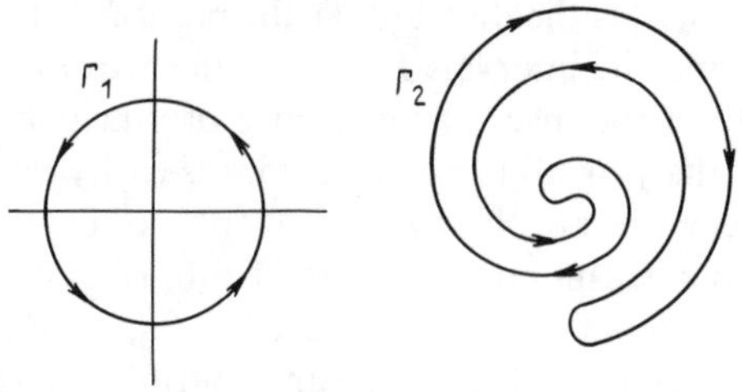

Fig. 14

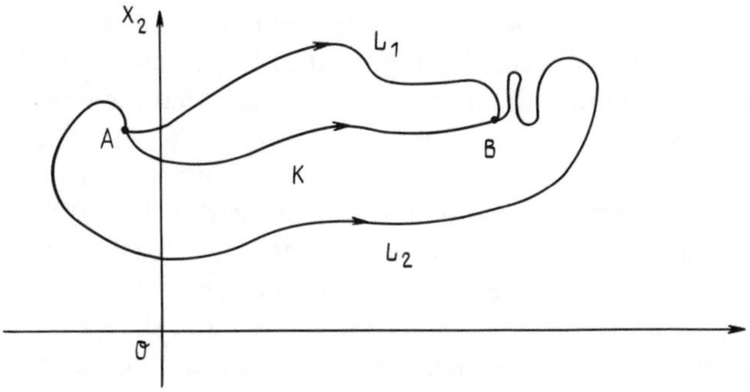

Fig. 15

Let K be a simple arc on the plane $\mathbb{R}^2$. We shall say that *the curve K is oriented* if one of its end-points is called the beginning. Let K and L be two oriented simple arcs with common beginning at a point A and common end B. We assume that K and L have no other common points. Then together they form a simple closed curve Γ. We orient Γ so that under a motion along K the point A precedes the point B. If Γ is positively oriented, we shall say that L is situated to the left of K. If the orientation of Γ is negative, we shall say that L is situated to the right of K. For example, the simple arc L_1 in Fig. 15 is situated to the left, and the simple arc L_2 to the right of K. It is not difficult to see that if a simple arc L lies to the left of K, then in turn K lies to the right of L.

What "right or left of a simple arc on a manifold" means we shall define later.

Without difficulty we can give an exact meaning to the intuitive definitions given above in the case when the arcs under consideration are sufficiently smooth. In the general case the reader can perceive in them a kind of vicious circle.

We now give the exact definitions. We first make some remarks.

Suppose that $z = (x, y) \in \mathbb{R}^2$, where $|z| = r \neq 0$. Then the polar coordinates are defined by the conditions

$$x = r\cos\theta, \quad y = r\sin\theta. \tag{3.1}$$

Any number θ for which the equalities (3.1) are satisfied is called the *polar angle* of the point z.

Let $z(t)$, $a \leqslant t \leqslant b$, be an arbitrary closed path on the plane $\mathbb{R}^2$, that is, such that $z(a) = z(b)$. We denote by L the set of points on the plane sketched out by the point $z(t)$ when t runs through the interval $[a, b]$, that is, $L = z([a, b])$. We take a point $c \notin L$ arbitrarily. Then we can define a continuous function $\varphi(t)$, $a \leqslant t \leqslant b$, such that for each $t \in [a, b]$ $\varphi(t)$ is the polar angle of the vector $z(t) - c$. If $\varphi_1(t)$ and $\varphi_2(t)$ are two functions satisfying this condition, then $\varphi_1(t) - \varphi_2(t) \equiv 2\pi m = \text{const}$, where m is an integer. The difference $\varphi(b) - \varphi(a)$ therefore does not depend on the choice of the function $\varphi(t)$. The number $v(c, z) = [\varphi(b) - \varphi(a)]/2\pi$ is an integer. The quantity $v(c, z)$ is called the *index of the point c with respect to the closed path $z(t)$, $a \leqslant t \leqslant b$*. The function $c \mapsto v(c, z)$ of the variable c is defined on the set $\mathbb{R}^2 \setminus L$ and is constant on each connected component of it.

Let Γ be a simple closed curve on the plane, $z(t)$, $a \leqslant t \leqslant b$, an arbitrary parametrization of it, that is, a path in $\mathbb{R}^2$ such that $z(a) = z(b)$, and suppose that the function z maps the half-open interval $(a, b]$ one-to-one onto the set Γ. We shall assume that the function $z(t)$ is extended periodically with period $T = b - a$ onto the whole set $\mathbb{R}$. (For any $t \in \mathbb{R}$ there is one and only one integer m such that $t - Tm \in (a, b]$. For this t we put $z(t) = z(t - Tm)$.) Henceforth we shall assume that this condition is automatically satisfied.

If the periodic functions z_1 with period T_1 and z_2 with period T_2 are parametrizations of a simple closed curve Γ, then $z_2(t) = z_1(\varphi(t))$, where $\varphi\colon \mathbb{R} \to \mathbb{R}$ is a continuous strictly monotonic function such that if $t_2 - t_1 = T_1$, then $\varphi(t_2) - \varphi(t_1) = \pm T_2$. We shall say that the parametrizations z_1 and z_2 are oriented in the same way if φ is an increasing function, and oriented in the opposite way if φ is a decreasing function. The set of all parametrizations of a simple closed curve is split into two disjoint classes so that two parametrizations of one class are oriented in the same way, and parametrizations belonging to different classes are oriented in the opposite way. We shall say that *the simple closed curve Γ is oriented* if the parametrizations of one class are called right, and the parametrizations of the other class are called left. Intuitively this is equivalent to the designation of a definite direction of going round the curve Γ.

Let Γ be an oriented simple closed curve on the plane $\mathbb{R}^2$, $z(t)$ a right parametrization of this curve, and $T > 0$ the period of the function $z(t)$. We have a closed path $z(t)$, $0 \leqslant t \leqslant T$, and for any point $c \notin \Gamma$ there is defined a number $v(c, z)$. This number does not depend on the choice of the parametrization of Γ. The curve Γ splits $\mathbb{R}^2$ into domains G and G', one of which is bounded. We shall assume that this is the domain G. For any point $c \in G'$, $v(c, z) = 0$, and for $c \in G$, $v(c, z) = \pm 1$. The curve Γ is said to be *positively oriented* if $v(c, z) = 1$ for all $c \in G$, and *negatively oriented* if $v(c, z) = -1$ for $c \in G$.

The following assertion is true.

Theorem 3.4.1. *Let U be a domain in $\mathbb{R}^2$, and $\varphi\colon U \to \mathbb{R}^2$ a topological map. Then either for any oriented simple closed curve $\Gamma \subset U$ the curve Γ and $\varphi(\Gamma)$ are oriented in the same way, or for any such curve the curves Γ and $\varphi(\Gamma)$ are oriented in the opposite way.*

If the first of the two possibilities holds, we shall say that the map φ preserves the orientation. In the second case we shall say that φ changes the orientation.

Let K be a simple arc on a plane. We shall say that *the arc K is oriented* if one of its end-points (we denote it by A) is called the beginning and the other (we denote it by B) is called the end of the arc. We shall call the parametrization $z(t)$, $a \leqslant t \leqslant b$, of the oriented simple arc K *right* if $z(a) = A$, $z(b) = B$, and *left* if $z(a) = B$, $z(b) = A$.

Let K and L be two oriented simple arcs on a plane such that the beginning of each of them is the point A, and the end is the point B. The curves K and L together form a simple closed curve Γ. We construct a parametrization of Γ. Namely, we specify a, b, c arbitrarily such that $a < c < b$, and let $z_1(t)$ be a parametrization of the arc K such that $z_1(a) = A$, $z_1(c) = B$, and $z_2(t)$ a parametrization of the arc L such that $z_2(c) = B$, $z_2(b) = A$, and put $z(t) = z_1(t)$ when $a \leqslant t \leqslant c$, and $z(t) = z_2(t)$ when $c \leqslant t \leqslant b$. It is not difficult to see that all the parametrizations of Γ that can be obtained in this way are of the same name. Calling them right, we obtain an orientation of Γ. We shall say that Γ lies to the *left* of K if Γ is positively oriented. If Γ is negatively oriented, we say that L is situated to the *right* of K.

Thus, we have given an exact meaning to the assertion "the simple arc L lies to the left (right)" of the simple arc K for the case of a plane. As we did above, we can now define the concept of left and right semineighbourhood of the oriented simple arc K on the plane.

Let us consider the case of a simple arc on an arbitrary two-dimensional manifold. Let M be a two-dimensional manifold with boundary, and K an arbitrary simple arc in M not containing boundary points of M. We shall call any open set $U \supset K$ such that the closure of U is homeomorphic to a closed disc on the plane $\mathbb{R}^2$ a *canonical neighbourhood of the arc K* (the existence of an open set U satisfying these conditions follows immediately from Theorem 3.1.3). Let $\varphi\colon U \to \mathbb{R}^2$ and $\psi\colon U \to \mathbb{R}^2$ be topological maps of the domain U into $\mathbb{R}^2$. The sets $G = \varphi(U)$ and $H = \psi(U)$ are open, and $\sigma = \psi \circ \varphi^{-1}$ is a topological map of G onto H. We shall say that φ and ψ are of the same name if the map σ preserves the orientation, and of different names if σ changes the orientation. The set of all homeomorphisms of U into $\mathbb{R}^2$ splits into two classes in such a way that two homeomorphisms belonging to one class have the same name, and homeomorphisms belonging to different classes have different names. We shall say that the domain U is oriented, or in other words that a definite orientation of U is specified, if the homeomorphisms of one class are called *right*, and homeomorphisms of the other class are called *left*.

Let U_1 and U_2 be two canonical neighbourhoods of a simple arc K. We denote by V the connected component of the set $U_1 \cap U_2$ that belongs to K. We assume that U_1 and U_2 are oriented, and let $\varphi\colon U_1 \to \mathbb{R}^2$ and $\psi\colon U_2 \to \mathbb{R}^2$ be right homeomorphisms of these neighbourhoods. Let $G = \varphi(V)$, $H = \psi(V)$. Then the homeomorphism $\sigma = \psi \circ \varphi^{-1}\colon G \to H$ is defined. The sets G and H are connected. We shall say that U_1 and U_2 are oriented in the same way, or in other words that their orientations are consistent, if σ preserves the orientation. Otherwise we shall say that U_1 and U_2 are oriented in the opposite way. The set of canonical neighbourhoods of the simple arc K thus splits into two classes so that two neighbourhoods of one class are oriented in the same way, and neighbourhoods of different classes are oriented in the opposite way. We shall say that along K there is specified a definite orientation of the manifold M if the elements of one of these two classes are called right neighbourhoods of K (then the elements of the other class are called left neighbourhoods of K).

Thus, let K be an oriented simple arc in a two-dimensional manifold M. We assume that an orientation is specified along K. Let U be a right canonical neighbourhood of K, and φ a topological map of the closure of U onto the disc $\bar{B}(0, 1)$ such that the beginning A of the simple arc K is transformed by this map into the point $P = (-\frac{1}{2}, 0) \in \mathbb{R}^2$, and the end B of K is mapped into the point $Q = (\frac{1}{2}, 0)$, and the simple arc itself is transformed by a given homeomorphism into the interval $[PQ]$. The existence of such a homeomorphism is easily established by applying Schoenflies's theorem. Obviously we can assume that this homeomorphism φ is right. Let L be an oriented simple arc contained in U, where A is its beginning and B its end, and there are no other points common to K and L. We shall say that L lies to the left (right) of the curve K if $\varphi(L)$ lies to the left (right) of the interval $[PQ]$ on the plane. Let $G \subset U$ be the open domain bounded by the simple arcs K and L. Then we shall call G a *left semineighbourhood* of K if L lies to the left of K, and a *right semineighbourhood* if L is situated to the right of K. Let (X_ν), $\nu = 1, 2, \ldots$, be an arbitrary sequence of points of M that converges to some interior point of K. Then there is a ν_0 such that when $\nu \geqslant \nu_0$ the point X_ν belongs to U. When $\nu \geqslant \nu_0$ there is defined a point $z_\nu = \varphi(X_\nu)$, $z_\nu = (x_\nu, y_\nu)$. We shall say that X_ν converges to a point X_0 to the right (left) with respect to K if there is a $\nu_1 \geqslant \nu_0$ such that $y_\nu > 0$ ($y_\nu < 0$) for all $\nu \geqslant \nu_1$. It is easy to prove that the property that the sequence (X_ν), $\nu = 1, 2, \ldots$, converges to the right (left) to an arbitrary interior point X_0 does not depend on the choice of the right neighbourhood U of K nor on the choice of the homeomorphism φ.

Let us define what is meant by "a sequence of simple arcs converges on the left (right) to a simple arc K on a manifold M". We first introduce a concept that refers to plane curves. Suppose we are given simple arcs K and K_ν, $\nu = 1, 2, \ldots$, on the plane $\mathbb{R}^2$. Then we shall say that K_ν converges to K as $\nu \to \infty$ if the arcs K and K_ν admit the parametrizations $X(t)$, $a \leqslant t \leqslant b$, and $X_\nu(t)$, $a \leqslant t \leqslant b$, $\nu = 1, 2, \ldots$, such that $|X_\nu(t) - X(t)| \to 0$ as $\nu \to \infty$ uniformly on $[a, b]$.

Let K be a simple arc in the manifold M that does not contain boundary points of M. We assume that K itself is oriented and along it there is specified an orientation of M. We specify arbitrarily a right neighbourhood U of the arc K and construct a homeomorphism φ of the set $\bar{U}$ into the plane that satisfies all the conditions listed above (that is, $\varphi(\bar{U})$ is the disc $\bar{B}(0, 1)$, and $\varphi(K)$ is the interval $[PQ]$, where $P = (-\frac{1}{2}, 0) = \varphi(A)$, $Q = (\frac{1}{2}, 0) = \varphi(B)$, A is the beginning and B the end of the arc K; in addition, we require that φ is a right homeomorphism). Let (K_ν), $\nu = 1, 2, \ldots$, be a sequence of oriented simple arcs in M, each of which has its beginning at A and its end at B. We shall say that the simple arcs K_ν converge to the simple arc K on the left (right) if there is a number ν_0 such that $K_\nu \subset U$ for all $\nu \geqslant \nu_0$ and K_ν lies to the left (right) of K, and as $\nu \to \infty$ the arc $\varphi(K_\nu)$ converges to the interval $[PQ]$.

§4. Two-Dimensional Riemannian Geometry

4.1. Differentiable Two-Dimensional Manifolds. The concept of a two-dimensional manifold with boundary has been defined in 3.1. Here we shall consider only manifolds whose boundary is empty, and we shall use the term "two-dimensional manifold" for their designation.

Let M be an arbitrary two-dimensional manifold. A *chart* or *local coordinate system* in M is any topological map $\varphi: U \to \mathbb{R}^2$, where U is an open set of the space M. If $\varphi: U \to \mathbb{R}^2$ is a chart in a two-dimensional manifold M, then $V = \varphi(U)$ is an open set in $\mathbb{R}^2$. U is called the *domain of definition of the chart* φ, and V is called its *range*. Suppose that $p \in U$ and $\varphi(p) = (t_1, t_2)$. The numbers t_1, t_2 are called the *coordinates* of the point $p \in U$ with respect to the given chart.

The charts $\varphi_1: U_1 \to \mathbb{R}^2$, $\varphi_2: U_2 \to \mathbb{R}^2$ of a two-dimensional manifold M are said to be *overlapping* if $U_1 \cap U_2$ is not empty. In this case there are defined open sets in $\mathbb{R}^2$: $G_1 = \varphi_1(U_1 \cap U_2)$ and $G_2 = \varphi_2(U_1 \cap U_2)$, and topological maps $\sigma = \varphi_2 \circ \varphi_1^{-1}: G_1 \to G_2$, $\tau = \varphi_1 \circ \varphi_2^{-1}: G_2 \to G_1$. Obviously $\sigma = \tau^{-1}$. We take a point $p \in U_1 \cap U_2$ arbitrarily. Let $\varphi_1(p) = (t_1, t_2)$, $\varphi_2(p) = (u_1, u_2)$. The numbers t_1, t_2 are the coordinates of the point p with respect to the chart φ_1, and u_1, u_2 are its coordinates with respect to the chart φ_2. We have $\sigma(t_1, t_2) = (u_1, u_2)$, so the function σ enables us to calculate the coordinates of a point $p \in U_1 \cap U_2$ with respect to the chart φ_2 from its coordinates in the chart φ_1. Similarly, $(t_1, t_2) = \tau(u_1, u_2)$. We shall call σ and τ the *transition functions* for the given charts φ_1 and φ_2.

Let $\varphi_1: U_1 \to \mathbb{R}^2$ and $\varphi_2: U_2 \to \mathbb{R}^2$ be two overlapping charts in a manifold M. We shall say that the charts φ_1 and φ_2 are *smoothly compatible* if the transition functions $\varphi_1 \circ \varphi_2^{-1}$ and $\varphi_2 \circ \varphi_1^{-1}$ corresponding to them have all the partial derivatives of any order, and these derivatives are continuous.

Let $\mathfrak{A}$ be a set of charts of a two-dimensional manifold M. Then we shall say that $\mathfrak{A}$ is an *atlas of class* C^∞ if the domains of definition of the charts belonging to the atlas $\mathfrak{A}$ cover M, and any two overlapping charts of $\mathfrak{A}$ are smoothly compatible.

A differentiable two-dimensional manifold is a two-dimensional manifold on which an atlas $\mathfrak{A}$ of class C^∞ is specified. We shall call the charts belonging to the atlas $\mathfrak{A}$ *basic charts*. A chart $\varphi: U \to \mathbb{R}^2$ in a manifold M, if it is smoothly compatible with any basic chart $\psi: G \to \mathbb{R}^2$ such that $U \cap G$ is non-empty, is said to be *admissible*. The set of all admissible charts in a two-dimensional manifold forms an atlas of class C^∞.

Let us define the concept of an orientable two-dimensional manifold and the concept of an orientation in it.

Let M be an arbitrary differentiable two-dimensional manifold and let $\varphi_1\colon U_1 \to \mathbb{R}^2$ and $\varphi_2\colon U_2 \to \mathbb{R}^2$ be two overlapping admissible charts in M. Let $\sigma = \varphi_2 \circ \varphi_1^{-1}$ and $\tau = \varphi_1 \circ \varphi_2^{-1} = \sigma^{-1}$ be the transition functions corresponding to them. If the Jacobian of the map σ is positive everywhere (obviously in this case the Jacobian of τ is also positive everywhere), we shall say that *the given charts are oriented in the same way*. If the Jacobian of σ is negative everywhere (then the Jacobian of τ is negative at every point of the domain of definition), then the charts φ_1 and φ_2 are said to be *oriented in the opposite way*.

A differentiable two-dimensional manifold M is said to be *orientable* if the set of all its admissible charts can be split into two classes in such a way that any two overlapping charts belonging to one class are oriented in the same way, and charts taken from different classes are oriented in the opposite way. We say that a definite orientation of M is specified, or in other words that M is oriented, if all the charts belonging to one of these classes are called right, and all the charts of the other class are called left.

Let M be a differentiable two-dimensional manifold, and $G \subset M$ an open subset of M. Then G as a topological space with the topology induced from M is obviously a two-dimensional manifold. We assume that $\varphi\colon U \to \mathbb{R}^2$ is an admissible chart in M such that $U \cap G$ is non-empty. Then the restriction of φ to the set $U \cap G$ is a chart in the two-dimensional manifold G. The totality of all charts in G that can be obtained in such a way is an atlas of class C^∞. We shall say that this atlas determines in G the structure of a differentiable manifold induced from the manifold M.

Any connected component of the two-dimensional manifold M is a two-dimensional manifold. If M is a differentiable manifold, then by the construction indicated above any connected component of M is turned into a differentiable two-dimensional manifold.

4.2. The Concept of a Two-Dimensional Riemannian Manifold. From an elementary course of differential geometry the reader will be familiar with such concepts as the first and second fundamental forms of a surface. The first fundamental form gives the possibility of calculating the lengths of curves lying on a surface, and the second fundamental form characterizes the bending of the surface in space. All the concepts relating to a surface that can be defined by means of the first fundamental form, and those properties of a surface that can be established by means of this fundamental form, determine the topic of the intrinsic geometry of a surface.

The concepts of a smooth and piecewise smooth path in a differentiable manifold. A two-dimensional Riemannian manifold is a differentiable manifold in which the length of a curve is defined. If the curve is sufficiently smooth, then its length can be calculated, starting from some differential quadratic form, as in the case of surfaces in three-dimensional Euclidean space.

Let $x(t) = (x_1(t), x_2(t))$, $a \leqslant t \leqslant b$, be a path on the plane $\mathbb{R}^2$. Then we shall say that x is a *smooth path of class C^r*, where $r \geqslant 1$ is an integer, if the functions

$x_1(t)$ and $x_2(t)$ have continuous derivatives $\frac{d^r x_1}{dt^r}(t)$ and $\frac{d^r x_2}{dt^r}(t)$ in $[a, b]$, and the vector $x'(t) = (x_1'(t), x_2'(t))$ is non-zero for all $t \in [a, b]$. We shall call the path $x(t)$, $a \leqslant t \leqslant b$, *piecewise smooth* of class C^r if we can find a finite sequence of points $a = t_0 < t_1 < \cdots < t_m = b$ such that the restriction of x to each of the intervals $[t_{i-1}, t_i]$ is a smooth path of class C^r. This means that the function $x(t)$ in each of the intervals (t_{i-1}, t_i) has a continuous derivative $x^{(r)}(t)$ and at the points t_i the function $x^{(r)}(t)$ has finite limits when t tends to t_i from the left or right. Hence it follows, in particular, that the first derivative $x'(t)$ also has limits to the left and right at these points. For the function $x'(t)$ these limits must be non-zero.

Let M be a differentiable two-dimensional manifold, and $\xi\colon [a, b] \to M$ a parametrized curve in M. We assume that there is an admissible chart $\varphi\colon U \to \mathbb{R}^2$ such that $\xi(t) \in U$ for all $t \in [a, b]$. We shall say that ξ is a *smooth* (*piecewise smooth*) path of class C^r if the path $x(t) = \varphi[\xi(t)]$, $a \leqslant t \leqslant b$, on the plane $\mathbb{R}^2$ is smooth (piecewise smooth) of class C^r. Let $\psi\colon V \to \mathbb{R}^2$ be any other admissible chart such that the point $\xi(t) \in V$ for all $t \in [a, b]$. Then the function $y(t) = \psi[\xi(t)]$ is defined. We obviously have $y(t) = \sigma[x(t)]$, where $\sigma = \varphi \circ \psi^{-1}$ is the transition function for the given charts φ and ψ. The function σ has all the partial derivatives of any order, and these derivatives are continuous. Hence it follows that $y(t)$ is a piecewise smooth path of class C^r. We should observe that the first derivatives of the components $y_1(t)$, $y_2(t)$ of the function y are expressed as follows:

$$y_1'(t) = \frac{\partial y_1}{\partial x_1} x_1'(t) + \frac{\partial y_1}{\partial x_2} x_2'(t), \quad y_2'(t) = \frac{\partial y_2}{\partial x_1} x_1'(t) + \frac{\partial y_2}{\partial x_2} x_2'(t).$$

Here the symbols y_1 and y_2 denote the components of the function σ. If $x'(t) \neq 0$, then since the Jacobian of the function σ is non-zero it follows from the equalities indicated that $y'(t) \neq 0$.

Definition of a two-dimensional Riemannian manifold. Let M be an arbitrary differentiable two-dimensional manifold. Then we shall say that a Riemannian geometry is specified in M, or briefly that M is a Riemannian manifold, if with any piecewise smooth path $\xi\colon [a, b] \to M$ of class C^1 there is associated a number $l(\xi; a, b)$ such that the following conditions are satisfied.

R1. For any $c \in (a, b)$ the following equality is satisfied:

$$l(\xi; a, b) = l(\xi; a, c) + l(\xi; c, b). \tag{4.1}$$

R2. For any admissible coordinate system $\varphi\colon U \to \mathbb{R}^2$ continuous functions $g_{ij}(x_1, x_2)$, $i, j = 1, 2$, are defined such that $g_{12}(x_1, x_2) = g_{21}(x_1, x_2)$ and for any $x \in G = \varphi(U)$ the quadratic form

$$g(x; z) \equiv \sum_{i=1}^{2} \sum_{j=1}^{2} g_{ij}(x_1, x_2) z_i z_j \tag{4.2}$$

in the variables z_1, z_2 is positive definite and for any piecewise smooth path $\xi\colon [a, b] \to M$ of class C^1 passing into the domain of definition of the given chart the following equality is satisfied:

$$l(\xi; a, b) = \int_a^b \sqrt{\sum_{i=1}^{2} \sum_{j=1}^{2} g_{ij}[x(t)]x_i'(t)x_j'(t)}\, dt. \tag{4.3}$$

The quantity $l(\xi; a, b)$ is called the *length of the path* $\xi\colon [a, b] \to M$ in the Riemannian manifold M. The value of $l(\xi; a, b)$ is completely determined by the conditions R1 and R2 for any piecewise smooth path ξ in the manifold M.

We assume that the manifold M is connected. Then for any two points $p, q \in M$ there is a piecewise smooth path in M that joins these points. The greatest lower bound of the lengths of such paths is denoted by $\rho_M(p, q)$ and is called the *distance* between the points p and q in the Riemannian space M. The function ρ_M of a pair of points of the two-dimensional manifold M defined in this way is a *metric*. We shall call it the *natural metric* of the given Riemannian manifold.

It is convenient to write the quadratic form (4.2) in the form

$$ds^2 = \sum_{i=1}^{2} \sum_{j=1}^{2} g_{ij}(x_1, x_2)\, dx_i\, dx_j. \tag{4.4}$$

This notation goes back to the infinitesimal representations of the time of the creation of mathematical analysis. We can give it a completely modern meaning by interpreting dx_i as a linear function such that for the vector $z = (z_1, z_2)$ we have $dx_i(z) = z_i$. We shall call the quadratic form (4.4) a representation of the line element of the manifold M by means of the chart φ. We shall call the quantity ds, that is, the square root of the quadratic form on the right-hand side of (4.4), the line element of the Riemannian geometry in the manifold M. We shall also call the quadratic form (4.4) the *metric tensor* of the Riemannian manifold M.

The ordinary Euclidean plane is a Riemannian manifold. In the orthogonal Cartesian coordinate system its line element is defined by $ds^2 = dx_1^2 + dx_2^2$. It is easy to construct other examples of two-dimensional Riemannian manifolds. For example, a sphere in $\mathbb{R}^3$, the surface of an infinite circular cylinder, and the Lobachevskij plane are two-dimensional Riemannian manifolds. Any two-dimensional surface in $\mathbb{R}^3$ that satisfies the regularity conditions adopted in differential geometry is turned into a Riemannian manifold if we introduce on it the metric induced from $\mathbb{R}^3$.

Properties of the natural metric of a Riemannian manifold. We show that the metric ρ_M in a Riemannian manifold M is compatible with the topology of this space. For this we first prove the following proposition.

Lemma 4.2.1. *Let $\varphi\colon U \to \mathbb{R}^2$ be an arbitrary admissible chart in the Riemannian manifold M, $p_0 \in U$, $x_0 = \varphi(p_0)$, $V = \varphi(U)$. Let B_r denote the disc $B(x_0, r)$ on the plane $\mathbb{R}^2$, and let $G_r = \varphi^{-1}(B_r)$. Then there are numbers $\delta > 0$ and $L < \infty$, $L \geqslant 1$ such that if $0 < r \leqslant \delta$, then*

$$B_M(p_0, r/L) \subset G_r \subset B_M(p_0, rL)$$

(the symbol $B_M(p_0, r)$ denotes a ball in the sense of the metric ρ_M in the manifold M).

Proof. Let $\delta > 0$ be such that the disc $B(x_0, 2\delta) \subset V$. Then the closed disc $\bar{B}(x_0, \delta)$ is contained in V. Since the disc $\bar{B}(x_0, \delta)$ is compact and the coefficients of the quadratic form $g(x, h)$ by means of which the functional $l(p; a, b)$ is defined are continuous, there is a number $L < \infty$ such that for any vector $h \in \mathbb{R}^2$ and any point $x \in \bar{B}(x_0, \delta)$ we have

$$\frac{1}{L^2}|h|^2 \leqslant g(x, h) \leqslant L^2|h|^2.$$

We show that the given L and δ are the required quantities.

Assume that $0 < r \leqslant \delta$. We take a point $p \in G_r$ arbitrarily. Let $x = \varphi(p)$, $x(t) = (1 - t)x_0 + tx$, $0 \leqslant t \leqslant 1$, $p(t) = \varphi^{-1}[x(t)]$. Suppose that $x(t)$ is a parametrized interval in $\mathbb{R}^2$ with ends x_0 and x. We have

$$\rho_M(p_0, p) \leqslant l(p; 0, 1) = \int_0^1 \sqrt{g[x(t), x'(t)]}\, dt$$

$$\leqslant L \int_0^1 |x'(t)|\, dt = L|x - x_0| < Lr.$$

Since $p \in G_r$ is taken arbitrarily, we have thus proved that $G_r \subset B_M(p_0, Lr)$.

Let $\xi(t)$, $a \leqslant t \leqslant b$, be an arbitrary piecewise smooth path such that $p(a) = p_0$, $q = p(b) \notin G$. There is a $c \leqslant b$ such that when $a \leqslant t \leqslant c$ the point $\xi(t) \in G_r$ and $\xi(c) \notin G_r$. Then, putting $x(t) = \varphi(\xi(t))$, we obtain

$$l(\xi; a, b) \geqslant l(p; a, c) = \int_a^c \sqrt{g[x(t), x'(t)]}\, dt$$

$$\geqslant \frac{1}{L}\int_a^c |x'(t)|\, dt \geqslant \frac{1}{L}|x(c) - x(a)| = \frac{r}{L}.$$

Hence it follows that $\rho_M(p, q) \geqslant r/L$ for any point $q \notin G_r$. Hence, if $\rho_M(p, q) < r/L$, we have $q \in G_r$, that is,

$$G_r \supset B_M(p_0, r/L).$$

This proves the lemma.

Corollary. *The natural metric ρ_M of the two-dimensional Riemannian manifold M is compatible with the natural topology of M.*

In order to prove the given assertion we first need to prove that any neighbourhood of an arbitrary point p_0 of the manifold M contains some disc $B_M(p_0, \varepsilon)$. The truth of this follows from the first inclusion of the lemma. Secondly, we need to prove that any disc, in the sense of the metric ρ_M, with centre p_0 contains some neighbourhood of the point p_0. This follows from the second inclusion, and thus the corollary is proved.

The following assertion is true.

Theorem 4.2.1. *The metric ρ_M on the two-dimensional Riemannian manifold M is the intrinsic metric. For any piecewise smooth path ξ: $[a, b] \to M$ of class C^1 in the manifold M, $l(\xi; a, b)$ is its length with respect to the metric ρ_M.*

We shall omit the proof of the theorem, as it is rather cumbersome (see Kobayashi and Nomizu (1963), Ch. IV).

The rule for transforming the coefficients of the metric tensor on going over to another coordinate system. In the classical handbooks on Riemannian geometry the rule for transforming the coefficients of the metric tensor is included in the definition of a Riemannian manifold. We show how to derive this rule, starting from the definition given here.

Let $\varphi: U \to \mathbb{R}^2$ and $\psi: U \to \mathbb{R}^2$ be two admissible coordinate systems with a common domain of definition U. Suppose that the differential quadratic forms

$$\sum_{i=1}^{2} \sum_{j=1}^{2} g_{ij}(x_1, x_2)\, dx_i\, dx_j, \quad \sum_{k=1}^{2} \sum_{l=1}^{2} h_{kl}(y_1, y_2)\, dy_k\, dy_l,$$

are representations of the line element of the Riemannian manifold by means of the charts φ and ψ respectively. Let us explain how the functions g_{ij} and h_{kl} are connected. Let $G_1 = \varphi(U)$, $G_2 = \psi(U)$. We take arbitrarily a point $y \in G_2$ and a vector $z \neq 0$ on the plane $\mathbb{R}^2$. We put $y(t) = y + tz$, $0 \leqslant t \leqslant u$. We shall assume that u is chosen so that $x(t) \in G_1$ for all $t \in [0, u]$. Suppose that $\xi(t) = \psi^{-1}[y(t)]$, $y(t) = \varphi[\xi(t)] = u[y(t)]$, where $u = \varphi \circ \psi^{-1}$ is the transition function for the given charts φ and ψ. Obviously $\xi(t)$, $0 \leqslant t \leqslant u$, is a piecewise smooth path of class C^1 lying in the domain U. We have

$$l(p; 0, u) = \int_0^u \sqrt{g[x(t), x'(t)]}\, dt,$$

on the other hand,

$$l(p; 0, u) = \int_0^u \sqrt{h[y(t), y'(t)]}\, dt.$$

Differentiating the given equalities with respect to u and putting $u = 0$, we obtain

$$g[x(0), x'(0)] = h(y, z).$$

We have $x = x(0) = \tau(y)$. The vector $\zeta = x'(0)$ is expressed in terms of the vector z by the formula $\zeta = d\tau(y, z)$. Here $d\tau(y)$ is a linear map – the differential of the function τ at the point y. Written in coordinates, the last equality takes the form

$$\zeta_1 = \frac{\partial x_1}{\partial y_1} \cdot z_1 + \frac{\partial x_1}{\partial y_2} \cdot z_2, \quad \zeta_2 = \frac{\partial x_2}{\partial y_1} \cdot z_1 + \frac{\partial x_2}{\partial y_2} \cdot z_2.$$

As a result we obtain

$$h(y, z) = g(x, d\tau(y, z)). \tag{4.5}$$

In expanded form this equality is

$$\sum_{k=1}^{2} \sum_{l=1}^{2} h_{kl}(y) z_k \cdot z_l = \sum_{k=1}^{2} \sum_{l=1}^{2} \sum_{i=1}^{2} \sum_{j=1}^{2} g_{ij}(x) \cdot \frac{\partial x_i}{\partial y_k} \frac{\partial x_j}{\partial y_l} \cdot z_k \cdot z_l.$$

Since the vector z is arbitrary, we conclude that

$$h_{kl}(y) = \sum_{i=1}^{2} \sum_{j=1}^{2} g_{ij}(x) \frac{\partial x_i}{\partial y_k} \cdot \frac{\partial x_j}{\partial y_l}, \tag{4.6}$$

where $x = \tau(y)$. This is the required representation of the coefficients of the quadratic form h_{kl}.

From (4.5) and (4.6), we obtain in turn

$$g(x, z) = h(y, d\tau^{-1}(y, z)), \tag{4.7}$$

where $\tau^{-1} = \psi \circ \varphi^{-1}$, $y = \tau^{-1}(x)$ and furthermore

$$g_{ij}(x) = \sum_{k=1}^{2} \sum_{l=1}^{2} h_{kl}(y) \cdot \frac{\partial y_k}{\partial x_i} \cdot \frac{\partial y_l}{\partial x_j}. \tag{4.8}$$

Properties of being approximately Euclidean in the small. Let p_0 be an arbitrary point of a two-dimensional Riemannian manifold. Then in a neighbourhood of this point we can define an admissible chart $\varphi: U \to \mathbb{R}^2$ such that the coefficients of the differential quadratic form that determines the line element of the manifold at the point $x_0 = \varphi(p_0)$ take the values $g_{11}(x_0) = 1$, $g_{12}(x_0) = g_{21}(x_0) = 0$, $g_{22}(x_0) = 1$ so at the point x_0 this quadratic form is $ds^2 = dx_1^2 + dx_2^2$.

The coefficients g_{ij} of the metric tensor of the manifold are continuous, and as $x \to x_0$, $g_{11}(x) \to 1$, $g_{22}(x) \to 1$, $g_{12}(x) = g_{21}(x) \to 0$. The Riemannian geometry determined by the line element $dx_1^2 + dx_2^2$ on the plane $\mathbb{R}^2$ is the usual Euclidean geometry. Therefore the result established here can be interpreted in the following way. In an infinitely small neighbourhood the Riemannian metric is Euclidean.

Area as a function of sets in a Riemannian manifold. On any differentiable two-dimensional Riemannian manifold we can define a totally additive set function S, called the area.

A set $A \subset M$ will be called a *Borel set* if for any admissible chart $\varphi: U \to \mathbb{R}^2$ in the manifold M the set $\varphi(A \cap U)$ is a Borel set.

A set A in a Riemannian manifold M is said to be *bounded* if the closure of A is compact.

Let $\mathfrak{B}(M)$ denote the totality of all Borel sets of the two-dimensional manifold M. The symbol $\mathfrak{B}_0(M)$ will denote the totality of all bounded Borel sets in M.

Let A be an arbitrary Borel set in the two-dimensional Riemannian manifold M. We assume that A is contained in the domain of definition of some admissible chart $\varphi: U \to \mathbb{R}^2$. Let $g_{11}, g_{12} = g_{21}, g_{22}$ be the coefficients of the differential quadratic form ds^2 with respect to this chart. Since the quadratic form $g(x, z) = g_{11}(x) \cdot z_1^2 + 2g_{12}(x) z_1 \cdot z_2 + g_{22}(x) \cdot z_2^2$ is positive definite, the quantity $g(x) = g_{11}(x) \cdot g_{22}(x) - [g_{12}(x)]^2$ is positive for all $x = (x_1, x_2)$ for which it is defined.

We put

$$S(A) = \iint_{\varphi(A)} \sqrt{g(x)}\, dx_1\, dx_2.$$

From the classical formula for changing the variables in a multiple integral and the rule for transforming the coefficients of a quadratic form on transition from one coordinate system to another it follows that the quantity $S(A)$ does not depend on the choice of local coordinate system in whose domain of definition A is contained.

If A is an arbitrary Borel set in M, then since M is a space with denumerable base A can be represented in the form

$$A = \bigcup_{m=1}^{\infty} A_m, \tag{4.9}$$

where each of the sets A_m is a Borel set and is contained in the domain of definition of some admissible chart, and the sets A_m are pairwise disjoint. We put

$$S(A) = \sum_{m=1}^{\infty} S(A_m).$$

It is easy to establish that the sum on the right-hand side does not depend on the choice of the representation (4.9) of the set A, so the given definition is reasonable.

We shall call the quantity $S(A)$, where A is a Borel set in a Riemannian manifold M, the *area of the set* A. A function of the set S defined here is completely additive, that is, for any sequence (A_m), $m = 1, 2, \ldots$, of pairwise disjoint Borel sets in M we have

$$S\left(\bigcup_{m=1}^{\infty} A_m\right) = \sum_{m=1}^{\infty} S(A_m).$$

The purely formal definition of the area of a set given here is justified by the fact that in the case when the Riemannian manifold is a smooth surface in the space $\mathbb{E}^3$ with the metric induced from $\mathbb{E}^3$, $S(A)$ exactly coincides with what we call the area of a set on a surface in differential geometry. We note that other more natural geometrical definitions of area in a Riemannian manifold lead to the same quantity $S(A)$.

4.3. The Curvature of a Curve in a Riemannian Manifold. Integral Curvature. The Gauss-Bonnet Formula. In the definition of a Riemannian manifold given in 4.2, for the coefficients g_{ij} in the representation of the line element we assume only that the coefficients are continuous. The main content of Riemannian geometry is the theory of curvature of Riemannian manifolds. The latter can be constructed only on the assumption that the functions g_{ij} have all the partial derivatives of not less than the second order, and all these derivatives are continuous. For simplicity we shall assume henceforth that the functions g_{ij} belong to the class C^∞, that is, each of them has all the partial derivatives of any order and these derivatives are continuous. If this condition is satisfied, we shall say that M is a Riemannian manifold of class C^∞.

Let M be a two-dimensional Riemannian manifold of class C^∞. We assume that M is connected. We shall denote the natural metric of M by ρ.

Theorem 4.3.1 (Kobayashi and Nomizu (1963), Ch. IV). *Any point X of a Riemannian manifold of class C^∞ has a neighbourhood U such that for any two points $Y, Z \in U$ there is a unique shortest curve of the manifold M that joins these points. Any shortest curve in a Riemannian manifold of class C^∞ is a curve of class C^∞.*

Curvature and turn of a simple arc. Let K be a smooth simple arc of class C^2 in a two-dimensional Riemannian manifold M. We assume that the arc K is oriented and along it there is specified a definite orientation of M. Then for each point X of the curve K we can define the number $k_l(X)$, which we shall call the *left curvature* of K at X. We put $k_r(X) = -k_l(X)$. The number $k_r(X)$ is called the *right curvature* of K at X. We do not give a formal definition of $k_l(X)$. From those properties of the curvature at a point of a curve that we shall give later there actually follows a way of calculating the curvature (the main property is the one contained in the Gauss-Bonnet theorem stated later). We note that if we change the orientation of the simple arc K, then $k_l(X)$ changes sign. In the same way, if under a fixed orientation of the simple arc we change the orientation of the manifold M along this arc, then $k_l(X)$ is also multiplied by -1.

If K is a shortest curve, then its curvature at each point is zero.

We assume that K is an oriented simple arc of class C^2 in a Riemannian manifold and that along K there is specified an orientation of M. Let $x(s)$, $0 \leqslant s \leqslant l$, be a parametrization of K, where the parameter s is arc length. We put

$$\kappa_l(K) = \int_0^l k_l[x(s)]\, ds, \quad \kappa_r(K) = -\kappa_l(K).$$

We shall call $\kappa_l(K)$ the *left turn* of K, and $\kappa_r(K)$ the *right turn* of K. All that we have said above about the behaviour of the curvature under a change of orientation of the curve K and a change of orientation of the manifold along K remains valid in connection with a turn.

The introduction of the two quantities $\kappa_l(K)$ and $\kappa_r(K)$, which differ only in sign, may seem odd, and it is usually not done in the standard textbooks on differential geometry. We shall define the left and right turns later for simple arcs in an arbitrary two-dimensional manifold of bounded curvature, and there the relation $\kappa_r(K) = -\kappa_l(K)$ may not be satisfied, generally speaking.

The Gaussian Curvature at a Point of a Manifold. For any point X of a two-dimensional Riemannian manifold M we can define a number $\mathcal{K}(X)$, the Gaussian curvature of the manifold M at the point X. Later we shall give formulae for calculating $\mathscr{K}(X)$ in some special local coordinate systems on the manifold. We note, however, that those properties of the curvature that we give later can be used, in principle, for its definition. (In the given case the main property is the Gauss-Bonnet formula.) With the help of $\mathscr{K}(X)$ we can introduce some additive set functions. Namely, let E be an arbitrary Borel set in M. Then if E is bounded (that is, the closure of E is compact), there is defined the integral

$$\iint_E \mathscr{K}(p)\, dS(p) = \omega(E).$$

We shall call $\omega(E)$ the *integral curvature* of the set E. In addition, we introduce the following quantities:

$$|\omega|(E) = \iint_E |\mathscr{K}(p)|\, dS(p),$$

$$\omega^+(E) = \iint_E \mathscr{K}^+(p)\, dS(p), \quad \omega^-(E) = \iint_E \mathscr{K}^-(p)\, dS(p).$$

Here $\mathscr{K}^+(p) = \max\{\mathscr{K}(p), 0\} = (|\mathscr{K}(p)| + \mathscr{K}(p))/2$, $\mathscr{K}^-(p) = \max\{-\mathscr{K}(p), 0\} = (|\mathscr{K}(p)| - \mathscr{K}(p))/2$. The quantities $|\omega|(E)$, $\omega^+(E)$ and $\omega^-(E)$ are defined for any Borel set E in the manifold M (the boundedness of E is not required). Infinite values of the integrals are allowed. The quantity $|\omega|(E)$ is called the *absolute integral curvature* of the set E, and $\omega^+(E)$ and $\omega^-(E)$ respectively the positive and negative parts of the curvature of E. For any E for which $\omega(E)$ is defined we have

$$\omega(E) = \omega^+(E) - \omega^-(E),$$
$$|\omega|(E) = \omega^+(E) + \omega^-(E).$$

The Gauss-Bonnet formula (Alekseevskij, Vinogradov and Lychagin (1988)). Here we give a relation connecting a turn of a curve and the integral curvature of a set.

We shall call a set G in a two-dimensional manifold M a *simple domain* if G is homeomorphic to the closed disc $\{(x, y) | x^2 + y^2 \leqslant 1\}$ on the plane $\mathbb{R}^2$. If G is a simple domain, then its boundary Γ is a simple closed curve.

Let G be a simple domain in a Riemannian manifold M, and Γ the boundary of G. We assume that the curve Γ is piecewise smooth of class C^2, that is, that Γ can be split into finitely many simple arcs that are smooth of class C^2. Let X_1, $X_2, \ldots, X_m, X_{m+1} = X_1$ be the points of Γ that bring about this splitting. Let Γ_j be the arc $[X_j, X_{j+1}]$ of this splitting, where $j = 1, 2, \ldots, m$. We specify arbitrarily a topological map φ of the domain G onto the disc $\bar{B}(0, 1)$ in $\mathbb{R}^2$. We orient Γ so that the following condition is satisfied. If $p(t)$, $a \leqslant t \leqslant b$, is a right parametrization of Γ, then the function $x(t) = \varphi[p(t)]$, $a \leqslant t \leqslant b$, defines in $\mathbb{R}^2$ the positive orientation of the boundary of the disc $\bar{B}(0, 1)$. In accordance with this, each of the arcs Γ_j is oriented. Along each of the arcs Γ_j we specify an orientation of M by means of the following agreement. Let U be a canonical neighbourhood of Γ_j and $\psi: U \to \mathbb{R}^2$ a right homeomorphism of U into $\mathbb{R}^2$. Let H be the connected component of the set $G \cap U$ containing Γ_j. Then $\psi \circ \varphi^{-1}$ is a topological map of the set $\varphi(M)$ that preserves the orientation. The orientation of M along Γ_j is completely determined by this condition. We shall say that it is an orientation of the arc Γ_j compatible with the specified homeomorphism $\varphi: G \to \mathbb{R}^2$. We note that the construction described here will be necessary when we consider general two-dimensional manifolds of bounded curvature.

For each j there is defined the left turn of the arc Γ_j. We shall also denote it by $\kappa_G(\Gamma_j)$ and call it the turn of the arc Γ_j on the side of the domain G. For each of the points X_j we define a number α_j, the *angle of the domain* G_j at this point. To this end, for each $j = 1, 2, \ldots, m$ we define a local coordinate system $\varphi_j: U_j \to \mathbb{R}^2$ such that $\varphi_j(X_j) = 0$ and the metric differential form of the manifold at the point $(0, 0)$ takes the form $dx_1^2 + dx_2^2$. We shall assume that U_j is sufficiently small and that $\varphi_j(U_j)$ is the disc $B(0, \delta)$. Then the intersection $U_j \cap \Gamma$ consists of two arcs contained in Γ_{j-1} and Γ_j, and the images of these arcs are smooth simple arcs starting from the point O and splitting the disc $B(0, \delta)$ into two domains B_1 and B_2. One of them (suppose it is B_1) is the image of $U_j \cap G$ under the map φ_j. We put α_j equal to the angle of the domain B_1 at the point O (see Fig. 16). We have $0 \leqslant \alpha_j \leqslant 2\pi$. The quantity α_j thus defined does not depend on the choice of special coordinate system.

Theorem 4.3.2 (Alekseevskij, Vinogradov and Lychagin (1988)). *Let G be a simple domain in a two-dimensional Riemannian manifold M of class C^∞. We assume that the boundary Γ of G is split by points $X_1, X_2, \ldots, X_m$ into smooth arcs of class C^2. Let α_j be the angle of G (defined as mentioned above) at the point X_j, and $\kappa_G(\Gamma_j)$ the turn on the side of G of the simple arc $\Gamma_j = [X_{j-1}X_j]$. Then we have*

$$\sum_{j=1}^{m} \kappa_G(\Gamma_j) + \sum_{j=1}^{m} (\pi - \alpha_j) = 2\pi - \omega(G)$$

(*the Gauss-Bonnet formula*)[1].

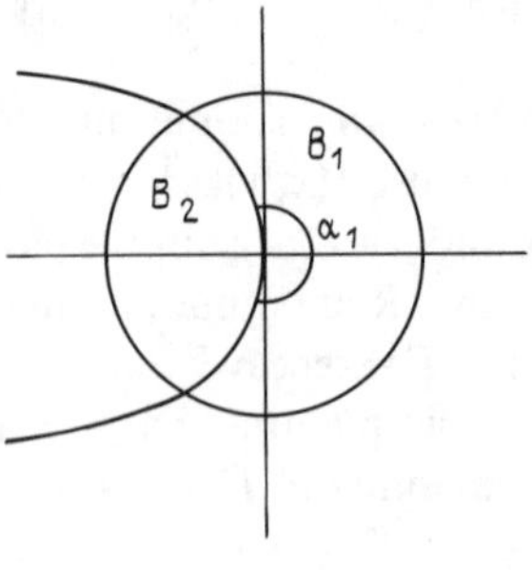

Fig. 16

[1] Encyclopaedia of Mathematical Sciences, vol. 28 (1991), p. 36

A *geodesic triangle* in the Riemannian manifold M is any domain $T \subset U$ whose boundary consists of three shortest curves successively joining the points A, B and C on the boundary of T and having no common points except the ends. These shortest curves form a simple closed curve that is piecewise smooth of class C^2. We denote by α, β and γ the angles of the boundary of the triangle T at the points A, B and C respectively measured from the side of T. The quantity $\alpha + \beta + \gamma - \pi = \delta(T)$ is called the *excess of the geodesic triangle T.*

Corollary 1. *For any geodesic triangle T in the Riemannian manifold M we have*

$$\delta(T) = \omega(A).$$

This result, in particular, enables us to find a way of calculating the Gaussian curvature at an arbitrary point of a two-dimensional Riemannian manifold M.

Corollary 2. *Let X be an arbitrary point in a Riemannian manifold M, and X_m, Y_m, Z_m, $m = 1, 2, \ldots$, sequences of points in M such that $X_m \to X$, $Y_m \to X$ and $Z_m \to X$ as $m \to \infty$ and for each m the points X_m, Y_m and Z_m do not lie on one shortest curve. Let T_m be the geodesic triangle bounded by the shortest curves joining the sequences X_m and Y_m, Y_m and Z_m, and finally Z_m and X_m. Then we have*

$$\mathscr{K}(X) = \lim_{m \to \infty} \delta(T_m)/\sigma(T_m).$$

Let us show how, by using the Gauss-Bonnet formula, we can define a turn of a simple arc in a Riemannian manifold. Let K be a simple arc in the manifold M. We assume that K is oriented and that along K there is specified an orientation of M. Let A be the beginning and B the end of K. We assume that we can find a finite sequence $X_0 = A, X_1, X_2, \ldots, X_m = B$ of points of the curve K, numbered in the order of disposition on K, such that each of the arcs $[X_{i-1}, X_i]$, $i = 1, 2, \ldots, m$, of K is a shortest curve. In this case we shall call K a *geodesic polygonal line* in the manifold M. The points X_i, $i = 0, 1, \ldots, m$, are called its *vertices*. For the polygonal line K we define two numbers $\kappa_l(K)$ and $\kappa_r(K)$, which we shall call the *left and right turns* of K respectively. Namely, G and H are arbitrary left and right semineighbourhoods of M. We denote by α_i, $i = 1, 2, \ldots, m - 1$, the angle at the point X_i of the domain G, and by β_i the angle of the domain H at the same point (see Fig. 17). Obviously $\alpha_i + \beta_i = 2\pi$. We put

$$\kappa_l(K) = \sum_{i=1}^{m-1} (\pi - \alpha_i), \quad \kappa_r(K) = \sum_{i=1}^{m-1} (\pi - \beta_i).$$

We have

$$\kappa_l(K) + \kappa_r(K) = 0.$$

The next assertion gives the answer to the question of how to define a turn of an arbitrary simple arc of class C^2.

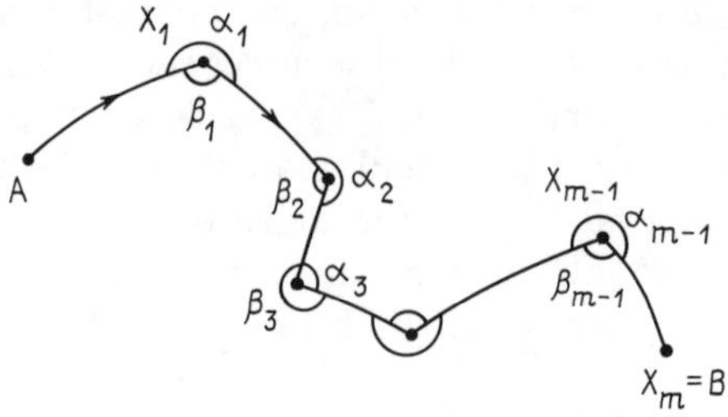

Fig. 17

Corollary 3. *Let K be a simple arc of class C^2 in a two-dimensional Riemannian manifold M. We assume that the arc K is oriented, A and B are the beginning and end of K, and along K there is specified an orientation of M. Let (L_m), $m = 1, 2, \ldots$, be a sequence of geodesic polygonal lines joining the end-points of K, converging to K on the left (right). Let α_m and β_m be the angles at A and B of the domain contained between L_m and K. If $\alpha_m \to 0$ and $\beta_m \to 0$ as $m \to \infty$, then*

$$\kappa_l(L_m) \to \kappa_l(K)$$

(respectively, $\kappa_r(L_m) \to \kappa_r(K)$ as $m \to \infty$).

Proof. We confine ourselves to the case when the polygonal lines L_m converge to K on the left. Let G_m be the domain included between K and L_m. We denotę by $\gamma_1, \gamma_2, \ldots, \gamma_n$ the angles of G_m at the vertices of L_m. The turn of each of the arcs into which L_m is split by its vertices is equal to zero, and applying the Gauss-Bonnet formula we obtain

$$\kappa_l(K) + (\pi - \alpha_m) + (\pi - \beta_m) + \sum_{i=1}^{n} (\pi - \gamma_i) = 2\pi - \omega(G_m).$$

Since G_m lies to the right of L_m,

$$\sum_{i=1}^{n} (\pi - \gamma_i) = \kappa_r(L_m) = -\kappa_l(L_m)$$

and as a result we obtain

$$\kappa_l(K) = \alpha_m + \beta_m - \omega(G_m) + \kappa_l(L_m).$$

By hypothesis $\alpha_m \to 0$, $\beta_m \to 0$ as $m \to \infty$. Since the L_m converge to K as $m \to \infty$, it follows that $\omega(G_m) \to 0$ as $m \to \infty$, and so $\kappa_l(K) = \lim_{m \to \infty} \kappa_l(L_m)$, as required.

4.4. Isothermal Coordinates in Two-Dimensional Riemannian Manifolds of Bounded Curvature. Let M be a two-dimensional Riemannian manifold, and $\varphi: U \to \mathbb{R}^2$ an admissible chart in M. We say that φ is an *isothermal coordinate system* in M if the line element of the manifold in this coordinate system has the form

$$ds^2 = \lambda(x, y)(dx^2 + dy^2).$$

We give an expression for some quantities connected with a two-dimensional Riemannian manifold in the isothermal coordinate system.

We have the following expression for the Gaussian curvature:

$$\mathscr{K} = -\frac{1}{2\lambda}\cdot \Delta \ln \lambda = -\frac{1}{2\lambda}\cdot\left(\frac{\partial^2 \ln \lambda}{\partial x^2} + \frac{\partial^2 \ln \lambda}{\partial y^2}\right).$$

Expressing $\ln \lambda$ in terms of K by the formula for solving the Poisson equation, we obtain

$$\ln \lambda(z) = \frac{1}{\pi}\iint_G \ln\frac{1}{|z-\zeta|}\cdot \mathscr{K}(\zeta)\cdot\lambda(\zeta)\,d\xi\,d\eta + h(z),$$

where $G = \varphi(U)$, and $h(z)$ is a harmonic function of the variable $z = x + iy$. The last equality can be rewritten as follows:

$$\ln \lambda(z) = \frac{1}{\pi}\cdot\iint_G \ln\frac{1}{|z-\zeta|}\,d\omega(\zeta) + h(z), \tag{4.10}$$

where $\omega(E)$ is a totally additive function, the *integral curvature of a set*.

The equality (4.10) is interesting in that it suggests an analytic approach to the solution of the problem of introducing two-dimensional manifolds of bounded curvature. For these the integral curvature is an arbitrary totally additive set function. It is natural to consider two-dimensional Riemannian manifolds with a line element which in a neighbourhood of each point admits the representation

$$ds^2 = \lambda(x, y)(dx^2 + dy^2),$$

where the function $\lambda(z)$, $z = x + iy$, admits a representation of the form (4.10) in which $\omega(E)$ is an arbitrary set function. It turns out that the general concept of a two-dimensional manifold of bounded curvature can actually be defined in this way. More details on this will be given below.

§ 5. Manifolds with Polyhedral Metric

5.1. Cone and Angular Domain. We first describe a general construction relating to arbitrary metric spaces. Suppose we are given a metric space (M, ρ). We denote the ray $\{x \in \mathbb{R} | x > 0\}$ of the number line $\mathbb{R}$ by $\mathbb{R}_0^+$. We construct the direct product $M \times \mathbb{R}_0^+$. Formally $M \times \mathbb{R}_0^+$ is the set of all pairs (x, t), where $x \in M, t > 0$. For the element $z = (x, t)$ of the set $M \times \mathbb{R}_0^+$ we put $r(z) = t$. To the set $M \times \mathbb{R}_0^+$ we adjoin an element O, for which we shall assume that $r(O)$ is equal to zero. The set $M \times \mathbb{R}_0^+$, augmented by the element O, will be denoted by Q_M. It is convenient to identify O with the set of all pairs of the form $(x, 0)$. For $z \in Q_M$ the number $r(z)$ is called the *polar radius* of the point z.

We now introduce in Q_M a metric ρ^*. Let z_1 and z_2 be two arbitrary elements of Q_M. If $z_1 = z_2$, we put $\rho^*(z, z) = 0$. If $z_1 = 0$, $z_2 \neq 0$, we put $\rho^*(z_1, z_2) = \rho^*(z_2, z_1) = r(z_2)$. Suppose that $z_1 = (x_1, r_1)$, $z_2 = (x_2, r_2)$, where $r_1 > 0$, $r_2 > 0$. If $\rho(x_1, x_2) \geqslant \pi$, we put $\rho^*(z_1, z_2) = r_1 + r_2$. If $\rho(x_1, x_2) < \pi$, then $\rho^*(z_1, z_2)$ is defined by the formula

$$\rho^*(z_1, z_2) = \sqrt{r_1^2 + r_2^2 - 2r_1 r_2 \cos \rho(x_1, x_2)}.$$

Obviously $\rho^*(z_1, z_2)$ in this case is equal to the side BC of the planar triangle ABC, for which $|AB| = r_1$, $|AC| = r_2$, $\angle BAC = \rho(x_1, x_2)$. It is easy to verify that $\rho^*(y_1, y_2)$ is the metric in Q_M. The metric space Q_M defined in this way, and also any space isometric to it, will be called a *cone over the metric space* (M, ρ). The point O is called the *vertex* of the cone Q_M.

Let a be an arbitrary point of the metric space (M, ρ). The totality of all points of the cone Q_M of the form $z = (a, r)$, where $r \geqslant 0$, will be called its *generator* corresponding to the point $a \in M$ and denoted by $\gamma(a)$. Let $(a, r_1) = z_1$ and $(a, r_2) = z_2$ be two arbitrary points of the generator $\gamma(a)$ of the cone Q_M. The set of all points $z \in Q_M$ of the form (a, r), where r lies between r_1 and r_2, is called the *interval of the generator* $\gamma(a)$ with ends z_1 and z_2. For any two points $z_1 = (a, r_1)$ and $z_2 = (a, r_2)$ we have $\rho^*(z_1, z_2) = |r_1 - r_2|$. Hence it follows that the length of the interval of the generator of Q_M with ends at the points $z_1 = (a, r_1)$ and $z_2 = (a, r_2)$ is equal to $|r_1 - r_2|$, and so any such interval is a shortest curve in the space Q_M. Let O be the vertex of the cone Q_M. For any point $z = (x, r) \in Q_M$ we have, by definition, $\rho^*(O, z) = r$. Let us specify arbitrarily a number $h > 0$. The set $Q_M(h)$ of all points $z = (x, r) \in Q_M$ for which $r = \rho^*(O, z) < h$ will be called a *finite cone* over the space (M, ρ) with length of generator equal to h. The set Γ_h of all points $z = (x, r)$ for which $r = \rho^*(O, z) = h$ is later called the *cross-section of the cone* Q_M at a distance h from its vertex.

If the metric of the space (M, ρ) is intrinsic, then (Q_M, ρ^*) is also a space with intrinsic metric.

If the metric spaces (M_1, ρ_1) and (M_2, ρ_2) are isometric, then the cones Q_{M_1} and Q_{M_2} are isometric spaces.

Later we shall require the following two special cases of the general construction described here.

We obtain the first case by choosing for M the interval $[a, b]$, $a < b$, of the number line $\mathbb{R}$. In this case for $x, y \in [a, b]$ we put $\rho(x, y) = |x - y|$. The cone Q_M corresponding to the case when M is the interval $[a, b]$ is called an *angular domain*, and the number $\theta = b - a$ is called the *magnitude* of this angular domain. If the intervals $M_1 = [a_1, b_1]$ and $M_2 = [a_2, b_2]$ have equal lengths, then the cones Q_{M_1} and Q_{M_2} are isometric. The angular domain whose magnitude is equal to θ will be denoted by $A(\theta)$. Obviously for any $\theta > 0$ the space $A(\theta)$ is homeomorphic to a half-plane. Let us define a map j of the cone $A(\theta) = Q_{[a,b]}$ into the plane $\mathbb{R}^2$ by putting for $(x, r) \in Q_{[a,b]}$

$$j(x, r) = (r \cos(x - a), r \sin(x - a)).$$

In a neighbourhood of any point $y = (x, r)$ of the cone $A(\theta)$ such that $r > 0$ the map j is topological. It is easy to verify that each curve in $A(\theta)$ is transformed by j into a curve of the same length, so it follows, in particular, that j is isometric in a neighbourhood of any point $y \in A(\theta)$ other than the vertex of the cone $A(\theta)$. In the case when $\theta < 2\pi$ the map j is one-to-one, and $j(A(\theta))$ is a planar angular domain, which in polar coordinates (r, φ), where r is the distance from the origin and φ is the polar angle, is defined by $0 \leqslant \varphi \leqslant \theta$. The generators $\gamma(a)$ and $\gamma(b)$ of $A(\theta)$ are called its *boundary rays*. If $\theta = \pi$, the cone $A(\theta)$ is isometric to a closed half-plane.

We obtain the second (special) case by choosing for M a circle in the plane $\mathbb{R}^2$. Let us specify a number $\theta > 0$ arbitrarily, and let C_θ be a circle in the plane E^2 with centre at the point $(0, 0)$ and such that its length is equal to θ. For arbitrary points $x, y \in C_\theta$ we put $\rho(x, y) = 0$ if $x = y$, and in the case when $x \neq y$ let $\rho(x, y)$ be the length of the shortest arc into which C_θ is split by the points x and y. Obviously $\rho(x, y)$ is the metric on C_θ. This metric is intrinsic. We shall call the cone over the metric space (C_θ, ρ) a *circular cone* and denote it by $Q(\theta)$. (In the case when $\theta \leqslant 2\pi$ the cone $Q(\theta)$ is isometric to the lateral surface of an infinite right-circular cone in the space $\mathbb{E}^3$.) The number θ is called the *total angle of the circular cone* $Q(\theta)$, and the difference $2\pi - \theta = \omega[Q(\theta)]$ is called its *curvature*. Any point of the cone $Q(\theta)$ other than its vertex has a neighbourhood isometric to a piece of the plane. If $\omega = 0$, then some neighbourhood of the vertex of the cone $Q(\theta)$ is also isometric to a flat domain and the space $Q(\theta)$ is isometric to the plane $\mathbb{E}^2$. If ω is non-zero, then no neighbourhood of the vertex of $Q(\theta)$ is isometric to a flat domain. To verify this let us consider the cross-section Γ_h of the cone $Q(\theta)$ at a distance h from its vertex. The set Γ_h is a simple closed curve whose length is equal to θh. If some neighbourhood of the point Q, the vertex of the cone $Q(\theta)$, were isometric to a flat domain, then for sufficiently small h the length of Γ_h would be equal to $2\pi h \neq \theta h$.

Suppose we are given a number $h > 0$. The set of all points x of the cone $A(\theta)$ at a distance less than h from its vertex will be called a *circular sector* of radius h with angle θ and denoted by $A(\theta, h)$. The set of all points of the circular cone $Q(\theta)$ at a distance less than h from its vertex will be called a *finite circular cone* with total angle θ and length of generator h and denoted by $Q(\theta, h)$.

Any metric space isometric to a circular (finite circular) cone will also be called a circular cone (finite circular cone respectively).

Shortest curves in the cones $A(\theta)$ and $Q(\theta)$. Let M be an arbitrary metric space with intrinsic metric. We construct a cone Q_M and take an arbitrary generator $\gamma(a)$ of it, where $a \in M$. Let $X = (a, r_1)$ and $Y = (a, r_2)$ be two arbitrary points of $\gamma(a)$, where $r_1 < r_2$. The set of all points $Z = (a, r)$, where $r_1 \leqslant r \leqslant r_2$, is a simple arc $L = [X, Y]$, which we shall call an *interval* of the given generator. It is easy to verify that if $Z \notin [X, Y]$, then $\rho^*(X, Y) < \rho^*(X, Z) + \rho^*(Z, Y)$ (a strict inequality). Hence it follows that if the path $X(t)$, $p \leqslant t \leqslant q$, in the space Q_M, joining the points X and Y, is not a parametrization of the interval $[X, Y]$, then its length $l(X; p, q) > \rho^*(X, Y) = r_2 - r_1$. This enables us to conclude that if the points X and Y in the cone Q_M lie on one generator of Q_M, then they are joined in Q_M by a unique shortest curve – the interval $[X, Y]$.

We now consider a special case of the cone $Q(\theta)$. Let X and Y be two arbitrary points of it. If one of them is the vertex of $Q(\theta)$, then X and Y lie on one generator, and so in the given case they are joined by a unique shortest curve – an interval of this generator. Let $X = (a, r_1)$, $Y = (b, r_2)$, where a and b are points of the circle C_θ. We shall assume that $a \neq b$, since otherwise X and Y lie on one generator of $Q(\theta)$. The points a and b split C_θ into simple arcs Γ_1 and Γ_2. Let θ_1 and θ_2 be the lengths of these arcs. We denote by Q_1 the set of all points (x, r) of $Q(\theta)$ such that $x \in \Gamma_1, r \geqslant 0$, and by Q_2 the totality of all $(x, r) \in Q(\theta)$ for which $x \in \Gamma_2$. In the induced intrinsic metric the domain Q_1 is isometric to the angular domain $A(\theta_1)$, and the domain Q_2 is isometric to $A(\theta_2)$. Under the isometry, to the vertex of $Q(\theta)$ there correspond the vertices of $A(\theta_1)$ and $A(\theta_2)$. The distance between the points a and b on C_θ is equal to $\min\{\theta_1, \theta_2\}$, and if $\theta_1 \geqslant \pi$, $\theta_2 \geqslant \pi$, then $\rho^*(X, Y) = r_1 + r_2$. The simple arc composed of the intervals $[X, O]$ and $[O, Y]$ of generators of $Q(\theta)$ has length equal to $r_1 + r_2 = \rho^*(X, Y)$, and so it is the shortest curve joining X and Y. It is easy to verify that for any point $Z \neq O$ we have $\rho^*(X, Z) + \rho^*(Z, Y) > r_1 + r_2$ (a strict inequality). Hence it follows that the shortest curve joining X and Y is unique. In the case when $\min\{\theta_1, \theta_2\} < \pi$ we have $\rho^*(X, Y) < r_1 + r_2$, so in this case the simple arc composed of the intervals $[X, O]$ and $[O, Y]$ is not a shortest curve. Suppose, for definiteness, that $\min\{\theta_1, \theta_2\} = \theta_1$. The set Q_1 is isometric to an angular domain on the plane equal to θ_1. Let us map Q_1 isometrically onto this angular domain. To the points X and Y there correspond points X' and Y' on the sides of the angular domain. It is obvious that the simple arc that goes over to the interval joining X' and Y' under this map is a shortest curve in the cone $Q(\theta)$, joining X and Y. If $\theta_2 = \theta_1 < \pi$, then applying our arguments to the domain Q_2 we deduce that in $Q(\theta)$ there are two distinct shortest curves joining X and Y. One of them goes into Q_1 and the other into Q_2 and the vertex of the cone lies inside the domain bounded by these shortest curves.

From what we have said here about shortest curves in $Q(\theta)$ there follows a circumstance in relation to which the case $\theta < 2\pi$ is qualitatively different from the case $\theta \geqslant 2\pi$. Namely, if $\theta < 2\pi$, then a shortest curve in the cone $Q(\theta)$ cannot pass through its vertex (that is, the vertex of the cone cannot be an interior point of the shortest curve). If $\theta \geqslant 2\pi$, then in the cone there are shortest curves passing through its vertex.

Let us investigate the arbitrariness with which a shortest curve in $Q(\theta)$, where $\theta \geqslant 2\pi$, that ends at the vertex of the cone, can be extended beyond this vertex so that the resulting curve remains a shortest curve. In the case $\theta = 2\pi$ the cone $Q(\theta)$ is isometric to a plane and here everything is clear. We shall assume that $\theta > 2\pi$. Let $X = (a, r)$, where $r > 0$, be an arbitrary point of $Q(\theta)$. On the circle C_θ we lay off from the point a on different sides of it the arcs $[a, b]$ and $[a, c]$ such that the length of each of them is equal to π, and let Γ be the arc $[b, c]$ of C_θ, $a \notin [b, c]$. If $Y = (p, s)$, where $p \in \Gamma$, $s > 0$, then the simple arc composed of intervals of the generators $[X, O]$ and $[O, Y]$ is a shortest curve. If $p \notin \Gamma$, then this simple arc is not a shortest curve. Thus, moving from X to O along the generator and then from O along any generator $l(p)$, where $p \in \Gamma$, we obtain a shortest curve in $Q(\theta)$ passing through O, and all the shortest curves that are extensions of the shortest curve $[X, O]$ beyond O can be constructed in this way.

The cone $Q(\theta)$ can be obtained from $A(\theta)$ by pasting together the extreme rays of the cone $A(\theta)$, identifying points of these rays that are equidistant from the vertex of the cone.

We assume that there is specified a finite collection of cones $A(\theta_i)$, $A(\theta_i) = Q_{[a_i, b_i]}$, $i = 1, 2, \ldots, m$, where the intervals $[a_i, b_i]$ are pairwise disjoint. Pasting together the angular domains $A(\theta_i)$ along extreme rays, we obtain from them a metric space that is isometric to $A(\theta)$, where $\theta = \theta_1 + \theta_2 + \cdots + \theta_m$. Formally in a given case the pasting is carried out as follows. Let $L_i = l(a_i)$ and $L_i' = l(b_i)$ be boundary rays of the angular domain $A(\theta_i)$. For each $i = 1, 2, \ldots, m$ we paste the boundary ray L_i' of the cone $A(\theta_i)$ to the ray L_{i+1} of the cone $A(\theta_{i+1})$, identifying points of these rays that are equidistant from the origin.

From what we have said it follows, in particular, that in the general case both $A(\theta)$ and $Q(\theta)$ can be obtained by pasting together finitely many flat convex angular domains.

5.2. Definition of a Manifold with Polyhedral Metric. Let M be an arbitrary two-dimensional manifold with boundary (see Alekseevskij, Vinogradov and Lychagin (1988)). We assume that M is connected and an intrinsic metric $\rho(X, Y)$ is defined in it. Then we shall say that *the metric $\rho(X, Y)$ is polyhedral* and the metric space (M, ρ) is a two-dimensional polyhedron if the conditions A and B formulated below are satisfied. The first of these conditions concerns the structure of the metric space (M, ρ) close to its interior points, and the second close to the boundary points of M.

A. For any interior point X of the manifold M we can find a neighbourhood that admits an isometric map φ onto some circular cone $Q(\theta, h)$, where $\theta > 0$, $h > 0$, such that $\varphi(X)$ is the vertex of this cone (θ and h depend on the point X, of course).

Let X be an interior point of M, and V a neighbourhood of this point such that condition A is satisfied. The quantity θ does not depend on the choice of the neighbourhood V. This follows from the fact that it can be defined in a way that does not require consideration of any special neighbourhoods of X. Namely, let $S(X, r)$ be the circle of radius r in (M, ρ) with centre X; $l(X, r)$ is its length. Then for sufficiently small $r > 0$, namely for $r < h$, we have $l(X, r) = \theta r$ and so $\theta = \lim_{r \to 0} l(X, r)/r$. The quantity θ is called the *total angle* at the point X of the manifold M and will be denoted by $\theta(X)$. We put $\omega(X) = 2\pi - \theta(X)$. The quantity $\omega(X)$ is called the *curvature of* M at the point $X \in M$. The point X is called a *vertex of* M if its curvature is non-zero. We note that if a point $Y \in V$ is distinct from X, then by what we said above about the structure of cones the point Y has a neighbourhood isometric to a disc on the plane, that is, to a cone $Q(2\pi, \delta)$, and so the curvature at the point Y is zero. Hence it follows, in particular, that

V does not contain vertices of M other than X, so all the points of the set of vertices of M are isolated.

B. For any boundary point X of the manifold M we can find a neighbourhood V of this point that admits an isometric map φ onto some circular sector $A(\theta, h)$ (the numbers $\theta > 0$ and $h > 0$ depend on the point X) such that $\varphi(X)$ is a vertex of the sector $A(\theta, h)$.

Let X be a boundary point of the manifold M, and V a neighbourhood of it for which condition B is satisfied. The number θ does not depend on the choice of V; this is established in exactly the same way as in the case of interior points of M. We put $\theta(X) = \theta$, $\kappa(X) = \pi - \theta$. The quantity $\kappa(X)$ is called the *turn of the boundary of the polyhedron M* at the point X. $\theta(X)$ is the *total angle of the polyhedron* at the point X. The point X is called a *boundary vertex of the polyhedron* if $\kappa(X) \neq 0$. If a point $Y \in V$ is a boundary point and distinct from X, then some neighbourhood of Y is isometric to a half-disc on the plane, that is, to a circular sector $A(\pi, \delta)$, where $\delta > 0$ and so it is not a boundary vertex of M. We deduce that V does not contain boundary vertices other than X. If $Y \in V$ is an interior point of M, then some neighbourhood of Y is isometric to a disc, and hence Y is not a vertex of M, so the set V also does not contain vertices of the polyhedron M.

A neighbourhood V of a point X of a polyhedron M that satisfies condition A in the case when X is an interior point of M, and condition B in the case when X is a boundary point, will be called a *canonical neighbourhood* of X.

A remark about terminological character. The word "vertex", used without adjoining the word "boundary", implies that the point to which it is applied is an interior point of the given polyhedron.

A polyhedron M will be called *complete* if the metric space (M, ρ) is complete (that is, any fundamental sequence of points in it is convergent).

Any compact set on a polyhedron contains finitely many vertices. This follows from the fact that by a theorem of Borel a compact set is covered by finitely many canonical neighbourhoods.

Numerous examples of two-dimensional manifolds with polyhedral metric can easily be constructed. Suppose, for example, that M is the boundary of a bounded convex polyhedron D in the space $\mathbb{E}^3$, endowed with the metric induced from $\mathbb{E}^3$. As a topological space, M is homeomorphic to the sphere $S(0, 1)$ (a topological map of M onto the sphere can be obtained by choosing arbitrarily an interior point P of the polyhedron D and projecting its boundary onto the sphere $S(P, 1)$ by rays starting from P). If $X \in M$ is an interior point of some face of a convex polyhedron M, then some neighbourhood of it is isometric to a disc on the plane. If X is an interior point of an edge of the polyhedral surface M, then some neighbourhood of it in M splits into two parts, each of which is a flat half-disc, and they have a common diameter. This neighbourhood is obviously isometric to a disc on the plane, so in this case the point X is not a vertex of a manifold M with polyhedral metric. Let X be a vertex of the polyhedron D. Consider the ball $B(X, \varepsilon)$ in the space $\mathbb{E}^3$. For sufficiently small $\varepsilon > 0$ the intersection $B(X, \varepsilon) \cap M$ is the union of finitely many circular sectors that adjoin each other along radii starting from X. This neighbourhood is isometric to the circular cone $Q(\theta, \varepsilon)$, where θ is the sum of the angles of the sectors at the point X. By well-known results of the theory of convex bodies, $\theta < 2\pi$. Thus the vertices of the convex polyhedron D are vertices of M as a manifold with polyhedral metric. The curvature at all the vertices of the manifold is positive.

Here we mention the following fact. Let X be a vertex of a convex polyhedron D. A plane P passing through X will be called the *supporting plane of the polyhedron D at the point X* if D is contained in one of the two closed half-spaces for which P serves as the common boundary. A vector v will be called an *outward normal vector* of D at X if it is orthogonal to some supporting plane P at X, directed into the half-space bounded by this plane that does not contain D, and the length of the vector v is equal to 1. We shall assume that the outward normal vectors are laid off from the fixed point O in $\mathbb{E}^3$. The set of their ends lies on the sphere $S(0, 1)$ and is called the *spherical image of the point X*. The following assertion is true.

Theorem 5.2.1. *The spherical image of a vertex of a convex polyhedron is a convex spherical polygon on the sphere $S(0, 1)$ whose area is equal to the curvature of the vertex X.*

The polygon mentioned in Theorem 5.2.1 is obtained as follows.

Let $H_1, H_2, \ldots, H_m$ be all the faces of the polyhedron D that meet at the point X, and v_i the outward normal vector orthogonal to the plane of the face H_i. The points $v_1, v_2, \ldots, v_m$ on the sphere $S(0, 1)$ are the vertices of a polyhedron – the spherical image of the point X – and there are no other vertices of it.

A proof of the theorem is given in Aleksandrov (1950c). It is useful to compare this theorem with a classical result of differential geometry, according to which the area of the spherical image of a set on a surface is equal to the integral curvature of this set.

A cone and an angular domain are obviously manifolds with a polyhedral metric. An infinite right circular cylinder and a Möbius surface are polyhedra. In the last two cases the curvature at each point of the polyhedron is zero.

A theorem on the triangulation of a manifold with polyhedral metric. The concept of a manifold with polyhedral metric can also be defined in another way, namely, the following assertion is true.

Theorem 5.2.2. *Let M be a two-dimensional manifold with boundary. We assume that M is connected and is endowed with a polyhedral metric. Then M admits a triangulation $\mathbf{K}$ such that any triangle $T \in \mathbf{K}$ is isometric to a triangle on the plane $\mathbb{E}^2$.*

We just give an outline of the proof.

First of all we observe that for any interior point $X \in M$ we can find a domain G homeomorphic to a closed disc such that G is split by shortest curves starting from X into parts, each of which is isometric to a planar triangle. If X

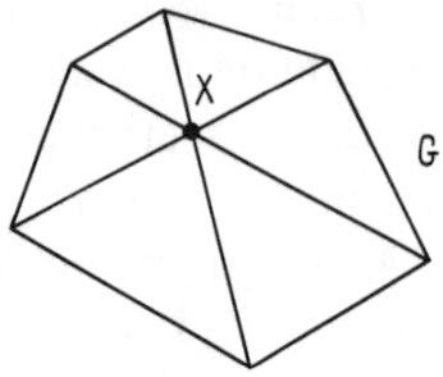

Fig. 18

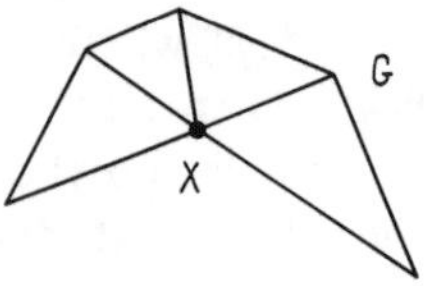

Fig. 19

is an interior point of M, then X is also an interior point of G, and if $X \in \partial M$, then X is a boundary point of G. (In the first case the splitting of G into triangles looks like that shown in Fig. 18, and in the second case in Fig. 19). A domain G satisfying these conditions will be called a *T-neighbourhood* of the point X. The existence of a T-neighbourhood of any point $X \in M$ is established by means of constructions relating to a cone.[2]

Next we construct a finite or denumerable sequence $G_1, G_2, \ldots, G_\nu, \ldots$ of subsets, each of which is a T-neighbourhood of some point $X \in M$ such that any compact subset of M intersects finitely many sets G_ν (in particular, if M is compact, the sequence G_ν is finite). We subdivide each of the domains G_ν into domains isometric to planar triangles. As a result we obtain a covering of M by some sequence of triangles T_μ, $\mu = 1, 2, \ldots$, such that any compact subset of M intersects finitely many triangles T_μ. For an arbitrary point $X \in M$ we denote by $Q(X)$ the intersection of all the triangles T_μ that contain X. The set $Q(X)$ is either a one-point set or a shortest curve or is isometric to a planar convex polygon. Subdividing into triangles all the domains $Q(X)$ that are isometric to polygons, we obtain the required triangulation of M.

Pasting together of manifolds with polyhedral metric. The following proposition is true.

Theorem 5.2.3. *Let* (M_ν), $\nu = 1, 2, \ldots$, *be a finite or denumerable set of two-dimensional manifolds with polyhedral metric. We assume that for this set there is specified a pasting rule that satisfies all the conditions of* 3.2, *and* M *is a manifold with intrinsic metric that arises as a result of the pasting. Then* M *is also a two-dimensional manifold with polyhedral metric.*

---

[2] Let X be an interior point of M, V a canonical neighbourhood of X, and $Q(\theta, h)$, $\theta = \theta(X)$, a finite circular cone to which V is isometric. The cone $Q(\theta, h)$ is split by its generators into finitely many circular sectors, so the angle of any of them is less than $\min(\pi, \theta/2)$. On each of these generators we specify some point. We obtain a finite set of points $Y_1, Y_2, \ldots, Y_m, Y_{m+1} = Y_1$. We shall assume that they are numbered so that Y_1 and Y_{i+1} lie on the sides of one circular sector. Let S_i be the sector bounded by the generators $\mathcal{O}Y_i$ and $\mathcal{O}Y_{i+1}$, and θ_i the angle of this sector. Since $\theta_i < \theta/2$, $\theta_i < \pi$, the shortest curve in the cone $Q(\theta, h)$ joining any two points on its sides goes inside the sector. Hence it is obvious that S_i is isometric to a sector of a disc on the usual plane $\mathbb{E}^2$ and the part T_i of it bounded by the shortest curves $[\mathcal{O}, Y_i]$, $[\mathcal{O}, Y_{i+1}]$ is isometric to a planar triangle. We thus obtain a finite collection of triangles $T_1, T_2, \ldots, T_m$, whose union G' constitutes a neighbourhood of the vertex of the cone $Q(\theta, h)$. Under an isometric map of $Q(\theta, h)$ onto V the set G' is transformed into a domain G having exactly those properties that we need. For the case when X is a boundary point of M the constructions are similar.

The proof of the theorem reduces to a consideration of neighbourhoods of points of M that arise as a result of pasting together boundary points of the manifolds M_ν. Suppose, say, that a point X is obtained by pasting together points $X_i \in M_{\nu_i}$, $i = 1, 2, \ldots, m$. In M_{ν_i} the point X_i has a neighbourhood isometric to some circular sector $A(\theta_i, h)$. By pasting together these neighbourhoods we obtain a neighbourhood of X. Since the pasting together of the circular sectors along their generators gives as a result either a circular sector or a circular cone, we deduce that X has a neighbourhood isometric to either a circular cone $K(\theta, h)$ or a circular sector $A(\theta, h)$. We have proved that M is a manifold with polyhedral metric.

Here we mention that, in particular, having pasted a manifold M with polyhedral metric to a second copy M' of it so that points of the boundary that correspond under an isometric map of M to M' are identified, we again obtain a two-dimensional polyhedron $\tilde{M}$ (in the terminology introduced above, this is a twice covered polyhedron M). The polyhedron $\tilde{M}$ does not have boundary points, and M is isometric to some subdomain of $\tilde{M}$ endowed with the induced metric. We assume that a point $X \in \tilde{M}$ is obtained by pasting together points $X' \in \partial M$ and $X'' \in \partial M'$. Then $\theta(X') = \theta(X'')$ and $\theta(X) = 2\theta(X')$, and $\omega(X) = 2\kappa(X')$.

Polygonal lines on a polyhedron. Let M be an arbitrary two-dimensional polyhedron. By Theorem 2.3.2 any point $X \in M$ has a neighbourhood, any two points of which can be joined by a shortest curve.

Let L be a shortest curve on M, and X an arbitrary interior point of L. If X is a vertex of M, then $\omega(X)$ is negative; this follows from what we said in 5.1 about shortest curves in a cone. If X is a boundary vertex, then the rotation of the boundary at this point is negative.

We shall call a simple arc L an *interval on the polyhedron M* if L is a shortest curve and no internal point of it is either a vertex or a boundary vertex of the polyhedron.

Any shortest curve on a polyhedron is a compact set and therefore contains finitely many vertices. Hence it follows that a shortest curve on a polyhedron consists of finitely many intervals.

A simple arc L on a manifold M is called a *polygonal line* if we can find on it a finite sequence of points $X_0, X_1, \ldots, X_m$, where X_0 and X_m are the end-points of L, numbered in the order of disposition on L, such that each of the arcs $[X_{i-1}, X_i]$ is a shortest curve. If L is a polygonal line, then, subdividing the shortest curves $[X_{i-1} X_i]$ into intervals, we deduce that L is the union of finitely many intervals.

We shall call a set K on a polyhedron M a *simple polygonal line* if K is a simple curve in the sense of the definition in 2.1 (that is, K is a closed subset of M and K is homeomorphic to either a circle or an interval of the number line) and any simple arc contained in K is a polygonal line.

The following assertion is true.

Theorem 5.2.4 (Aleksandrov and Zalgaller (1962)). *Let M be a two-dimensional manifold with polyhedral metric, and Q a connected closed subset of*

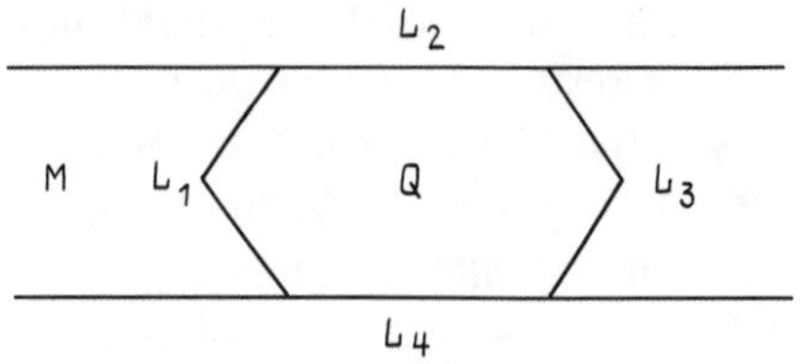

Fig. 20

M. We assume that Q is a two-dimensional manifold with boundary in the topology induced from M, and that any component of the boundary of Q is a simple polygonal line in M. Let ρ_Q be the induced metric in Q. Then the metric space (Q, ρ_Q) is a two-dimensional polyhedron.

Remark. Under the conditions of the theorem the boundary of Q, as a two-dimensional manifold, generally speaking does not coincide with the boundary of Q as a subset of M.

The proof of the theorem reduces to a consideration of the structure of neighbourhoods of the boundary points of Q (see Fig. 20).

The area of a set on a polyhedron. On any manifold with polyhedral metric there is defined a set function called the area or two-dimensional Lebesgue measure. Let us specify a triangulation **K** of a polyhedron M such that each of the triangles $T_1, T_2, \ldots, T_\nu, \ldots$ occurring in it is isometric to a planar triangle. A set $A \subset T_\nu$ is said to be *measurable* if the image of A is a measurable set under an isometric map of T_ν into a plane. We denote by $\sigma(A)$ the two-dimensional Lebesgue measure of the image of A under an isometric map of T_ν into a plane and call it the *area* or *two-dimensional Lebesgue measure* of the set A. Let $A \subset M$ be an arbitrary set. Then A is said to be measurable if $A \cap T_\nu$ is a measurable set for each ν and we put

$$\sigma(A) = \sum_{\nu} \sigma(A \cap T_\nu).$$

It is not difficult to show that the class of measurable sets and the function σ do not depend on the choice of triangulation **K** of the polyhedron M.

5.3. Curvature of a Set on a Polyhedron. Turn of the Boundary. The Gauss-Bonnet Theorem. For two-dimensional manifolds with polyhedral metric we define a set function that can be regarded as an analogue of the concept of the integral curvature of a set in a Riemannian manifold.

Let M be a two-dimensional manifold with polyhedral metric. A set $A \subset M$ is said to be *bounded* if the closure of A is compact. The totality of all bounded Borel sets on a polyhedron M will be denote by $\mathfrak{B}_0(M)$. If a set $A \subset M$ is bounded, then by a theorem of Borel its closure is covered by finitely many canonical neighbourhoods, and so A contains finitely many vertices. We shall denote the totality of all Borel sets in M by $\mathfrak{B}(M)$.

For an arbitrary number $x \in \mathbb{R}$ we put $x^+ = x$, $x^- = 0$ for $x \geqslant 0$ and $x^+ = 0$, $x^- = -x$ for $x < 0$. Obviously we always have $x^+ - x^- = x$, $x^+ + x^- = |x|$. Let X be a vertex of M. We put $\omega^+(X) = [\omega(X)]^+$, $\omega^-(X) = [\omega(X)]^-$, $|\omega|(X) = |\omega(X)|$. The numbers $\omega^+(X)$ and $\omega^-(X)$ are called the *positive and negative parts of the curvature of the point X.*

Let A be an arbitrary Borel set on a polyhedron. We put

$$\omega^+(A) = \sum_{X \in A} \omega^+(X), \quad \omega^-(A) = \sum_{X \in A} \omega^-(X),$$

$$|\omega|(X) = \sum_{X \in A} |\omega|(X) = \omega^+(A) + \omega^-(A).$$

The summation on both sides is carried out over the set of all vertices of the polyhedron that belong to A. Infinite values of the sum are allowed. The numbers $\omega^+(A)$ and $\omega^-(A)$ are called *the positive and negative parts of the curvature of the set A*, and $|\omega|(A)$ is called its *absolute curvature.*

If A is a bounded set in M, then, as we mentioned above, A contains finitely many vertices of M, and so the sum

$$\omega(A) = \sum_{X \in A} \omega(X)$$

is defined and finite. It is called the *curvature of the set A.*

Let us consider some examples. If a polyhedron M is a cone $Q(\theta)$, then it has a unique vertex and the curvature of a set $A \subset Q(\theta)$ is equal to zero if the vertex O of the cone $Q(\theta)$ does not belong to A, and is equal to $2\pi - \theta$ if $O \in A$. In particular, $\omega(M) = 2\pi - \theta$.

Let M be the surface of a cube. Then M has eight vertices. The total angle at each of them is equal to $3\pi/2$, so it follows that all the curvatures of the vertices of the cube are equal to $\pi/2$, and so $\omega(M) = 8(\pi/2) = 4\pi$. Similarly, if M is the surface of a regular tetrahedron, then M has four vertices and the total angle at each of them is equal to $3(\pi/3) = \pi$. In accordance with this the curvatures of all the vertices of the tetrahedron are equal to π and $\omega(M) = 4\pi$. We see that in both cases the curvature of the polyhedron M is equal to the same number, namely 4π. This observation is not accidental. Moreover, the following general assertion is true.

Theorem 5.3.1. *Let M be a compact two-dimensional manifold without boundary and with polyhedral metric. Then*

$$\omega(M) = 2\pi\chi(M),$$

where $\chi(M)$ is the Euler characteristic of M.

Proof. We specify arbitrarily a triangulation $\mathbf{K}$ of the polyhedron M such that each triangle of $\mathbf{K}$ is isometric to a planar triangle. The existence of such a triangulation follows from Theorem 5.2.2. Let n_0 be the number of vertices of $\mathbf{K}$, n_1 the number of edges, and n_2 the number of faces. Then according to Euler's formula $n_0 - n_1 + n_2 = \chi(M)$. We observe that $2n_1 = 3n_2$, from which we obtain $\chi(M) = n_0 - \frac{1}{2}n_2$. Let T_i, $i = 1, 2, \ldots, n_2$, be all the triangles that form the triangulation **K**. We denote the angles of the triangle T_i by α_i, β_i and γ_i. We have $\alpha_i + \beta_i + \gamma_i = \pi$ for any i. Hence

$$\pi n_2 = \sum_{i=1}^{n_2} (\alpha_i + \beta_i + \gamma_i).$$

We split the last sum into groups so that in each of these groups there occur the terms that are the angles at one vertex of the polyhedron. The sum of the angles corresponding to a vertex X of the triangulation **K** is equal to $\theta(X) = 2\pi - \omega(X)$. Hence we conclude that $\pi n_2 = 2\pi n_0 - \omega(M)$, and so $\omega(M) = 2\pi(n_0 - \frac{1}{2}n_2)$. As we showed above, $n_0 - \frac{1}{2}n_2 = \chi(M)$, so the theorem is proved.

To each boundary vertex X of M there corresponds a number $\kappa(X)$, the turn of the boundary at this point. This enables us to define some additive functions specified on the boundary of the polyhedron. Let $\mathfrak{B}(\partial M)$ denote the totality of all Borel subsets of ∂M, and $\mathfrak{B}_0(\partial M)$ the totality of all bounded Borel subsets of ∂M. For $E \in \mathfrak{B}_0(\partial M)$ we put

$$\kappa(E) = \sum_{X \in E} \kappa(X).$$

For $E \in \mathfrak{B}(\partial M)$ let

$$|\kappa|(E) = \sum_{X \in E} \kappa(X), \quad \kappa^+(E) = \sum_{X \in E} [\kappa(X)]^+, \quad \kappa^-(X) = \sum_{X \in E} [\kappa(X)]^-.$$

We shall call $\kappa(E)$ the *turn of the boundary on the set* E, and $|\kappa|(E)$, $\kappa^+(E)$ and $\kappa^-(E)$ the absolute turn, the positive part of the turn and the negative part of the turn of the boundary on the set E respectively.

If, in particular, the boundary of M is compact, then there is defined a number $\kappa(\partial M)$, the turn of the boundary of M. The number $\kappa(\partial M)$ is an analogue of integral geodesic curvature.

The next assertion follows from Theorem 5.3.1.

Theorem 5.3.2 (the Gauss-Bonnet theorem for polyhedra). *Let M be a two-dimensional polyhedron. We assume that M is a compact manifold and the boundary of M is not empty. Then*

$$\omega(M) + \kappa(\partial M) = 2\pi\chi(M),$$

where $\chi(M)$ is the Euler characteristic of M.

Proof. Let $\tilde{M}$ be a twice convered manifold M. Then $\chi(\tilde{M}) = 2\chi(M)$. In fact, we specify arbitrarily a triangulation **K** of M. The manifold $\tilde{M}$ is obtained by pasting together two different copies of M. From the triangulation of M there naturally arises a triangulation $\tilde{\mathbf{K}}$ of $\tilde{M}$. Let n_0 be the number of vertices, n_1 the number of edges, n_2 the number of faces of **K**, n_0' the number of vertices of **K** belonging to the boundary of M, and n_1' the number of edges of this triangulation contained in ∂M. Then obviously the number $\tilde{n}_0$ of vertices of $\tilde{\mathbf{K}}$ is equal to $2n_0 - n_0'$, the number $\tilde{n}_1$ of edges of this triangulation is equal to $2n_1 - n_1'$, and the number $\tilde{n}_2$ of faces of $\tilde{\mathbf{K}}$ is equal to $2n_2$. Each connected component of the boundary of M is a simple curve. Hence it follows that on it there lie as many edges of the triangulation as vertices, and so $n_0' = n_1'$. This enables us to conclude that

$$\chi(\tilde{M}) = \tilde{n}_0 - \tilde{n}_1 + \tilde{n}_2 = (2n_0 - n_0') - (2n_1 - n_1') + 2n_2$$

$$= 2(n_0 - n_1 + n_2) = 2\chi(M).$$

To any interior vertex X of a polyhedron M there correspond two vertices of $\tilde{M}$, and at each of them the curvature is equal to $\omega(X)$. To each boundary vertex X there corresponds one vertex of $\tilde{M}$, whose curvature is equal to $2\kappa(X)$. Hence it follows that $\omega(\tilde{M}) = 2\omega(M) + 2\kappa(\partial M)$. According to Theorem 5.3.1, $\omega(\tilde{M}) = 2\pi\chi(\tilde{M}) = 4\pi\chi(M)$, and so the theorem is proved.

5.4. A Turn of a Polygonal Line on a Polyhedron. Let us specify arbitrarily a two-dimensional manifold M with polyhedral metric.

Let L be a polygonal line on a polyhedron M; L is a simple arc. We assume that L does not contain boundary points of M. We orient L in an arbitrary way and specify along L a definite orientation of the manifold M. We have thus defined the concepts of the left-hand and right-hand sides of L. With each interior point X of L we associate two numbers, which we shall call the *left and right angles* at a point of L. Let us describe the construction by means of which they are determined. Let A be the beginning and B the end of L, and $X_0 = A$, $X_1, \ldots, X_{m-1}, X_m = B$ the points of L such that each of the arcs $[X_{i-1}X_i]$ is an interval. (The points X_i are assumed to be numbered in the order of disposition on L.) We assume that X lies inside one of the arcs $[X_{i-1}X_i]$, $i = 1, 2, \ldots, m$. Then some sufficiently small neighbourhood of the point X is isometric to a disc on the plane, and under an isometric map into the plane the part of L lying in this neighbourhood goes over to an interval. In this case we assume that the left and right angles at X are equal to π. We now assume that X is one of the points X_i, $0 < i < m$. Let V be a canonical neighbourhood of X_i. We assume that this neighbourhood is sufficiently small so that the intersection $V \cap L$ consists of intervals $(X'X_i]$, $[X_iX'')$, where X' lies between X_{i-1} and X_i, and X'' is between X_i and X_{i+1}. The polygonal line L splits V into two components, one of which lies to the left of L (that is, it is contained in some left half-neighbourhood of L). We denote this component by V_l. The other component $V \backslash L$, which we denote by V_r, lies to the right of L. Under an isometric map of V onto the circular cone $Q(\theta, \varepsilon)$, where $\theta = \theta(x)$, $\varepsilon > 0$, the arcs $[X'X_i]$, $[X_iX'']$ go into some generators of this cone, and the domains V_l and V_r are mapped into sectors of the cone $Q(\theta, \varepsilon)$. Let $\theta_l(X, L)$ be the angle of the sector corresponding to V_l, and $\theta_r(X, L)$ the angle of this sector corresponding to V_r. Obviously $\theta_l(X, L)$ and $\theta_r(X, L)$ do not depend on the choice of the neighbourhood V. We shall call $\theta_l(X, L)$ the angle at X on the left-hand side of L (or to the left of L), and $\theta_r(X, L)$ the angle at X on the right-hand side of L (or to the right of L). Obviously $\theta_l(X, L) + \theta_r(X, L) = \theta(X)$. We put $\kappa_l(X, L) = \pi - \theta_l(X, L)$, $\kappa_r(X, L) = \pi - \theta_r(X, L)$. We shall call $\kappa_l(X, L)$ and $\kappa_r(X, L)$ the left and right *turns of the polygonal line* L at the point X, respectively. We have $\kappa_l(X, L) + \kappa_r(X, L) = 2\pi - \theta(X) = \omega(X)$. In particular, we deduce that if the point $X = X_i$ is not a vertex of the polyhedron M, then $\kappa_r(X, L) = -\kappa_l(X, L)$. In the case when X does not coincide with one of the points X_i, $\kappa_l(X, L) = \kappa_r(X, L) = 0$. We put

$$\kappa_l(L) = \sum_{i=1}^{m-1} \kappa_l(X_i, L), \quad \kappa_r(L) = \sum_{i=1}^{m-1} \kappa_r(X_i, L).$$

We shall call $\kappa_l(L)$ and $\kappa_r(L)$ the *left and right turns* of L, respectively. The sum $\kappa_l(L) + \kappa_r(L)$ is equal to the sum of the curvatures at the vertices of the polyhedron that are interior points of L, that is, this sum is equal to the curvature of the open arc $(AB) = L \setminus \{A, B\}$. We thus have

$$\kappa_l(L) + \kappa_r(L) = \omega[(AB)]. \tag{5.1}$$

If L is a shortest curve, then it follows from what we said in 5.1 about shortest curves on a cone that $\theta_l(X, L) \geqslant \pi$, $\theta_r(X, L) \geqslant \pi$ at each point $X \in L$, and so $\kappa_l(X, L) \leqslant 0$, $\kappa_r(X, L) \leqslant 0$. In particular, we deduce that $\kappa_l(L) \leqslant 0$ and $\kappa_r(L) \leqslant 0$. We note that in the case of polyhedra, generally speaking, we cannot assert that a turn of a shortest curve is equal to zero, as this holds for shortest curves in a Riemannian manifold. It is easy to verify this by considering a shortest curve on a cone $K(\theta)$, where $\theta > 2\pi$, passing through its vertex O. It is obvious from the arguments of 5.1 that the right angle and also the left angle of the shortest curve at the vertex of the cone can take any values from π to $\theta - \pi$; according to this, the left and right turns of a shortest curve at a vertex of an open polygon can take any values from the interval $(2\pi - \theta, 0) = (\omega(0), 0)$.

Let us mention one consequence of Theorem 5.3.2. Let L be an oriented polygonal line on a polyhedron M, contained in an open set $U \subset M$ homeomorphic to an open disc in $\mathbb{R}^2$. We denote by A and B the beginning and end of L. Along the arcs L we specify a definite orientation of M. Let K be an oriented open polygon with beginning A and end B that does not have any other points in common with L and is contained in U. We assume that K lies to the left of L. Let G be a domain between K and L, homeomorphic to a closed disc, and G^0 the totality of all its interior points. Then

$$\kappa_l(L) + \kappa_r(K) = \alpha + \beta - \omega(G^0), \tag{5.2}$$

where α and β are the angles of G at A and B.

For the proof it is sufficient to consider the manifold obtained if we introduce the induced metric in G and apply Theorem 5.3.2. The Euler characteristic of a disc is equal to 1. The boundary of G consists of the arcs K and L. Since L lies to the right of K, at each boundary vertex of G lying inside K the turn of the boundary of G is equal to $\kappa_r(X, K)$. At each boundary vertex of G lying inside L the turn of the boundary of G is equal to $\kappa_l(X, L)$. At A and B the turns are equal to $\pi - \alpha$ and $\pi - \beta$. Hence we conclude that $\kappa(\partial G) = 2\pi - \alpha - \beta + \kappa_l(K) + \kappa_r(L)$. The equality (5.2) now follows directly from Theorem 5.3.2.

Let K and L be open polygonal lines satisfying all the conditions listed above. We have $\kappa_r(K) = \omega(K \setminus \{A, B\}) - \kappa_l(K)$. Observing that $(K \setminus \{A, B\}) \cup G^0 = G \setminus L$, we obtain the following relation:

$$\kappa_l(L) = \kappa_l(K) + \alpha + \beta - \omega(G \setminus L). \tag{5.3}$$

Similarly we obtain

$$\kappa_r(K) = \kappa_r(L) + \alpha + \beta - \omega(G \setminus K). \tag{5.4}$$

5.5. Characterization of the Intrinsic Geometry of Convex Polyhedra. Here we give a result that is in a certain sense the beginning of the whole problem area considered in the present article. We have in mind a theorem of A.D. Aleksandrov on the realization of a convex polyhedron with a given metric (see Aleksandrov (1948a)). In differential geometry there was known a problem posed by Weyl in 1918. If R is a closed convex surface in three-dimensional Euclidean space that satisfies the usual regularity conditions in differential geometry, then by introducing in R the metric induced from $\mathbb{E}^3$ we obtain a two-dimensional Riemannian manifold homeomorphic to a sphere and such that at each point of it the Gaussian curvature is non-negative. We now assume that we are given a two-dimensional Riemannian manifold M homeomorphic to a sphere and such that at each point of it the Gaussian curvature is non-negative. *Weyl's problem* is as follows: can we assert that there is a closed convex surface R such that M is isometric to this surface or, as we shall say later, on which there is realized a given Riemannian metric defined on a sphere? In Weyl's original statement analytic Riemannian manifolds were considered, that is, it was assumed that the transition functions for admissible charts are real-analytic, and the coefficients of the metric tensor are real-analytic functions. In addition, it was assumed that the Gaussian curvature is positive everywhere. In this statement, Weyl's problem was solved by H. Lewy (Lewy (1935)).

Aleksandrov (Aleksandrov (1948a)) took a purely geometrical approach to the solution of Weyl's problem. The main part of it forms an analogue of the Weyl-Lewy theorem for convex polyhedra. Namely, Aleksandrov proved the following proposition.

Theorem 5.5.1 (Aleksandrov (1948a)). *Let M be a two-dimensional manifold with polyhedral metric, homeomorphic to a sphere. Then for M to be isometric to the surface of a closed convex polyhedron in $\mathbb{E}^3$ it is necessary and sufficient that the curvature of each vertex of $\overline{M}$ should be non-negative.*

It is necessary to state the following remark making the formulation of the theorem more precise. To the list of convex polygons we have to add doubly covered flat convex polygons—polyhedra that are degenerate in a certain sense.

Now let ρ be an arbitrary Riemannian metric of positive curvature on the sphere. Then we can construct a sequence (ρ_n), $n = 1, 2, \ldots$, of polyhedral metrics of positive curvature that converges uniformly to ρ as $n \to \infty$. By Theorem 5.5.1, for each n there is a closed convex polyhedron R_n on whose surface the metric ρ_n is realized (that is, such that the sphere with metric ρ_n is isometric to the surface of R_n). From the sequence of polyhedra (R_n) we can extract a subsequence (R_{n_k}), $n_1 < n_2 < \cdots < n_k < \ldots$, that converges to some convex body R. The surface of this body is the required one: on it there is realized the specified Riemannian metric ρ.

To obtain a complete proof of the Weyl-Lewy theorem by the method described, however, some additional work is required. The fact is that in this theorem we consider analytic Riemannian metrics and prove that they are realized on convex surfaces that are also analytic. The limit of the sequence of convex polyhedra can be an arbitrary convex body. A priori such a situation is not excluded, when a given analytic Riemannian metric can be realized not only for the surface whose existence follows from Lewy's theorem but also on some other surfaces. The completion of the given approach to a solution of Weyl's problem is contained in deep research by Pogorelov on the unique determination of convex surfaces and on the degree of regularity of a convex surface depending on how regular its intrinsic metric is (Pogorelov (1969)).

In connection with the method adopted by Aleksandrov for the solution of Weyl's problem, there naturally arises the problem of examining all intrinsic metrics on a sphere, each of which is the limit of polyhedral metrics of positive curvature. The manifolds with intrinsic metric that arise in this way are special cases of two-dimensional manifolds of bounded curvature. Namely, they are manifolds whose curvature as a set function is non-negative.

5.6. An Extremal Property of a Convex Cone. The Method of Cutting and Pasting as a Means of Solving Extremal Problems for Polyhedra. Here we shall consider two-dimensional manifolds with polyhedral metric, homeomorphic to a closed disc. Later we shall state a theorem from which it follows that for a wide class of extremal problems about such polyhedra the solution is achieved on polyhedra of sufficiently special type. For the latter the corresponding extremal problem reduces essentially to a problem from elementary geometry.

A two-dimensional polyhedron R will be called a *convex cone* if it is homeomorphic to a closed disc, it has at most one vertex, and at each boundary vertex of R the turn is positive and the curvature of R is non-negative.

If a convex cone does not have vertices, which is allowed by the definition, then it is isometric to a convex polygon on a plane. It is clear that conversely any such polygon, in the sense of the given definition, is a convex cone. Let D be an arbitrary bounded convex polygon on a plane in $\mathbb{E}^3$, and O a point lying outside this plane and such that the foot of the perpendicular drawn from it to the plane of D belongs to the polygon. Then the lateral surface of the pyramid with vertex O and base D is a convex cone.

Let OAB and $O'A'B'$ be two equal triangles on a plane, where $|OA| = |O'A'|$, $|OB| = |O'B'|$ and $|AB| = |A'B'|$, and the angles at A and B are acute. Pasting these triangles together along the sides OA and $O'A'$, OB and $O'B'$ so that the points O and O' are identified, we obtain a polyhedron, which is a convex cone.

One can show that all possible types of convex cones are exhausted by the examples mentioned.

Let K_1 and K_2 be rectifiable simple closed curves in the metric spaces M_1 and M_2, and φ a map of K_1 onto K_2. We shall say that φ maps K_1 onto K_2 with the lengths of arcs preserved if φ maps K_1 onto K_2 one-to-one and takes any arc of K_1 into an arc of K_2 of the same length.

Suppose we are given metric spaces (M_1, ρ_1) and (M_2, ρ_2). We shall call a map $\varphi: M_1 \to M_2$ *contracting* if for any two points $X, Y \in M_1$

$$\rho_2[\varphi(X), \varphi[Y)] \leqslant \rho_1(X, Y).$$

Theorem 5.6.1. (Reshetnyak (1961a)). *Let R be a two-dimensional polyhedron homeomorphic to a closed disc and such that $\omega^+(R) < 2\pi$. Then there is a convex cone Q and a map $\varphi: Q \to R$ such that the following conditions are satisfied:*

1) *the curvature of Q does not exceed $\omega^+(R)$;*
2) *φ is a contracting map;*
3) *the boundary of the cone Q is transformed by the map φ into the boundary of R with the lengths of arcs preserved.*

The theorem shows that if $\omega^+(R) < 2\pi$, then a polyhedron of the type considered here can be deformed in such a way that the distances between its points are not decreased, the length of the boundary remains fixed, and the whole curvature proves to be collected at one point. The positive part of the curvature is not increased, and the negative part becomes zero.

The properties of the cone Q defined in Theorem 5.6.1 are described by the following theorem.

Theorem 5.6.2 (Reshetnyak (1961a)). *Let Q and R be polyhedra homeomorphic to a closed disc. We assume that there is a contracting map φ of Q onto R such that the boundary of Q is mapped onto the boundary of R with lengths preserved. Then*

a) *if some arc $K \subset \partial R$ is a shortest curve, then the arc $\varphi^{-1}(K)$ is also a shortest curve;*
b) *the lengths of the curves ∂R and ∂Q are the same;*
c) *the area of R does not exceed the area of Q;*
d) *$\kappa^+(K) \leqslant \kappa^+(\varphi(K))$ for any simple arc $K \subset \partial Q$.*

Theorems 5.6.1 and 5.6.2 turn out to be useful for solving different kinds of extremal problems. Here we give an example of a problem of this kind. The given theorems admit a generalization to the case of arbitrary two-dimensional manifolds of bounded curvature. Other examples in which they find application are more appropriately considered later in a section devoted to extremal problems in manifolds of bounded curvature.

Let R be a polyhedron homeomorphic to a disc, and $X_1, \ldots, X_n$ all the boundary vertices. Each of the arcs $[X_1X_2], [X_2X_3], \ldots, [X_nX_1]$ is an interval. Let $l_1, \ldots, l_n$ be the lengths of these intervals.

Theorem 5.6.3 (Aleksandrov (1945)). *In the set of all polyhedra R homeomorphic to a disc and such that $\omega^+(R) \leqslant \omega_0 < 2\pi$, the boundary of which consists of intervals whose lengths are $l_1, l_2, \ldots, l_n$, the greatest area is that of a polyhedron that is a convex cone whose curvature is equal to ω_0 and whose vertex is equidistant from all its boundary vertices.*

The proof of the theorem is as follows. Let R be a polyhedron satisfying the conditions of the theorem, Q the convex cone corresponding to R in accordance with Theorem 5.6.1, and $Y_1, Y_2, \ldots, Y_n$ the points that belong to the boundary of Q and are transformed into the boundary vertices $X_1, X_2, \ldots, X_n$ of R by the contracting map φ mentioned in Theorem 5.6.1. None of the arcs $[Y_1 Y_2], \ldots, [Y_{n-1} Y_n], [Y_n Y_1]$ contains boundary vertices of Q, because otherwise the rotation of this arc would be positive, which is impossible by virtue of assertion d) of Theorem 5.6.2, according to which $\kappa^+([Y_{i-1} Y_i]) \leqslant \kappa^+([X_{i-1} X_i]) = 0$. Thus the boundary of Q consists of intervals whose lengths are equal to $l_1, \ldots, l_n$.

We have $0 \leqslant \omega(Q) \leqslant \omega^+(R) \leqslant \omega_0$. By assertion c) of Theorem 5.6.2 the area of Q is not less than the area of R. Thus Theorem 5.6.3 reduces to the case when Q is a convex cone. For this case the proof is carried out by means of elementary geometry. Cutting the cone along the shortest curve joining its vertex to one of the boundary vertices, we obtain a polyhedron P isometric to a flat polyhedron $OZ_1 Z_2 \ldots Z_n Z_1'$ constructed as follows (see Fig. 21). The points Z_1 and Z_1' lie on the sides of the angle $Z_1 O Z_1'$ at an equal distance from the point O. The value of this angle is equal to the total angle at the vertex of the cone Q and so it is not less than $2\pi - \omega_0$, and the open polygon $Z_1 Z_2 \ldots Z_n Z_1'$ goes inside it. We have $|Z_1 Z_2| = l_1, |Z_2 Z_3| = l_2, \ldots, |Z_n Z_1'| = l_n$. The area of the polyhedron P is equal to the area of the cone Q. The problem reduces to that of finding, among all such polyhedra P, the one that has greatest area. This is a problem of elementary geometry. It is easy to establish that the area of P is maximal when the angle $Z_1 O Z_1'$ is equal to $2\pi - \omega_0$, and the points $Z_1, Z_2, \ldots, Z_n, Z_1'$ are at the same distance from the point O. Pasting the extremal polyhedron P along the sides OZ_1 and OZ_1', we obtain a convex cone constructed as described in Theorem 5.6.3, and the theorem is proved.

The proof of Theorem 5.6.1 uses the method of cutting and pasting suggested by Aleksandrov. In its main features it consists in using the following two devices.

We assume that some domain G on the polyhedron R is isometric in the induced metric to a planar non-convex quadrangle $H = ABCD$ with non-convex vertex D (see Fig. 22). We regard the vertex D as lying opposite to A and assume that the quadrangle H is endowed with the induced metric. On the plane we construct a triangle $T = A'B'C'$ with sides $A'B' = AB$, $A'C' = AC$, $B'C' = BD + CD$. Such a triangle exists because obviously $AB + BC > BD + DC$.

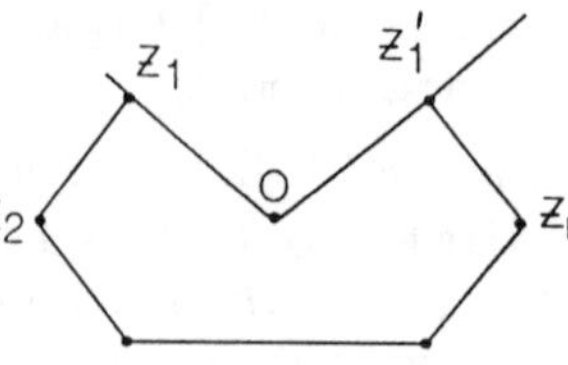

Fig. 21

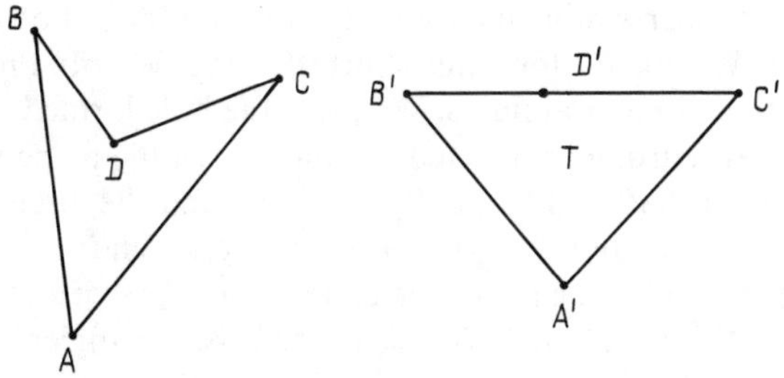

Fig. 22

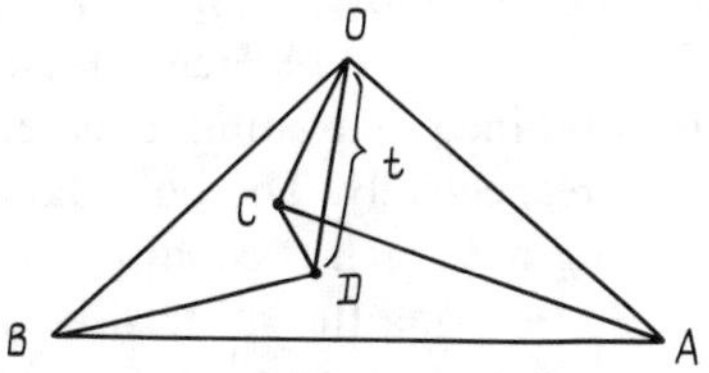

Fig. 23

The first device, which constitutes the method of cutting and pasting, consists in cutting out from R a domain G and instead of it pasting in the triangle T. The sides $[A'B']$, $[A'C']$ and $[B'C']$ of the triangle must be pasted to the arcs corresponding to the sides $[AB]$, $[AC]$ and the arc $[BC]$ on the boundary of H. In the polyhedron R' which is obtained as a result the total angle at the vertex corresponding to the point D turns out to be less than that of the original polyhedron R. By virtue of this the given method can be used to remove vertices with negative curvature. By means of this we also get rid of boundary points at which the turn of the boundary is negative.

The triangle T in the construction we have described has a contracting map onto H under which the boundary of T is mapped into the boundary of H with arc lengths preserved. The existence of such a map is established by the following construction. In the space $\mathbb{E}^3$ we construct a pyramid whose base is the quadrangle H, and the point D is the foot of the perpendicular drawn from the vertex of the pyramid to the plane of H (Fig. 23). Let t be the height of the pyramid, and $\theta(t)$ the total angle at its vertex. As $t \to 0$, $\theta(t)$ tends to a limit equal to $\pi + \alpha$, where $\alpha > \pi$ is the angle of H at D. Obviously $\theta(t) \to 0$ as $t \to \infty$. Since $\pi + \alpha > 2\pi$ and $\theta(t)$ is a continuous function of t, there is a t_0 such that $\theta(t_0) = 2\pi$. When $t = t_0$ the lateral surface of the given pyramid is obviously isometric to the triangle T. The map of the orthogonal projection of the lateral surface of the pyramid corresponding to $t = t_0$ is the required contracting map of T onto H.

The second device enables us to go over to the case when the positive part of the curvature of a polyhedron is concentrated at one point. It consists in the following. Let X and Y be two vertices of the polyhedron R such that at each of them the curvature is positive, and $\omega(X) + \omega(Y) < 2\pi$. We join X and Y by a shortest curve L. For simplicity we assume that L does not contain boundary points of R. We cut R along this shortest curve. We obtain a polyhedron homeomorphic to a circular annulus. We paste the hole formed in R as a result of the cut by a polyhedron constructed as follows. On the plane we construct a triangle $X'Y'Z'$ such that $|X'Y'| = \rho(X, Y) = l(L)$, and the angles at X' and Y' are equal to $\alpha = \frac{1}{2}\omega(X)$ and $\beta = \frac{1}{2}\omega(Y)$ respectively. Since $\alpha + \beta < \pi$, such a triangle exists. Let $X''Y''Z''$ be a triangle equal to $X'Y'Z'$ with $|X'Y'| = |X''Y''|$, $|X'Z'| = |X''Z''|$, $|Y'Z'| = |Y''Z''|$. We paste the given triangles along the sides starting from Z' and Z''. We paste $X'Z'$ to $X''Z''$ and $Y'Z'$ to $Y''Z''$ so that Z' and Z'' are identified. As a result we obtain a polyhedron Q with a single vertex Z obtained by pasting together Z' and Z''. The total angle at its vertex Z is equal to $2(\pi - \alpha - \beta) = 2\pi - \omega(X) - \omega(Y)$. Hence the curvature at Z is equal to $\omega(X) + \omega(Y)$. The points obtained by pasting together X' and X'', Y' and Y'' will be denoted by X and Y respectively. The boundary of Q is split into two parts by X and Y, and the length of each is equal to $\rho_R(X, Y) = l(L)$. Now we paste Q to the polyhedron R cut along the shortest curve L so that one of the arcs $[XY]$ forming the boundary of Q is pasted to one bank of the cut, and the second to the other bank. The point X on Q must be pasted to the point X in R, and Y on Q to the point Y in R. As a result we obtain a polyhedron R'. The total angle in R' at X will be equal to $2\pi - \omega(X) + 2\alpha = 2\pi$ and similarly the total angle in R' at Y is also equal to 2π. Thus the points X and Y have ceased to be vertices of the polyhedron. The curvature at Z is equal to $\omega(X) + \omega(Y)$. It is not difficult to see that no new vertices with positive curvature arise. Thus the curvature of X and Y turns out to be collected at one point – the point Z. The resulting polyhedron R' has a contracting map onto R. To obtain this it is sufficient to be able to construct a contracting map of Q onto an interval of length $\rho_R(X, Y)$ under which each arc contained in one of the arcs $[XY]$ into which the boundary of Q is divided by X and Y goes over to an arc of the same length. Without difficulty the given construction can be extended to the case when the shortest curve L joining the points X and Y contains boundary points of R.

5.7. The Concept of a *K*-Polyhedron. A polyhedron in the sense of the definition given above is a two-dimensional manifold with intrinsic metric, pasted together from some set of polyhedra taken on the usual Euclidean plane. In some questions it turns out to be useful to consider polyhedra pasted together from finitely many polygons lying on a surface of constant curvature.

Suppose we are given a real number K. The symbol Σ_K will denote a sphere in $\mathbb{E}^3$ of radius $r = 1/\sqrt{K}$ in the case $K > 0$, the usual Euclidean plane $\mathbb{E}^2$ in the case $K = 0$, and the Lobachevskij plane with Gaussian curvature K when $K < 0$.

We first define the concept of a cone of class K (K is an arbitrary real number) over a metric space (M, ρ). In the case $K = 0$ a *cone of class* K is a cone Q_M in the sense of the definition in 5.1.

Suppose that $K > 0$. We put $\sigma = \pi/\sqrt{K}$ and consider the direct product of the set M and the interval $I_\sigma = (0, \sigma)$, that is, the set of all pairs of the form (x, r), where $r \in (0, \sigma)$. For $z = (x, r) \in M \times I_\sigma$ we put $r(z) = r$. To the product $M \times I_\sigma$ we add two elements O and O', for which we put $r(O) = 0$, $r(O') = \sigma = \pi/\sqrt{K}$. Formally we can represent O as the totality of all pairs of the form $(x, 0)$, where $x \in M$, and O' as the set of pairs of the form (x, σ). We shall denote the set $(M \times I_\sigma) \cup \{O, O'\}$ by $Q_K(M)$. In the set $Q_K(M)$ we introduce a metric ρ_K^*. If $z_1 = z_2$, we put $\rho_K^*(z_1, z_2) = 0$. If $z_1 = O$, $z_2 \neq O$, then $\rho_K^*(z_1, z_2) = \rho_K^*(z_2, z_1) = r(z_2)$. If $z_1 = O'$, $z_2 \neq O'$, then $\rho_K^*(z_1, z_2) = \rho_K^*(z_2, z_1) = \sigma - r(z_2)$. In particular, $\rho_K^*(O, O') = \sigma$. Let $z_1 = (x_1, r_1)$, $z_2 = (x_2, r_2)$, where $0 < r_1 < \sigma$, $0 < r_2 < \sigma$. Then if $\rho(x_1, x_2) \geqslant \pi$, we put $\rho_K^*(z_1, z_2) = r_1 + r_2$. If $\rho(x_1, x_2) < \pi$, then $\rho_K^*(z_1, z_2)$ is defined as follows. On the sphere Σ_K we construct a spherical triangle ABC for which $|AB| = r_1$, $|AC| = r_2$, and $\angle ABC = \rho(x_1, x_2)$. In this case we put $\rho_K^*(z_1, z) = |BC|$. It is easy to verify that the function ρ_K^* of pairs of points of the set $Q_K(M)$ obtained in this way is a metric.

Suppose that $K < 0$. In this case the construction is exactly similar to the one we used in the case $K = 0$. We first construct the set Q_M, as we did in 5.1. We define a metric ρ_K^* in the set Q_M, but not the same as in 5.1. Namely, we put $\rho_K^*(z_1, z_2) = 0$ in the case $z_1 = z_2$. If $z_1 = O$, $z_2 \neq z_1$, we put $\rho_K^*(z_1, z_2) = \rho_K^*(z_2, z_1) = r(z_2)$. Finally, if $z_1 = (x_1, r_1)$, $z_2 = (x_2, r_2)$, where $r_1 > 0$, $r_2 > 0$, then $\rho_K^*(z_1, z_2)$ is defined as follows. On the Lobachevskij plane Σ_K we construct a triangle ABC with sides $|AB| = r_1$, $|AC| = r_2$, and $\angle BAC = \rho(x_1, x_2)$. We put $\rho_K^*(z_1, z_2) = |BC|$. We can show that in this case also ρ_K^* is a metric in Q_M. We shall also denote the metric space that arises in this way by $Q_K(M)$.

For a point $x \in M$ we shall denote the totality of all points $(x, r) \in Q_M$ by $\gamma(x)$. In the case $K > 0$ we also assume that the points O and O' belong to $\gamma(x)$. In the case $K < 0$ we assume that $O \in \gamma(x)$. We call the set $\gamma(x)$ the generator of the cone $Q_K(M)$ corresponding to the point x of the space M.

All that we said in 5.1 about intervals of generators extends automatically to the case of cones $Q_K(M)$ for arbitrary K.

In the case $K > 0$ the points O and O' are called vertices of the cone $Q_K(M)$. We shall call O the upper vertex and O' the lower vertex of $Q_K(M)$. In the case $K < 0$ the point O is called the vertex of $Q_K(M)$. Suppose we are given a number $h > 0$. We shall denote the set of all points $z \in Q_K(M)$ for which $r(z) < h$ by $Q_K(M, h)$ and the set of all points $z \in Q_K(M)$ such that $r(z) = h$ by $\Gamma_K(M, h)$.

Suppose that M is an interval $[a, b]$, where $a < b$, of the number line $\mathbb{R}$ and that $\rho(x, y) = |x - y|$ for $x, y \in M$. We put $\theta = b - a$. We shall denote the cone $Q_K(M)$ corresponding to the given space M by $A_K(\theta)$. The notation is reasonable, since the cones $Q_K([a_1, b_1])$ and $Q_K([a_2, b_2])$ for which $b_1 - a_1 = b_2 - a_2 = \theta$ are isometric. We shall call θ the total angle of the cone $A_K(\theta)$.

Suppose that $M = C_\theta$, where C_θ is a circle in the plane $\mathbb{E}^2$, and that the metric in C_θ is defined as in 5.1. The cone $Q_K(C_\theta)$ is denoted by $Q_K(\theta)$.

The results of 5.1 about the construction of cones $Q(\theta)$ and $A(\theta)$ by pasting together a finite collection of cones $A(\theta_i)$ extends in an obvious way to the case of cones $Q_K(\theta)$ and $A_K(\theta)$.

By analogy with what we did in 5.1, we define cones $Q_K(\theta, h)$, $A_K(\theta, h)$ and curves $\Gamma_K(\theta, h)$ in $Q_K(\theta)$ and $A_K(\theta)$. We have the relation

$$\frac{l[\Gamma_K(\theta, h)]}{h} \to \theta$$

as $h \to 0$.

In the case $\theta = 2\pi$ the cone $Q_K(\theta)$ is isometric to a plane Σ_K. Any point of the cone $Q_K(\theta)$ other than its vertices has a neighbourhood isometric to a disc on the plane Σ_K. Any point of the cone $A_K(\theta)$ that does not belong to its boundary rays also has a neighbourhood isometric to a disc on the plane Σ_K. If $X \in A_K(\theta)$ lies on one of the boundary rays of $A_K(\theta)$ and is distinct from its vertex, then some neighbourhood of X is isometric to a half-disc on the plane Σ_K.

All that we said in 5.1 about shortest curves on the cone Q extends in an obvious way to the case of the cones $Q_K(\theta)$ (with the stipulation that in the case $K > 0$, $\theta > 2\pi$ if we extend a shortest curve beyond the vertex of the cone we must require that the length of the resulting curve is not greater than $\sigma = \pi/\sqrt{K}$, otherwise it ceases to be a shortest curve).

Let M be a two-dimensional manifold with boundary, and K a real number. We assume that an intrinsic metric is specified in M. We shall say that the metric space (M, ρ) is a *polyhedron of class K*, or briefly a *K-polyhedron*, if the following conditions are satisfied:

a) each internal point X of M has a neighbourhood V that for some $\theta > 0$, $\varepsilon > 0$ admits an isometric map φ onto the cone $Q_K(\theta, \varepsilon)$ such that $\varphi(X)$ is a vertex of this cone;

b) each point $X \in \partial M$ has a neighbourhood V isometric to the cone $A_K(\theta, \varepsilon)$ for some $\theta > 0$, $\varepsilon > 0$. Under the isometry map of V onto $A_K(\theta, \varepsilon)$ the point X goes over to a vertex of this cone.

As in the case of ordinary polyhedra, the neighbourhood V of X that satisfies one of the conditions, a) or b), is called a *canonical neighbourhood of this point*. The quantity θ is called the total angle at X and is denoted by $\theta(X)$. If X is an interior point of M, we put $\omega(X) = 2\pi - \theta(X)$. We shall call $\omega(X)$ the curvature of the point X of the K-polyhedron M. If X is a boundary point of M, we put $\pi - \theta(X) = \kappa(X)$. The number $\kappa(X)$ is called the turn of the boundary at X. Then $X \in M$ is called a vertex of the K-polyhedron M if it is an interior point of M and its curvature is not equal to zero. A point $X \in M$ is called a boundary vertex of K if X belongs to the boundary of M and the turn of the boundary at X is not equal to zero.

All that we said in 5.2 about the structure of a polyhedron close to points that are not vertices or boundary vertices automatically extends to the case of K-polyhedra.

The theorem about triangulation of polyhedra is also true in the case of K-polyhedra, except that instead of triangles on a Euclidean plane we should take triangles on the plane Σ_K.

The theorem about pasting together manifolds with polyhedral metric (Theorem 5.2.3) also has a complete analogue for K-polyhedra. The concepts of a polygonal line on a K-polyhedron and the area are defined in complete analogy with the case considered in 5.2.

Let M be an arbitrary K-polyhedron and suppose that a set $A \subset M$ is such that its closure is compact (that is, in the terminology of 5.3, A is bounded). Then A contains finitely many vertices. We shall denote the sum of the curvatures at all the vertices of the polyhedron that belong to A by $\omega_K(A)$ and call it the *excess of the curvature* with respect to K. A complete analogue of the curvature of a set on a K-polyhedron is the set function ω defined by

$$\omega(A) = \omega_K(A) + K\sigma(A),$$

where $\sigma(A)$ is the area of the set.

Theorems on the total curvature of a compact polyhedron (Theorem 5.3.1), like the Gauss-Bonnet theorem (Theorem 5.3.2), remain valid for polyhedra of class K.

Aleksandrov's theorem on the characterization of closed convex polyhedra in Euclidean space has its analogue for arbitrary spaces of constant curvature. We do not give the exact statement here, since this result is contained in a general theorem of Pogorelov (Pogorelov (1957)).

The method of cutting and pasting can be successfully used also in the study of extremal problems for K-polyhedra. An analogue of Theorem 5.6.1 for the case $K < 0$ was established in Belinskij (1975).

Chapter 2
Different Ways of Defining Two-Dimensional Manifolds of Bounded Curvature

§6. Axioms of a Two-Dimensional Manifold of Bounded Curvature. Characterization of such Manifolds by Means of Approximation by Polyhedra

6.1. Axioms of a Two-Dimensional Manifold of Bounded Curvature.

A. Preliminary remarks. We first make some remarks about Riemannian manifolds. Let (R, ρ) be a two-dimensional manifold with intrinsic metric. We assume that its metric is Riemannian. Let us take a point $X_0 \in R$ arbitrarily. Then, as we said above, there is a neighbourhood U of X_0 such that any two points $X, Y \in U$ can be joined by a shortest curve, and this shortest curve is unique. We choose a neighbourhood U that satisfies this condition. We shall assume that U is homeomorphic to a disc and that the quantity

$$|\omega|(U) = \int_U |K(X)|\, d\sigma(X)$$

is finite. (We recall that $K(X)$ is the Gaussian curvature of the manifold at the point X, and that the additive set function $\sigma(E)$ is its area.) Obviously there is a neighbourhood U of the point X_0 that satisfies all these conditions.

Let A, B and C be three distinct points belonging to U, and K, L and M the shortest curves joining the points B and C, C and A, A and B respectively. We assume that K, L and M are contained in U. If A, B and C do not lie on one shortest curve, then K, L and M have no points in common other than their ends. In fact, suppose that any two of the given shortest curves have points in common other than A, B and C. Suppose, for example, that K and M have a point X in common, other than B. The shortest curves K and M are contained in U and since any two points of U are joined by a unique shortest curve the arcs $[BX]$ of K and M coincide. The shortest curves K and M have a common tangent at X. Since through a point of a Riemannian manifold there passes a unique shortest curve with a given tangent, it follows that one of the shortest curves K and M is contained in the other and we see that the points A, B and C lie on one shortest curve, contrary to our assumption. The shortest curves K, L and M form a simple closed curve, which bounds a domain T homeomorphic to a closed disc. We shall call this domain a geodesic triangle. Let α, β and γ be the angles of the triangle T at A, B and C respectively. Then from the Gauss-Bonnet formula it follows that

$$\delta(T) = \alpha + \beta + \gamma - \pi = \omega(T) = \iint_T K(X)\, d\sigma(X).$$

The quantity $\delta(T)$ is called the *excess of the triangle T*.

Let $T_1, T_2, \ldots, T_m$ be an arbitrary set of geodesic triangles contained in U and such that no two have interior points in common. Then

$$\sum_{i=1}^{m} \delta(T_i) = \sum_{i=1}^{m} \omega(T_i) \leqslant \sum_{i=1}^{m} \omega^+(T_i) \leqslant \omega^+(U) = \Omega_0 < \infty. \tag{6.1}$$

For any system of pairwise non-overlapping geodesic triangles contained in U the sum of their excesses is thus not greater than some constant $\Omega_0 < \infty$. We observe that we can make the sum on the left-hand side of (6.1) arbitrarily close to $\omega^+(U)$ by covering the set of those X where $K(X) > 0$ by sufficiently small geodesic triangles.

Two-dimensional manifolds of bounded curvature are defined as two-dimensional manifolds with intrinsic metric that satisfy an additional condition relating to excesses of geodesic triangles. The statement of this condition is suggested by the remark made above about two-dimensional Riemannian manifolds.

In order to give a definition of two-dimensional manifolds of bounded curvature it is necessary to give an exact meaning to the concept of the angle between curves. For a start it is sufficient to be able to do this when the curves are shortest curves starting from one point. Having the concept of angle, we can define the concept of the excess of a triangle.

B. Definition of the angle between curves in a space with intrinsic metric. Let (M, ρ) be an arbitrary metric space with intrinsic metric, and K and L simple arcs with a common origin O in the space M. We specify points $X \in K$ and $Y \in L$ arbitrarily. On the plane $\mathbb{E}^2$ we construct a triangle $O'X'Y'$ with sides

$$|O'X'| = \rho(O, X), |O'Y'| = \rho(O, Y), |X'Y'| = \rho(X, Y),$$

and let $\gamma(X, Y)$ be the angle at the vertex O' of this triangle. We put $\rho(O, X) = x$, $\rho(O, Y) = y$ and $\rho(X, Y) = z$. Then we have

$$\gamma(X, Y) = \arccos \frac{x^2 + y^2 - z^2}{2xy}.$$

Using the notation we have introduced, we can now state the definition.

Definition 6.1. The limit

$$\lim_{X \to O, Y \to O} \gamma(X, Y), \tag{6.2}$$

if it exists, is called the *angle between the curves K and L* at the point O.

The limit (6.2) may not exist in very simple cases, even if the curves K and L are sufficiently good from the viewpoint of the geometry of the space M (for example, they are shortest curves).

Suppose, for example, that (M, ρ) is a Banach space. Then in order that between any rectilinear intervals starting from an arbitrary point of this space there should exist an angle in the sense of the given definition it is necessary and sufficient that (M, ρ) should be a Hilbert space.

There naturally arises the idea, suggested by one of the conditions defining a manifold of bounded curvature, of introducing the requirement that between any two shortest curves starting from one point there should exist an angle in the sense of Definition 6.1. However, there is no need to do this. We introduce quantities that are always defined for an arbitrary pair of curves starting from one point.

Definition 6.2. Let K and L be simple arcs in a space with intrinsic metric, starting from some point O. Then in the notation introduced above we put

$$\bar{\gamma} = \overline{\lim_{X \to O, Y \to O}} \gamma(X, Y), \quad \underline{\gamma} = \underline{\lim_{X \to O, Y \to O}} \gamma(X, Y). \tag{6.3}$$

The quantities $\bar{\gamma}$ and $\underline{\gamma}$ are called the *upper and lower angles between the curves K and L* at the point O.

Formally $\bar{\gamma}$ and $\underline{\gamma}$ represent the following. For $h > 0$ we denote by $\bar{\gamma}(h)$ and $\underline{\gamma}(h)$ the least upper and greatest lower bounds of $\gamma(X, Y)$ on the set of all pairs (X, Y) such that $X \in K$, $Y \in L$, $0 < \rho(O, X) \leqslant h$, $0 < \rho(O, Y) \leqslant h$. The function $\bar{\gamma}(h)$ is non-decreasing, the function $\underline{\gamma}(h)$ is non-increasing, and

$$\bar{\gamma} = \lim_{h \to 0} \bar{\gamma}(h), \quad \underline{\gamma} = \lim_{h \to 0} \underline{\gamma}(h).$$

We have $0 \leqslant \underline{\gamma} \leqslant \bar{\gamma} \leqslant \pi$, and $\underline{\gamma} = \bar{\gamma}$ if and only if there exists a limit (6.2), which is then equal to $\underline{\gamma} = \bar{\gamma}$.

Definition 6.3. Let K be a simple arc starting from a point O of a space with intrinsic metric ρ. Then if there exists a limit

$$\lim_{X \to O, Y \to O} \gamma(X, Y)(X \in K, Y \in K), \tag{6.4}$$

we shall say that *the curve K has a direction at the point O.*

Obviously, if the limit (6.4) exists, then it is equal to zero. The assertion "the curve K has a direction at the point O" is equivalent to the following: "there exists an angle between the curves K and L at the point O in the case when $K = L$".

If K is a shortest curve and O is one of its end-points, then for any points $X, Y \in K$ either $\rho(O, X) = \rho(O, Y) + \rho(Y, X)$ or $\rho(O, Y) = \rho(O, X) + \rho(X, Y)$ and so $\gamma(X, Y) = 0$. Hence it follows that in this case $\lim_{X \to O, Y \to O} \gamma(X, Y)$ exists, so any shortest curve has a direction at each of its end-points.

Let L be a simple arc of the space $\mathbb{E}^3$, and O its origin. It is easy to verify that in order that L should have a direction at O in the sense of Definition 6.3 it is necessary and sufficient that it should have a tangent at this point, that is, that the ray OX with origin O passing through an arbitrary point $X \in K$ should converge to some limiting arc as $X \to O$.

The definitions of angle mentioned here have various modifications; see Aleksandrov and Zalgaller (1962).

C. Axioms of two-dimensional manifolds of bounded curvature. The main object of the investigation is defined as a metric space (R, ρ) satisfying Axioms A and B below.

Axiom A. (R, ρ) is a two-dimensional manifold without boundary and its metric ρ is intrinsic.

Henceforth we shall always assume that R is a space for which Axiom A is satisfied.

A *triangle* in the manifold R is a figure T consisting of three points A, B, C and three shortest curves K, L, M that join the points B and C, C and A, A and B, respectively. The points A, B, C are called the *vertices of the triangle T*, and the shortest curves $K = [BC]$, $L = [CA]$, $M = [AB]$ its *sides*. A triangle with vertices A, B, C will be denoted by ABC. We note that since two points in a space with intrinsic metric can be joined by several and even infinitely many shortest curves, a triangle is not uniquely determined by specifying its vertices. Henceforth, from the context it will always be clear whether ABC means a specific triangle with vertices A, B, C or an arbitrary one of such triangles.

Let $G \subset R$ be a domain in R that is homeomorphic to an open disc on a plane. Then we shall say that the triangle T is contained in the domain G if all its sides lie in G.

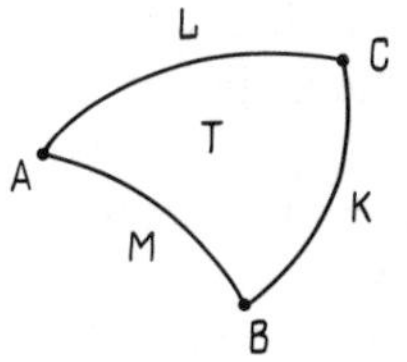

Fig. 24

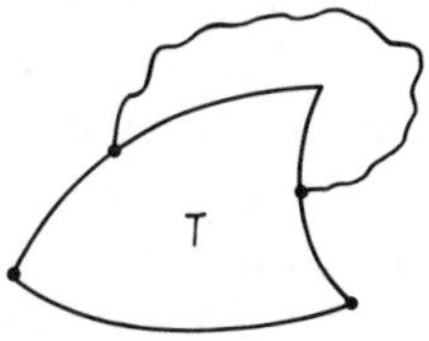

Fig. 25

A triangle T is said to be *homeomorphic to a disc* if its sides form a simple closed curve that bounds a domain homeomorphic to a disc (Fig. 24). We shall identify this domain with the triangle T. In accordance with this we shall say that its points are points of the triangle T, the interior points of the domain are interior points of the triangle, and its boundary points are boundary points of the triangle T.

A triangle T is said to be *simple* if it is homeomorphic to a disc and satisfies the following condition. No two boundary points of T can be joined by a curve that lies outside the triangle and is shorter than a simple arc with ends at these points that forms that part of the boundary of the triangle enveloped by the given curve (see Fig. 25).

Simple triangles T_1 and T_2 contained in G will be said to be *non-overlapping* if they do not have interior points in common.

Suppose we are given a triangle $T = ABC$, and that $K = [BC]$, $L = [CA]$, $M = [AB]$ are its sides. Let $\bar{\alpha}$ be the upper angle between the simple arcs L and M at the point A, $\bar{\beta}$ the upper angle between K and M at the point B, and $\bar{\gamma}$ the upper angle between L and K at the point C. The quantity

$$\bar{\delta}(ABC) = \bar{\delta}(T) = \bar{\alpha} + \bar{\beta} + \bar{\gamma} - \pi$$

is called the *upper excess* of the triangle T.

Axiom B. For any point $X \in R$ there is a neighbourhood G homeomorphic to an open disc in $\mathbb{R}^2$ and a constant $\Omega(G) < \infty$ such that for any system T_1, T_2, $\ldots$, T_m of pairwise non-overlapping simple triangles contained in G we have the inequality

$$\sum_{i=1}^{m} \bar{\delta}(T_i) \leqslant \Omega(G) < \infty. \tag{6.5}$$

Definition 6.4. A metric space satifying Axioms A and B is called a *two-dimensional manifold of bounded curvature.*

Henceforth Axiom B will be called the *axiom of boundedness of curvature.*

We note that in the given statement of Axiom B we have considered sums of excesses of triangles, and not the sums of their absolute values, as was done in the original version of Axiom B given in Aleksandrov (1948b), in which it was required that the sums $\sum_{i=1}^{m} |\bar{\delta}(T_i)|$ should be bounded. The boundedness of the sums of the absolute values of the excesses in a suitable neighbourhood of a point X follows from Axioms A and B. The proof of this is obtained, however, as a result of rather deep constructions that form the main part of the theory of two-dimensional manifolds of bounded curvature.

We note that between any two shortest curves in a two-dimensional manifold of bounded curvature starting from one point there exists an angle in the sense of Definition 6.1. The proof of this is also obtained at a rather late stage of the constructions (see 7.1).

In order that we can conclude from Definition 6.4 that a two-dimensional Riemannian manifold is a manifold of bounded curvature it is necessary to prove that for arbitrary shortest curves K and L starting from one point O the quantity $\bar{\gamma}$ defined by (6.3) is equal to the angle between K and L in the sense of the definition given previously. This is actually so, but the proof of the given assertion is not trivial. It can be obtained by an application of the classical apparatus of differential geometry. Later we shall give its proof, which relies on certain results relating to two-dimensional manifolds of bounded curvature (see 7.1).

6.2. Theorems on the Approximation of Two-Dimensional Manifolds of Bounded Curvature by Manifolds with Polyhedral and Riemannian Metric. A number of the main results of the theory of two-dimensional manifolds of bounded curvature are theorems on the approximation of two-dimensional manifolds of bounded curvature by manifolds with polyhedral metric and two-dimensional Riemannian manifolds. These theorems give a complete characterization of two-dimensional manifolds of bounded curvature. By means of them we can define the class of manifolds of the type under consideration as a result of closing the class of Riemannian manifolds (or manifolds with polyhedral metric) with respect to a suitable limiting process.

Let R be an arbitrary two-dimensional manifold, and U a domain in R homeomorphic to a closed disc. A metric ρ defined in U will be said to be *polyhedral* if this metric is intrinsic and the metric space (U, ρ) is polyhedral in the sense of the definitions in 5.2. Similarly, if (U, ρ) is a two-dimensional Riemannian manifold with boundary, we shall say that ρ is a *Riemannian metric* in U. If ρ is either a polyhedral or a Riemannian metric defined in U, then the symbols ω_ρ, $|\omega_\rho|$, ω_ρ^+ denote, respectively, the integral curvature, the absolute curvature and the positive part of the curvature as set functions in the manifold (U, ρ). The symbol $|\kappa_\rho|(\partial U)$ denotes the absolute turn of the boundary of the polyhedron (U, ρ). In the case when (U, ρ) is a Riemannian manifold, $|\kappa_\rho|(\partial U)$ denotes the integral of the modulus of the geodesic curvature of the boundary with respect to arc length.

Let (ρ_n), $n = 1, 2, \ldots$, be an arbitrary sequence of metrics defined in an open set U of a two-dimensional manifold R. We shall say that as $n \to \infty$ the metrics ρ_n converge to a metric ρ specified in U if $\rho_n(X, Y) \to \rho(X, Y)$ as $n \to \infty$ for any $X, Y \in U$, and the convergence is uniform.

We recall that a set U in a topological space is called a *neighbourhood of a point* X of this space if X is an interior point of U.

We shall call the following theorem the *first theorem on approximation*.

Theorem 6.2.1. *Let (R, ρ) be a two-dimensional manifold of bounded curvature. For any point X_0 of it we can construct a neighbourhood U of this point that is homeomorphic to a closed disc and a sequence of metrics (ρ_n), $n = 1, 2, \ldots$, defined in U such that (U, ρ_n) is a polyhedron homeomorphic to a disc for each n, $\rho_n(X, Y) \to \rho(X, Y)$ as $n \to \infty$ in the domain U, and there is a constant $C < \infty$ such that for all n*

$$|\omega_{\rho_n}|(U) + |\kappa_{\rho_n}|(\partial U) \leqslant C.$$

A proof of Theorem 6.2.1 is given in Aleksandrov and Zalgaller (1962). We note that in that work the theorem is stated in a weaker form than the one here (boundedness is guaranteed only for absolute curvatures), but from the arguments given there it follows that the given statement of the theorem is true.

Remark. Theorem 6.2.1 (Aleksandrov and Zalgaller (1962)) remains true if in its statement instead of polyhedral metrics we consider Riemannian metrics.

The given theorem establishes a necessary condition that is satisfied by any two-dimensional manifold of bounded curvature. This condition is also sufficient. In the part that touches on the sufficiency, the conditions imposed on the sequence of polyhedral metrics ρ_n can be weakened. Namely, the following theorem, which we shall call the *second theorem on approximation*, is true.

Theorem 6.2.2. *Let (R, ρ) be a two-dimensional manifold without boundary, endowed with an intrinsic metric ρ. We assume that for any point $X_0 \in R$ we can find a neighbourhood U of this point and a sequence of polyhedral metrics ρ_n, $n = 1, 2, \ldots$, specified in U that converges to the metric ρ and is such that the sequence $(\omega^+_{\rho_n}(U))$, $n = 1, 2, \ldots$, is bounded. Then (R, ρ) is a two-dimensional manifold of bounded curvature.*

Theorem 6.2.1 establishes the possibility of a local approximation of the metric of a two-dimensional manifold of bounded curvature by polyhedral or Riemannian metrics in a neighbourhood of an arbitrary point of the manifold. There arises the question, is a similar kind of theorem true about the approximation of a metric by polyhedral or Riemannian metrics defined on the whole manifold? The answer to this question is positive. The required approximations are easily constructed by means of theorems on the analytic representation of a two-dimensional manifold of bounded curvature, to which § 7 of this article is devoted.

6.3. Proof of the First Theorem on Approximation. The complete proof of Theorem 6.2.1 presented in Aleksandrov and Zalgaller (1962) turns out to be rather extensive, and here we give only an outline of it, mentioning some of the most important parts of it. The arguments on which the proof of Theorem 6.2.1 is based are conceptually a generalization of the technique worked out by Aleksandrov in his investigation of two-dimensional manifolds of non-negative curvature. We derive an analogue of Theorem 6.2.1, touching on approximation by Riemannian metrics, from a theorem on the analytic representation of a two-dimensional manifold of bounded curvature presented in § 7.

A. Estimate of the distortion of the angles of a triangle under transition to a development of it.* Henceforth R denotes a metric space with intrinsic metric ρ. We shall assume that R is locally compact and that all the later constructions refer to a small domain G of R in which any two points can be joined by a shortest curve.

For points $X, Y \in G$ instead of $\rho(X, Y) = l$ we shall simply write $|XY| = l$.

In the arguments on which the proof of Theorem 6.2.1 relies, an essential role is played by a method that consists in comparing a triangle in the space (R, ρ) with a flat triangle having the same lengths of sides. Let T be a triangle in the space (R, ρ), A, B, C its vertices, and the shortest curves $K = [BC]$, $L = [CA]$, $M = [AB]$ its sides. Let $a = |BC|$, $b = |CA|$, $c = |AB|$ be the lengths of the sides of T. On the plane $\mathbb{E}^2$ we construct a flat triangle $A'B'C'$, the lengths of whose sides are $|B'C'| = a$, $|C'A'| = b$, $|A'B'| = c$. Such a triangle can certainly be constructed, since we only need it to satisfy the inequalities $a \leqslant b + c$, $b \leqslant c + a$, $c \leqslant a + b$. These inequalities obviously hold. We shall call $A'B'C'$ a *development of the triangle* T. Henceforth we shall assume that to each vertex and each side of T there corresponds a vertex or side of the same name of the triangle $A'B'C'$, that is, the vertex A corresponds to the vertex A', the side $[BC]$ to $[B'C']$, and so on. Let α_0, β_0, γ_0 be the angles at the vertices A', B', C' of the triangle $A'B'C'$. We observe that $A'B'C'$ may degenerate into an interval. In this case one of the numbers α_0, β_0, γ_0 is equal to π and the other two are equal to zero. Let $\bar{\alpha}$, $\bar{\beta}$, $\bar{\gamma}$ be the upper angles between the sides of the triangle T at its vertices A, B, C. Then there is defined the number $\bar{\delta}(T) = \bar{\alpha} + \bar{\beta} + \bar{\gamma} - \pi = (\bar{\alpha} - \alpha_0) + (\bar{\beta} - \beta_0) + (\bar{\gamma} - \gamma_0)$, the upper excess of the triangle T.

Suppose we are given a triangle T with vertices A, B, C in the space R. We arbitrarily take points X and Y on the sides AB and AC of the given triangle, other than A. We shall join X and Y by a shortest curve XY. We assume that T is contained in a domain, any two points of which can be joined by a shortest curve, and so such shortest curves exist. Let $\bar{\delta}(AXY)$ be the upper excess of the triangle AXY. We put

$$\bar{\nu}_A^+(T) = \sup_{X \in AB, Y \in AC} \left[\inf_{[XY]} \bar{\delta}(AXY) \right].$$

The greatest lower bound (in square brackets) is taken with respect to the set of all triangles whose sides are the intervals $[AX]$ and $[AY]$ of the sides AB and AC of T and an arbitrary shortest curve joining X and Y (there may be many such shortest curves). Similarly we define the quantities $\bar{\nu}_B^+(T)$ and $\bar{\nu}_C^+(T)$. We put $\bar{\nu}^+(T) = \max\{\bar{\nu}_A^+(T), \bar{\nu}_B^+(T), \bar{\nu}_C^+(T)\}$.

Lemma 6.3.1. *In the notation introduced above, we have*

$$\bar{\alpha} - \alpha_0 \leqslant \bar{\nu}_A^+(T), \quad \bar{\beta} - \beta_0 \leqslant \bar{\nu}_B^+(T), \quad \bar{\gamma} - \gamma_0 \leqslant \bar{\nu}_C^+(T).$$

We note that the lemma refers to an arbitrary metric space and in its formulation it is not assumed that the given manifolds are manifolds of bounded curvature.

The proof of Lemma 6.3.1 will be carried out in 6.4.

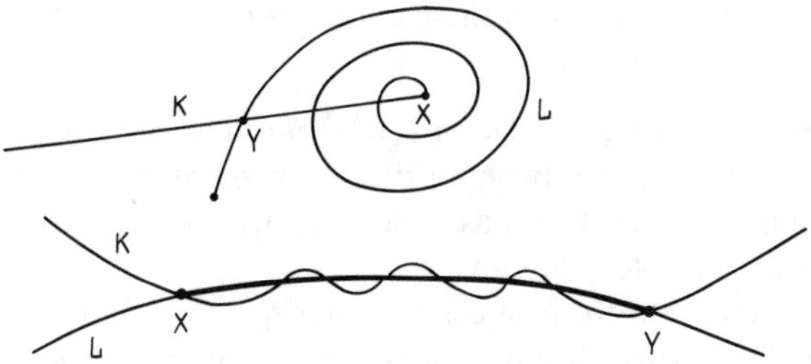

Fig. 26

B. Some methods of transforming shortest curves.* In the constructions described later, two methods are of essential use. The first is called the removal of superfluous intersections and consists in the following. We assume that the shortest curves K and L have at least two points in common. Let X and Y be the leftmost and rightmost of these points on the curve K. The arcs $[XY]$ on K and L have the same length, because otherwise by replacing of the arc $[XY]$ on one of them by the arc $[XY]$ taken on the other we could reduce the length of one of the curves K and L without changing its ends, which contradicts the fact that they are shortest curves. Replacing the arc $[XY]$ of K by the arc $[XY]$ of L, we obtain a shortest curve K'. The common points of K' and L form a simple arc lying on each of them. Replacing the pair of shortest curves K, L by the pair K', L, we can remove the singularities in a mutual arrangement of the type of those shown in Fig. 26.

We assume that there is a finite collection of shortest curves, no two of which have common intersections, that is, the common part of two of them is either an empty set, or it consists of one point, or it is a simple arc. If to this collection we add a new shortest curve, then by replacing some arcs of it we can arrange that the resulting collection of shortest curves does not have superfluous intersections.

We assume that R is a two-dimensional manifold with intrinsic metric. Let L be a shortest curve in R joining the points $A, B \in R$. We orient L so that A is its beginning and B is its end. We shall assume that L lies in a domain G of R homeomorphic to an open disc, and for any point $X \in L$ the distance to the boundary is greater than the length of L. We assume that G is oriented. Then we can define what is meant by L being left or right. We shall say that L is the leftmost (rightmost) shortest curve joining the points A and B if any simple arc whose ends are arbitrary points $X, Y \in L$ and which lies to the left (respectively, right) of L has length greater than that of the arc $[XY]$ of L. For any points A, $B \in G$, each of which is distant from the boundary of G by a distance greater than $\rho(A, B)$, there are leftmost and rightmost shortest curves joining A and B. The leftmost shortest curve joining A and B is unique. The same is true for the rightmost shortest curve.

The concepts of leftmost and rightmost shortest curves make it possible to select uniquely, in the set of all shortest curves joining given points, one definite shortest curve. The use of these concepts constitutes the second method of transforming shortest curves.

C^*. A lemma about partitioning on a triangle. The next step in the proof of Theorem 6.2.1 consists in establishing that each point X of a two-dimensional manifold of bounded curvature has a neighbourhood that can be partitioned into arbitrarily fine simple triangles.

As a preliminary we introduce certain concepts. Next we consider domains homeomorphic to a closed disc in a two-dimensional manifold R with intrinsic metric of bounded curvature.

Let G be such a domain. We shall say that G is *convex* if for any two points $X, Y \in G$ there is a shortest curve contained in G that joins them. If any shortest curve joining two arbitrary points $X, Y \in G$ is contained in G, then G is said to be *completely convex*.

A domain G is said to be *boundedly convex* if its boundary is rectifiable and G can be enclosed in a domain U homeomorphic to an open disc so that the distance from the boundary of G to the boundary of the domain U exceeds the perimeter of G by more than four times, and the following condition is satisfied. For any two points $X, Y \in \partial G$ there is no simple arc with ends X and Y that goes outside G and is shorter than the part of the boundary of G bounded by this arc.

The domain G is said to be *absolutely convex* if it is completely convex and boundedly convex.

We shall call a point $X \in R$ a *transit point* if there is a shortest curve for which X is an interior point.

Transit points in a manifold R form an everywhere dense set. In fact, let $X \in R$. Then in any neighbourhood of X there is a shortest curve, and hence also transit points of the manifold. On any simple arc and any simple closed curve in R the transit points form an everywhere dense set. In fact, let L be a simple arc, and X an arbitrary interior point of L. Then for any neighbourhood U of X we can find a neighbourhood $V \subset U$ of this point such that L partitions V. Let W be a neighbourhood of X such that any two points $Y, Z \in W$ can be joined by a shortest curve, and any such shortest curve is contained in V. We choose points $Y, Z \in W$ that lie on different sides of L. The shortest curve joining them intersects L. The point of intersection of this shortest curve with L is a transit point contained in the neighbourhood U of the point X.

The result of this step is included in the following proposition.

Lemma 6.3.2. *Suppose that a domain G in a two-dimensional manifold of bounded curvature, homeomorphic to a closed disc, is absolutely convex. We assume that G is a polygon, that is, the boundary of G is a closed polygonal line. Then for any $\varepsilon > 0$ we can find a finite system of pairwise non-overlapping simple triangles such that G is their union, the diameter of any of these triangles does not exceed ε, and the vertices of all these triangles other than boundary vertices of G are transit points and at each of them any side is less than the sum of the other two.*

The proof of Lemma 6.3.2 in its technical aspect is rather cumbersome. It is essential to use the axiom of boundedness of curvature. To its full extent the reader can become acquainted with it in the monograph Aleksandrov and Zalgaller (1962). Here we just outline the proof of Lemma 6.3.2 in very general terms.

We first prove the following proposition.

Lemma 6.3.3. *For any point X of a two-dimensional manifold of bounded curvature, whatever the values of $\varepsilon > 0$, $\delta > 0$, there is a closed neighbourhood V of this point contained in the disc $B(X, \delta)$ that is absolutely convex and homeomorphic to a closed disc on a plane, where the boundary of V is a simple closed polygonal line whose length is less than ε, and all the vertices are transit points.*

In order to prove Lemma 6.3.3 it is first necessary to establish that for any $\varepsilon > 0$ and $\delta > 0$ we can construct a neighbourhood U of the point X lying in the disc $B(X, \delta)$ that is bounded by a simple closed curve whose length is less than ε. The proof of this assertion uses the axiom of boundedness of curvature, and if this axiom is not satisfied, then generally speaking the assertion is not true, as the following example shows.

On the plane $\mathbb{R}^2$ we consider the disc $\bar{B}(0, 1)$. Let M be the topological manifold obtained from $\mathbb{R}^2$ by identifying points of $\bar{B}(0, 1)$ so that to the disc $\bar{B}(0, 1)$ there corresponds one point of M, which we denote by O. Obviously M is a two-dimensional manifold homeomorphic to $\mathbb{R}^2$. In M we introduce an intrinsic metric ρ so that in a neighbourhood of any point $X \neq O$ the metric ρ coincides with the natural metric of a plane. The metric ρ is uniquely determined by the given condition. It is not difficult to see that in the two-dimensional manifold with intrinsic metric that we have constructed any closed curve inside which the point O lies has length not less than 2π.

After we have established that any point has a small neighbourhood with arbitrarily small perimeter, the validity of Lemma 6.3.3 is established as follows. Suppose we are given $\varepsilon > 0$ and $\delta > 0$. We first construct a neighbourhood U lying in the disc $B(X, \delta)$ and such that the boundary of U is a simple closed curve Γ whose length is less than ε_1. Here ε_1 is determined by ε and δ. Transit points lying on the curve Γ form on it an everywhere dense subset, and so it can be partitioned into finitely many arbitrarily small arcs whose ends are transit points. Let us specify such a partitioning arbitrarily. Joining the ends of the arcs of the partition by shortest curves so that no superfluous intersections arise (see B^*), we obtain a closed polygonal line Γ' that encircles the point X. Without loss of generality we can assume that Γ' is simple (if not, then by discarding loops that are part of Γ' it is converted into a simple closed polygonal line). If the points on Γ that were joined to obtain Γ' are situated sufficiently densely on Γ, then Γ' is contained in $B(X, \delta)$. All the vertices of Γ' are transit points. We shall consider closed curves in R that envelop Γ' and lie in $\bar{B}(X, \delta)$. Among them there are curves of shortest length. If ε_1 is sufficiently small, we can guarantee that such a "minimal" curve does not touch the boundary of $\bar{B}(X, \delta)$. In this case such a curve is a polygonal line, and its vertices are vertices of Γ'. Among the closed curves of shortest length that envelop Γ' there is an outermost one. It bounds the required neighbourhood of the point X.

The proof of Lemma 6.3.2 is obtained as follows. Suppose we are given $\varepsilon > 0$. For an arbitrary point $X \in \overline{G}$ we denote by $V(X)$ a neighbourhood of X that satisfies all the conditions of Lemma 6.3.3 and corresponds to the given ε and $\delta = \varepsilon/2$. By a theorem of Borel, G is covered by finitely many neighbourhoods $V(X)$. Let $V_1, V_2, \ldots, V_m$ be these neighbourhoods. All the possible intersections of the domains V_i with the set G form a finite collection of convex polygons whose union is G. Subdividing these polygons into triangles, we obtain the partition whose existence is asserted by Lemma 6.3.2.

In conclusion, a few words about the proof of Lemma 6.3.3. As we mentioned above, it reduces to proving that any point X of a two-dimensional manifold of bounded curvature has arbitrarily small neighbourhoods with an arbitrarily small perimeter. The existence of such a neighbourhood is proved as follows.

We first establish that for any $r > 0$ there is a finite collection of triangles T_1, $T_2, \ldots, T_m$ contained in the disc $B(X, r)$ that satisfies the following conditions. The point X is a vertex of each of the triangles $T_1, T_2, \ldots, T_m$, and the sides of a triangle T_i, $i = 1, 2, \ldots, m$, starting from X have no common points other than X. The common part of two adjacent triangles is an arc beginning at X and lying on a side of each of them (see Fig. 27). Finally, it is still required that the side of T_i opposite to X is shorter than each of the other two for any $i = 1, 2, \ldots, m$. From the fact that $T_i \subset B(X, r)$ for any i it follows that the lengths of the sides of the triangles T_i starting from X do not exceed r.

The construction of the collection of triangles $T_1, T_2, \ldots, T_m$ satisfying the conditions listed requires rather laborious arguments. However, we use only the fact that R is a two-dimensional manifold with intrinsic metric. The axiom of boundedness of curvature is not used.

We shall say that a triangle T_i is boundedly convex at a point X if there is a neighbourhood V of X such that any shortest curve joining points P and Q that belong to V and lie on sides of T_i is contained in T_i.

We assume that among the triangles $T_1, T_2, \ldots, T_m$ there is one that is not boundedly convex. Let T_i be this triangle, and Y and Z its vertices other than X. Then in any neighbourhood V of X there are points $P \in [XY]$ and $Q \in [XZ]$ that can be joined by a shortest curve lying outside the triangle T_i. This shortest curve together with the intervals $[PY]$ and $[QZ]$ on the sides of T_i and its side

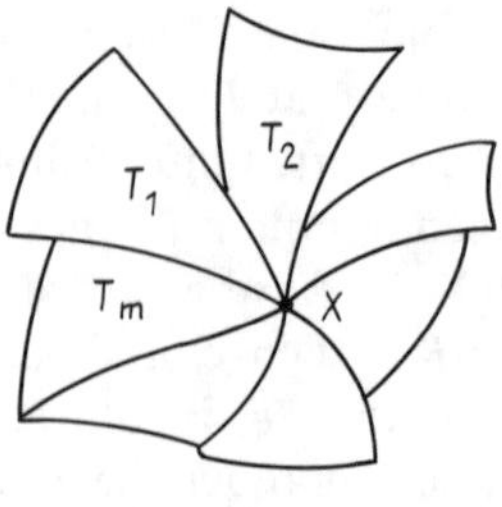

Fig. 27

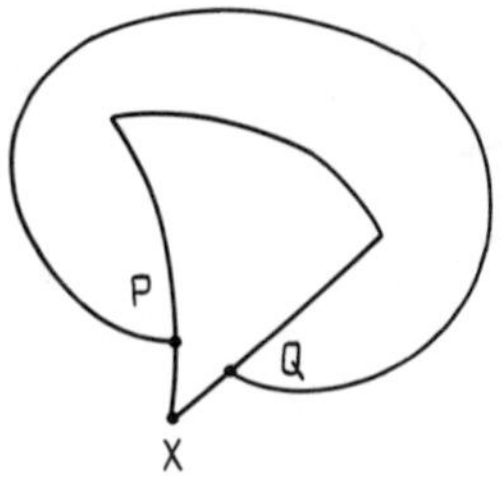

Fig. 28

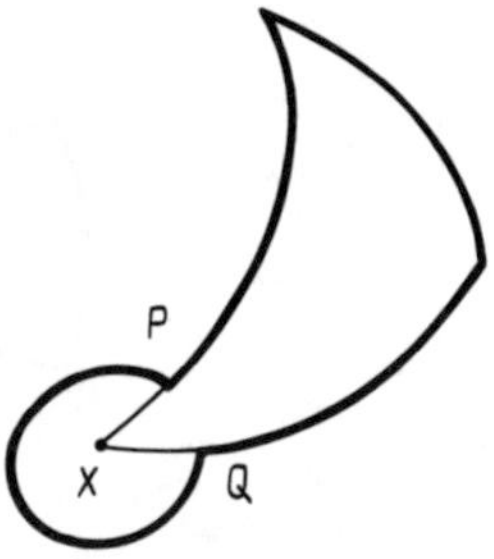

Fig. 29

[YZ] form a closed polygonal line Γ. If P and Q are sufficiently close to X, the situation represented in Fig. 28 cannot hold, and Γ encircles the point X, as represented in Fig. 29. (Let $\eta = \min\{|XP|, |XQ|\}$; in the case represented in Fig. 29 the shortest curve $[PQ]$ goes out from the disc of radius η and so its length is not less than $2\eta - |XP| - |XQ|$. If $|XP| < \eta/2$, $|XQ| < \eta/2$, then the situation becomes impossible, because otherwise it turns out that $S([PQ]) > |XP| + |XQ|$.) The length of Γ does not exceed the perimeter of the triangle T_i. This, in turn, does not exceed $3r$. Choosing $r = \varepsilon/3$, we deduce that Γ is the required polygonal line.

Henceforth we shall assume that each of the triangles $T_1, T_2, \ldots, T_m$ is boundedly convex at the point X. We assume that the upper angle of each of them at X is less than π. We specify a sufficiently small $h > 0$ and lay off on the sides of the triangles T_i starting from X intervals equal to h. Joining the ends of these intervals by shortest curves, we obtain a closed polygonal line. Thanks to the condition of bounded convexity at the point X for sufficiently small h, each of these shortest curves goes into the triangle T_i. By virtue of the fact that the upper angle of T_i is less than π for small h, this shortest curve does not go through the point X. It is not difficult to see that the polygonal line thus constructed is the required one if h is sufficiently small.

Difficulties arise when the triangle T_i, being boundedly convex at the point X, has angle at this point equal to π. We cannot guarantee that for points P and Q sufficiently close to X and Y, taken on the sides of T_i starting from X, the shortest curve $[PQ]$ does not go through X. The condition of boundedness of curvature enables us to prove that the number of such triangles T_i does not exceed some constant depending only on the point X. Detailed analysis of this requires a consideration of several possibilities. Let us restrict ourselves to the case that is the principal one. We observe that the angle of development of T_i corresponding to X does not exceed $\pi/3$, since this is the smallest of the angles of development. Since $\bar{\alpha} = \pi$, on the basis of Lemma 6.3.1 we deduce that $\bar{\nu}_X^+(T_i) \geqslant 2\pi/3$. Hence it follows that among the triangles XPQ, where P and Q are points on the sides of the triangle T_i starting from X, there is one whose excess is greater than $\pi/3$. In the basic case when the side PQ of this triangle goes inside the triangle T_i we obtain a simple triangle with excess greater than $\pi/3$.

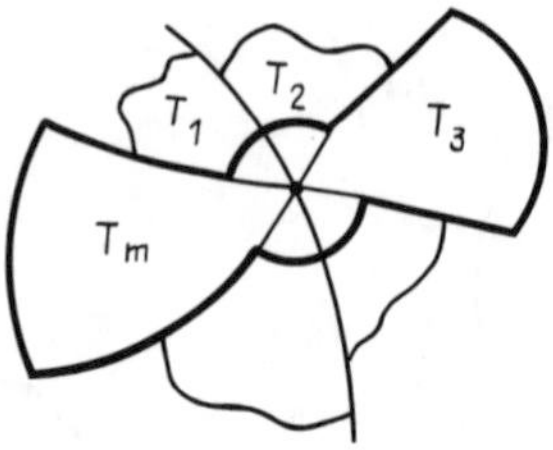

Fig. 30

We can assume that all the constructions are carried out in a neighbourhood G of the point X such that the sum of the excesses of any finitely many non-overlapping simple triangles lying in this neighbourhood does not exceed some constant $\Omega < \infty$. Hence it follows that the number of triangles T_i for which $\bar{\alpha} = \pi$ does not exceed the number $3\Omega/\pi$. The desired polygonal line is now constructed as follows (Fig. 30). Let us specify a sufficiently small number $h > 0$. For each $i = 1, 2, \ldots, m$ we define an arc L_i lying in the triangle T_i and not containing the point X. If the angle of T_i at X is different from π, then L_i is a shortest curve joining points on the sides of T_i lying at a distance h from the point X. If $\bar{\alpha}_i = \pi$, then L_i consists of the side of T_i lying opposite to X, and intervals of two other sides consisting of points P lying on these sides and such that $|PX| \geqslant h$. The arcs $L_1, L_2, \ldots, L_m$ form a closed polygonal line Γ^* encircling the point X. We note that $s(L_i) \leqslant 2h$ in the case $\bar{\alpha}_i < \pi$ and $s(L_i) \leqslant 3r$ in the case $\bar{\alpha}_i \geqslant \pi$ (we recall that the lengths of the sides of T_i starting from X do not exceed r, and the side lying opposite to X is very short). Hence we conclude that $s(L) \leqslant 2mh + 9\Omega r/\pi$. Since $h > 0$ and $r > 0$ are arbitrary, it follows that $s(L)$ can be made less than any preassigned $\varepsilon > 0$, as required.

D^*. Construction of an approximating sequence of polyhedra. We now describe the final part of the proof of Theorem 6.2.1. Since we intend to explain only the idea of the proof, some details of a technical character are omitted here.

Let G be a neighbourhood of the point X homeomorphic to a closed disc, such that its boundary is a simple closed polygonal line. We shall assume that G is absolutely convex and such that for any system of pairwise non-overlapping simple triangles contained in G the sum of their excesses does not exceed some constant $\Omega < \infty$. The existence of a neighbourhood G satisfying all these conditions follows from Lemma 6.3.3.

Let $A_1, A_2, \ldots, A_m$ be all the vertices of the closed polygonal line Γ that bounds the domain G.

Let Δ be a partition of G into pairwise non-overlapping simple triangles $T_1, T_2, \ldots, T_r$ such that each of the points A_i is a vertex of one of these triangles, and the vertices of the triangles T_i other than those that lie on the boundary of G are transit points, and in each of the triangles any side is less than the sum of the other two. Let $d(\Delta)$ be the largest diameter of the triangles T_i, $i = 1, 2, \ldots, r$. According to Lemma 6.3.2, whatever the number $\varepsilon > 0$ the partition Δ can be chosen so that $d(\Delta)$ is less than ε. Let T_i^* be a development of the triangle T_i. Thanks to the condition that in T_i each side is less than the sum of the other two, the triangle T_i^* does not degenerate into an interval. Pasting together the triangles T_i^* in the order in which the triangles T_i adjoin each other in the composition of the partition Δ, we obtain a two-dimensional polyhedron, which we denote by G_Δ. We map the polyhedron G_Δ topologically onto G so that the image of the triangle T_i^*, $i = 1, 2, \ldots, r$, is the triangle T_i and the sides of T_i^* are mapped isometrically onto the corresponding sides of T_i. For arbitrary points X, $Y \in G$ we denote by $\rho_\Delta(X, Y)$ the distance on the polyhedron G_Δ between the points that go into X and Y under the given topological map.

Specifying a sequence $\Delta_1, \Delta_2, \ldots, \Delta_\nu$ of partitions of G such that $d(\Delta_\nu) \to 0$ as $\nu \to \infty$, by means of the construction we have described we obtain a sequence of polyhedral metrics $\rho_\nu = \rho_{\Delta_\nu}$ defined in G. As $\nu \to \infty$ these metrics converge to the metric ρ of the domain G. The proof of the last assertion uses the axiom of boundedness of curvature. Since this assertion by itself is quite likely, we shall not dwell on how it can be proved.

The partition Δ of G into triangles obtained by applying Lemma 6.3.2 cannot, generally speaking, be a triangulation of G in the sense of the definition in 3.1, since the vertices of the triangles defined may be interior points of the sides of other triangles (see Fig. 31). (In the given case our terminology differs from that used in Aleksandrov and Zalgaller (1962).) In this connection we need to define more exactly what such a vertex of the partition Δ is. Let S be the union of the sides of the triangles forming the partition. A point $X \in S$ will be called a *vertex* of Δ if it is either a vertex of one of the triangles forming Δ or such that no neighbourhood of it with respect to S is a simple arc.

To complete the proof of Theorem 6.2.1 it is sufficient to establish that the absolute curvature and absolute turn of the boundary of G_Δ are bounded above by a constant that depends only on G. The main instrument for constructing the required estimate is the Gauss-Bonnet formula for polyhedra. In addition, we use the inequalities of Lemma 6.3.1 and the singularities of the partition Δ of G, namely the fact that all the vertices of the triangles that constitute Δ and are not boundary points of G are transit points.

Namely, the following propositions are true.

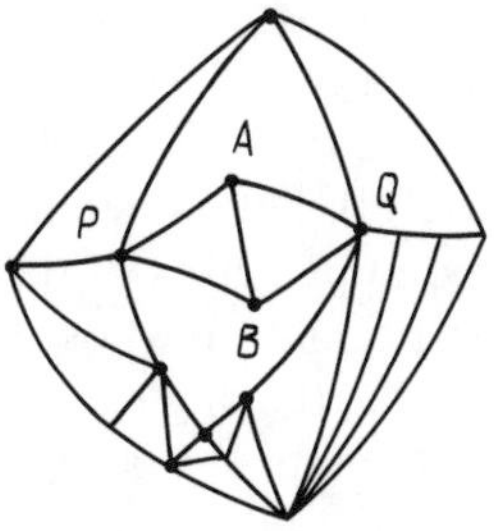

Fig. 31

I. For the triangles $T_1, T_2, \ldots, T_m$ that constitute Δ we have

$$\sum_{i=1}^{m} \bar{\nu}^+(T_i) \leqslant \Omega_0. \tag{6.6}$$

II. Let X be a vertex of one of the triangles of the partition Δ, and $\bar{\alpha}_1, \bar{\alpha}_2, \ldots, \bar{\alpha}_n$ the angles at X of all the triangles of the partition that meet at X. If for some i the point X is an interior point of a side of a triangle T_i of the given partition, we put $\bar{\alpha}_i = \pi$. Let

$$\theta_\Delta(X) = \bar{\alpha}_1 + \bar{\alpha}_2 + \cdots + \bar{\alpha}_n. \tag{6.7}$$

If X is an interior point of G, then

$$\theta_\Delta(X) \geqslant 2\pi. \tag{6.8}$$

If X belongs to the boundary of G, and X is distinct from each of the points $A_1, A_2, \ldots, A_m$, then

$$\theta_\Delta(X) \geqslant \pi. \tag{6.9}$$

We shall carry out the proof of Propositions I and II later. For now we just mention that the fulfilment of the inequalities (6.8) and (6.9) follows from the fact that the points X to which they refer are transit points.

Using Propositions I and II, we now show how to obtain the required estimates for $|\omega|(G_\Delta) + |\kappa|(\partial G_\Delta)$.

We use the Gauss-Bonnet formula for polyhedra. Let us apply it to the polyhedron G_Δ, which is homeomorphic to a disc, and so its Euler characteristic is equal to 1. On the basis of Theorem 5.3.2 we have

$$\omega(G_\Delta) + \kappa(\partial G_\Delta) = 2\pi.$$

Moreover,

$$\omega(G_\Delta) = \omega^+(G_\Delta) - \omega^-(G_\Delta),$$
$$\kappa(\partial G_\Delta) = \kappa^+(\partial G_\Delta) - \kappa^-(\partial G_\Delta).$$

Hence

$$\omega^-(G_\Delta) + \kappa^-(\partial G_\Delta) = \omega^+(G_\Delta) + \kappa^+(\partial G_\Delta) - 2\pi. \tag{6.10}$$

To obtain upper bounds for $\omega^-(G_\Delta)$ and $\kappa^-(\partial G_\Delta)$ it is sufficient to obtain upper bounds for $\omega^+(G_\Delta)$ and $\kappa^+(\partial G_\Delta)$. These bounds are established by means of Propositions I and II stated above.

Let us find a bound for $\omega^+(G_\Delta)$. Let X be an arbitrary vertex of the polyhedron G_Δ, and $\bar{X}$ the corresponding vertex of the partition Δ. Let $T_{l_1}, T_{l_2}, \ldots, T_{l_k}$ be all the triangles of the partition Δ on whose boundary $\bar{X}$ lies. Let $\bar{\alpha}_i$ be the angle at $\bar{X}$ of the triangle T_{l_i}, and α_i the angle at X of the triangle $T_{l_i}^*$. If $\bar{X}$ is an interior point of a side of $T_{l_i}^*$, we assume that $\bar{\alpha}_i = \alpha_i = \pi$. In this case X is an interior point of a side of T_{l_i} and $\bar{\alpha}_i - \alpha_i = 0$. If $\bar{X}$ is a vertex of T_{l_i}, then it follows from Lemma 6.3.1 that $\bar{\alpha}_i - \alpha_i \leqslant \bar{\nu}^+(T_i)$. Summing over $i = 1, 2, \ldots, k$, we obtain

$$\theta_\Delta(\overline{X}) - \theta(X) \leqslant \sum_i \bar{\nu}^+(T_i), \tag{6.11}$$

where in the sum on the right we take only those triangles T_i for which the point X is a vertex. If $\overline{X}$ is an interior vertex of the paritition Δ, then by Proposition II we have $\theta_\Delta(\overline{X}) \geqslant 2\pi$, from which it follows that in this case

$$\omega(X) = 2\pi - \theta(X) \leqslant \theta_\Delta(\overline{X}) - \theta(X) \leqslant \sum_i \bar{\nu}^+(T_i).$$

Since the right-hand side of the last inequality is non-negative, this enables us to conclude that

$$\omega^+(X) \leqslant \sum_i \bar{\nu}^+(T_i).$$

Let us sum the given inequality over the set of all interior vertices of the polyhedron G_Δ. For each $i = 1, 2, \ldots, r$ the triangle T_i in the sum obtained on the right occurs as many times as it has vertices lying inside G, that is, at most three times, and as a result we arrive at the following inequality:

$$\omega^+(G_\Delta) \leqslant 3 \sum_{i=1}^{r} \bar{\nu}^+(T_i).$$

By Proposition I the sum on the right does not exceed Ω_0, and we obtain the following bound for $\omega^+(G_\Delta)$:

$$\omega^+(G_\Delta) \leqslant 3\Omega_0. \tag{6.12}$$

Let $\overline{X}$ be an arbitrary boundary vertex of the partition Δ, and X, as above, the point corresponding to it on the polyhedron G_Δ. We again use the inequality (6.11). If $\overline{X}$ is not one of the points A_i, then by Proposition II we have $\theta_\Delta(\overline{X}) \geqslant \pi$, and so in this case we have

$$\sum \bar{\nu}^+(T_i) \geqslant \theta_\Delta(\overline{X}) - \theta(X) \geqslant \pi - \theta(X) = \kappa(X) \tag{6.13}$$

(the summation is taken over all triangles T_i for which X is a vertex). If $\overline{X} = A_i$, then $\theta_\Delta(\overline{X}) \geqslant 0$ and we obtain the bound

$$\kappa(X) = \pi - \theta(X) \leqslant \pi + \theta_\Delta(\overline{X}) - \theta(X) \leqslant \sum_i \bar{\nu}^+(T_i) + \pi. \tag{6.14}$$

The right-hand sides of (6.13) and (6.14) are non-negative, so it follows that these inequalities remain true if the expression on the left is replaced by $[\kappa(X)]^+$. Summing over the set of all boundary vertices of G_Δ, we obtain

$$\kappa^+(\partial G_\Delta) = \sum_i [\kappa(X)]^+ \leqslant m\pi + \sum_i \bar{\nu}^+(T_i) \leqslant m\pi + 3\Omega_0 \tag{6.15}$$

and the required bound for $\kappa^+(\partial G_\Delta)$ is established. From the equality (6.10) and the inequalities (6.12) and (6.15) it then follows that

$$\omega^-(G_\Delta) + \kappa^-(\partial G_\Delta) \leqslant 6\Omega_0 + (m-2)\pi.$$

Hence, finally, we obtain the required bound

$$|\omega|(G_\Delta) + |\kappa|(\partial G_\Delta) \leqslant 12\Omega_0 + (2m-2)\pi.$$

To complete the proof of Theorem 6.2.1 it remains to establish the validity of Propositions I and II.

Proof of Proposition I. Let $T = ABC$ be a simple triangle in a two-dimensional manifold. We assume that $\bar{\nu}_A^+(T) > 0$. We shall consider all possible pairs of points $X \in [AB]$, $Y \in [AC]$, $X \neq A$, $Y \neq A$, for which $\inf\limits_{[XY]} \delta(AXY)$ is positive. To find the quantity $\nu_A^+(T)$ only such pairs need to be taken into consideration. Let X and Y be such that $\delta(AXY) > 0$ for any shortest curve $[XY]$. Then none of these shortest curves passes through A. By virtue of the bounded convexity of the triangle T (we recall that this condition occurs in the definition of a simple triangle), there is a shortest curve that joins the points X and Y and lies in T. Let us find the rightmost of such shortest curves (the rightmost in relation to the arc $[XA] \cup [AY]$). It is not difficult to show that the triangle AXY, which has as sides this shortest curve and the intervals $[AX]$ and $[AY]$ of sides of the triangle T, is simple. We have $\delta(AXY) \geqslant \inf\limits_{[XY]} \delta(AXY)$. This enables us to conclude that $\bar{\nu}_A^+(T)$ does not exceed the least upper bound of the excesses of the simple triangles AXY contained in T. Moreover, it follows that $\bar{\nu}^+(T)$ does not exceed the least upper bound of the excesses of the simple triangles contained in T. From what we have said it obviously follows that for the triangles $T_1, T_2, \ldots, T_r$ that constitute the partition Δ not only the sum of their excesses, but also the sum

$$\sum_{i=1}^{r} \bar{\nu}^+(T_i)$$

does not exceed the constant Ω_0. Thus Proposition I is proved.

Proof of Proposition II. Let us restrict ourselves to an assertion relating to the case when X is an interior point of a side of the polyhedron G. Let $[XX_0]$, $[XX_1], \ldots, [XX_m]$ be the sides of triangles of the partition Δ for which X is a common vertex. Let $\bar{\alpha}_i$ be the angle between the shortest curves $[XX_{i-1}]$ and $[XX_i]$ at the point X. On the plane we specify a point O arbitrarily and construct non-overlapping triangles $A_{i-1}OA_i$, $i = 1, 2, \ldots, m$, such that $|OA_i| = |XX_i|$, $i = 0, 1, \ldots, m$, and $\angle A_{i-1}OA_i = \bar{\alpha}_i$ (Figs. 32, 33). We assume that $\bar{\alpha}_1 + \bar{\alpha}_2 + \cdots + \bar{\alpha}_m < \pi$. We denote by δ the smallest of the lengths of the intervals $[OA_i]$. Let B_0 and B_m be the points on the rays OA_0 and OA_m that are at a distance δ from the point O. The rays OA_i, $i = 1, 2, \ldots, m-1$, lie inside the angle A_0OA_m. By the assumption we have made, the value of this angle is less than π, and so it is convex. Consequently, the interval joining B_0 and B_m intersects each of these arcs. Let B_i be the point of the arc OA_i that lies on the interval $[B_0B_m]$. We put $d_i = |OB_i|$, $i = 0, 1, 2, \ldots, m$, $d_0 = d_m = \delta$. Let $l = |B_0B_m|$, $l_i = |B_{i-1}B_i|$, $l = l_1 + l_2 + \cdots + l_m$. Obviously we have $l < d_0 + d_m$ (the inequality is strict!). We put $d_0 + d_m - l = \varepsilon$. On the shortest curve $[XX_i]$, $i = 0, 1, 2, \ldots, m$,

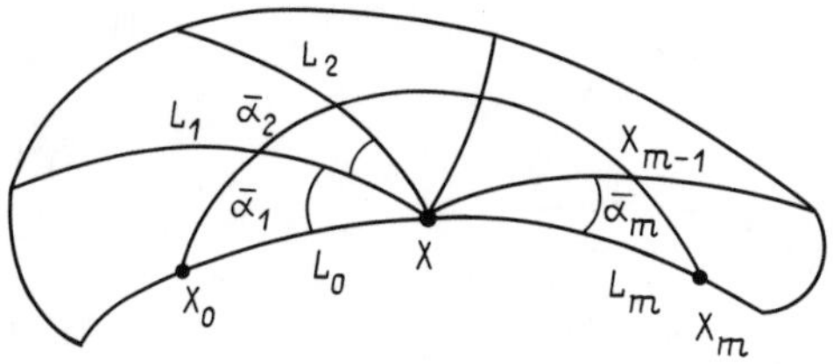

Fig. 32

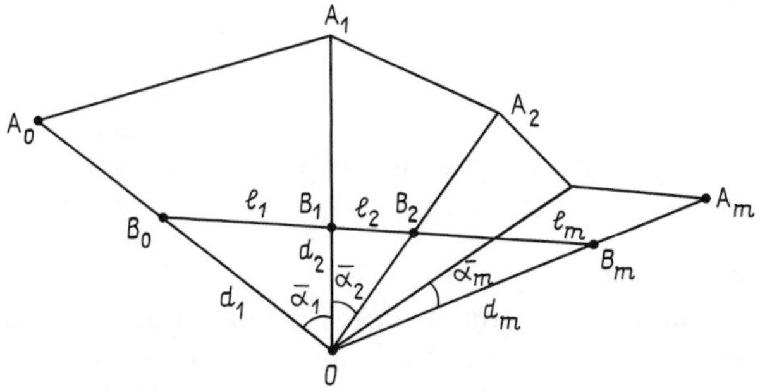

Fig. 33

we mark the point $Y_{i,n}$ such that $|XY_{i,n}| = d_i/n$ (n is a natural number). Let $\lambda_{i,n}$ be the length of the shortest curve $[Y_{i-1,n}Y_{i,n}]$. We construct a development of the triangle $XY_{i-1,n}Y_{i,n}$. Let $\gamma_{i,n}$ be the angle of this development at the vertex corresponding to the point X. We have

$$\varlimsup_{n\to\infty} \gamma_{i,n} \leqslant \bar{\alpha}_i. \tag{6.16}$$

We construct the simple triangle obtained from the development of the triangle $XY_{i-1,n}Y_{i,n}$ by a similarity transformation with coefficient of expansion n. Two sides of this triangle are equal to d_{i-1} and d_i, and the third is equal to $n\lambda_{i,n}$ and the angle opposite it is equal to $\gamma_{i,n}$. In the triangle $OB_{i-1}B_i$ the sides $[OB_{i-1}]$ and $[OB_i]$ are equal to d_{i-1} and d_i respectively, and the angle between them is equal to $\bar{\alpha}_i$. The inequality (6.16) enables us to conclude that $l_i = |B_{i-1}B_i| \geqslant \varlimsup_{n\to\infty} n\lambda_{i,n}$. So there is a number n_0 such that $n\lambda_{i,n} < l_i + (\varepsilon/m)$ when $n \geqslant n_0$ for each $i = 1, \ldots, m$. Hence we conclude that when $n \geqslant n_0$

$$|Y_{0,n}Y_{m,n}| \leqslant \sum_{i=1}^{m} \lambda_{i,n} < \sum_{i=1}^{m} \frac{l_i}{n} + \frac{\varepsilon}{n} = \frac{1}{n}(l + \varepsilon) = \frac{1}{n}(d_0 + d_m) = |Y_{0,n}O| + |OY_{m,n}|.$$

We obtain a contradiction with the fact that the arc $[X_0X] \cup [XX_m]$ is a shortest curve. For the case when X is a boundary vertex Proposition II is proved.

In the case when X is an interior vertex the triangles T_i are situated as shown in Fig. 34. The dotted line denotes the shortest curve passing through the point X (X is a transit point!). This shortest curve splits the neighbourhood of X into

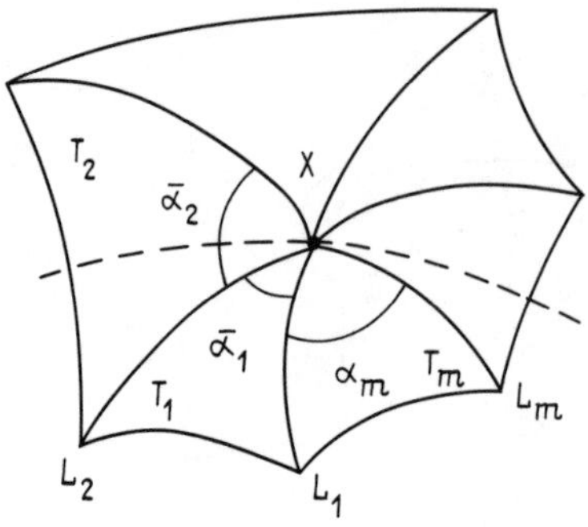

Fig. 34

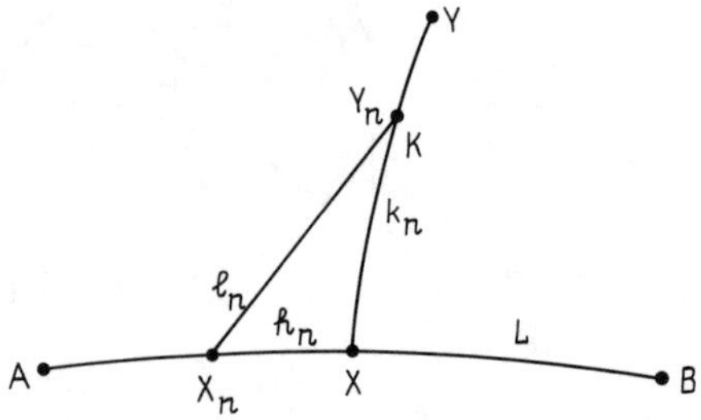

Fig. 35

two semicircles. Applying the inequality we have proved to each of these semicircles, we conclude that in the given case $\bar{\alpha}_1 + \bar{\alpha}_2 + \cdots + \bar{\alpha}_m \geqslant 2\pi$.

6.4. Proof of Lemma 6.3.1.

Lemma 6.3.1 contains an estimate of the difference between the angles of a triangle in a space with intrinsic metric and the corresponding angles of its development. It is obvious from the preceding text that this estimate is used substantially in the proof of the first approximation theorem.

Next, R denotes a locally compact metric space with intrinsic metric, and ρ denotes this metric. All the arguments relate to a domain of the space R in which any two points can be joined by a shortest curve.

Lemma A. *Let $L = [AB]$ be a shortest curve in R, $X(s)$, $0 \leqslant s \leqslant l$, its parametrization, where the parameter s is arc length, and Y a point outside L. For $s \in [0, l]$ we put $z(s) = |YX(s)|$. We fix arbitrarily a point $X = X(s)$, where $0 < s \leqslant l$, and a shortest curve L, and join X and Y by a shortest curve. We denote the greatest lower bound of the upper angles at X between the arc $[AX]$ and all possible shortest curves $[XY]$ by $\bar{\xi}(s)$. For any $s \in [0, l]$ we have*

$$\left(\frac{dz}{ds}\right)_{l.l.^1}(s) = \varliminf_{h \to +0} \frac{z(s) - z(s-h)}{h} \geqslant \cos \bar{\xi}(s).$$

Proof (see Fig. 35). We fix arbitrarily the value $s \in (0, l]$ and the shortest curve $K = [XY]$, where $X = X(s)$. Let (s_n), $n = 1, 2, \ldots$, be an arbitrary sequence of points of the interval $(0, s)$ such that $s_n \to s$ as $n \to \infty$. We put $s - s_n = h_n$, $X(s_n) = X_n$ and on the shortest curve K we specify a sequence of points (Y_n) such that $Y_n \to X$ as $n \to \infty$ and at the same time $h_n/|XY_n| \to 0$ as $n \to \infty$. We put $|XY_n| = k_n$, $|X_nY_n| = l_n$. By the triangle inequality we have

$$|YY_n| \geqslant |YX_n| - |X_nY_n|,$$

that is, $z(s) - k_n \geqslant z(s_n) - l_n$. Hence

$$\frac{z(s) - z(s_n)}{h_n} \geqslant \frac{k_n - l_n}{h_n}.$$

[1] short for: left lower

Let ξ_n be the angle of the development of the triangle X_nY_nX at the vertex corresponding to the point X. We have

$$\cos \xi_n = \frac{h_n^2 + k_n^2 - l_n^2}{2h_n k_n} = \frac{k_n - l_n}{h_n} + \frac{h_n^2 - (k_n - l_n)^2}{2h_n k_n}.$$

Furthermore, $|k_n - l_n| \leqslant h_n$, and so

$$\frac{k_n - l_n}{h_n} \geqslant \cos \xi_n - \frac{h_n}{2k_n}.$$

From the condition $h_n/k_n \to 0$ as $n \to \infty$ we obtain

$$\varliminf_{n\to\infty} \frac{z(s) - z(s_n)}{s - s_n} \geqslant \varliminf_{n\to\infty} \frac{k_n - l_n}{h_n} \geqslant \varliminf_{n\to\infty} \cos \xi_n.$$

We have

$$\varliminf_{n\to\infty} \cos \xi_n = \cos\left(\varlimsup_{n\to\infty} \xi_n\right) \geqslant \cos \bar{\xi},$$

where $\bar{\xi}$ is the upper angle between the shortest curves K and L at the point X. Since the sequence $s_n \to s - 0$ is taken arbitrarily, it follows that

$$\left(\frac{dz}{ds}(s)\right)_{l.l.} \geqslant \cos \bar{\xi}.$$

The shortest curve $K = [XY]$ was chosen arbitrarily. The left-hand side of the last inequality does not depend on the choice of K. Going over to the greatest lower bound over the set of all shortest curves joining Y and X, we obtain the required result.

Lemma B. *In the notation of Lemma A let $\alpha(s)$ and $\xi_0(s)$ be the angles of the development of the triangle $AX(s)Y$ at the vertices corresponding to the points A and $X = X(s)$ respectively. If* $\sin \xi_0(s) \neq 0$, *then*

$$\left(\frac{d\alpha}{ds}\right)_{l.l.}(s) = \varliminf_{h\to+0} \frac{\alpha(s) - \alpha(s-h)}{h} \geqslant \frac{1}{s}\cdot\frac{\cos \bar{\xi}(s) - \cos \xi_0(s)}{\sin \xi_0(s)}.$$

Proof. Suppose we are given a planar triangle with sides x, y, z. Let α be the angle opposite the side z. We have

$$z = \sqrt{x^2 + y^2 - 2xy \cos \alpha}.$$

Fixing y, we regard z as a function of the variables x and α. Obviously z is a differentiable function of x and α. We have $dz = \dfrac{x - y\cos\alpha}{z}dx + \dfrac{xy}{z}\sin\alpha\, d\alpha$.

Let ξ_0 denote the angle opposite the side y. Then we obtain $\dfrac{\sin\alpha}{z} = \dfrac{\sin\xi_0}{y}$. Moreover, $x - y\cos\alpha = z\cos\xi_0$ and finally $dz = \cos\xi_0\,dx + x\sin\xi_0\,d\alpha$. Hence it follows that if we give x and α increments Δx and $\Delta\alpha$, then the increment of z

can be represented as

$$\Delta z = \cos \xi_0 \Delta x + x \sin \xi_0 \Delta \alpha + \theta(\Delta x, \Delta \alpha)(|\Delta x| + |\Delta \alpha|),$$

where $\theta(\Delta x, \Delta \alpha) \to 0$ as $\Delta x \to 0$ and $\Delta \alpha \to 0$. We put $x = s$, $z = z(s)$, $y = z(0)$, $\alpha = \alpha(s)$. Obviously $\alpha(s)$ is a continuous function of s. We put $\Delta s = -h$, $\Delta z = z(s-h) - z(s)$, $\Delta \alpha = \alpha(s-h) - \alpha(s)$. After obvious transformations we obtain

$$\frac{z(s) - z(s-h)}{h} = \cos \xi_0 + [s \sin \xi_0 - \sigma\theta(h, \Delta\alpha)] \frac{\alpha(s) - \alpha(s-h)}{h} - \theta(h, \Delta\alpha). \tag{6.17}$$

Here $\sigma = \operatorname{sign} \Delta\alpha$. As $h \to 0$, $\alpha(s) - \alpha(s-h) \to 0$ and so $\theta(h, \Delta\alpha) \to 0$. We choose a sequence of quantities $h_m \to +0$, $m = 1, 2, \ldots$, such that $\dfrac{\alpha(s) - \alpha(s-h_m)}{h_m} \to \left(\dfrac{d\alpha}{ds}\right)_{l.l.}(s)$ as $m \to \infty$. Then, putting $h = h_m$ in (6.17) and proceeding to the limit, we obtain

$$\begin{aligned}&\left(\frac{d\alpha}{ds}\right)_{l.l.}(s) s \sin \xi_0 + \cos \xi_0 \\ &\quad = \lim_{m\to\infty} \frac{z(s) - z(s-h_m)}{h_m} \geqslant \varliminf_{h\to+0} \frac{z(s) - z(s-h)}{h} \geqslant \cos \bar{\xi}(s)\end{aligned}$$

and so the lemma is proved.

Relying on Lemmas A and B, we can now prove Lemma 6.3.1.

Despite what we have proved, we assume that

$$\bar{\alpha} - \alpha_0 \geqslant \bar{v}_A^+ + 3\varepsilon, \tag{6.18}$$

where $\varepsilon > 0$. Since ABC is one of the triangles AXY considered in the definition of $\bar{v}_A^+$, we can choose the shortest curve BC so that

$$(\bar{\alpha} - \alpha_0) + (\bar{\beta} - \beta_0) + (\bar{\gamma} - \gamma_0) \leqslant \bar{v}_A^+ + \varepsilon. \tag{6.19}$$

In view of (6.18), it follows from (6.19) that at least one of the quantities $\beta_0 - \bar{\beta}$ and $\gamma_0 - \bar{\gamma}$ is not less than ε.

We put $|AB| = x_0$, $|AC| = y_0$, $|BC| = z_0$. For points $X \in [AB]$ and $Y \in [AC]$ such that $|AX| = x > 0$, $|AY| = y > 0$ we denote by $\alpha(x, y)$ the angle of the development of the triangle AXY at its vertex corresponding to A. We prove that there are numbers $x_1 > 0$, $y_1 > 0$ such that $x_1 \leqslant x_0$, $y_1 \leqslant y_0$, $x_1 y_1 < x_0 y_0$ and

$$\alpha(x_0, y_0) - \alpha(x_1, y_1) > M \ln \frac{x_0 y_0}{x_1 y_1} \tag{6.20}$$

where $M = \sin^2 \dfrac{\varepsilon}{2} > 0$. We assume that $\beta_0 - \bar{\beta} \geqslant \varepsilon$. Since $\beta_0 - \bar{\beta} > 0$, we have $\beta_0 > 0$. Moreover, in this case $\beta_0 \neq \pi$, because otherwise the angle $\bar{\beta}$ would be equal to π. We now use the inequality of Lemma B, putting $s = x_0$, $\xi_0(s) = \beta$, $\bar{\xi}(s) = \bar{\beta}$ in it. We obtain

$$\left(\frac{\partial\alpha}{\partial x}(x_0, y)\right)_{l.l.} \geqslant \frac{1}{x_0}\cdot\frac{\cos\bar{\beta} - \cos\beta_0}{\sin\beta_0} \geqslant \frac{2}{x_0}\cdot\sin\frac{\bar{\beta}+\beta_0}{2}\cdot\sin\frac{\beta_0-\bar{\beta}}{2}.$$

By hypothesis, $\pi > \beta_0 \geqslant \bar{\beta} + \varepsilon$, $\bar{\beta} > 0$. Hence it follows that $\pi - \varepsilon \geqslant \bar{\beta}$, $\beta_0 \geqslant \varepsilon$ and so $\pi - \frac{\varepsilon}{2} \geqslant \frac{\bar{\beta}+\beta_0}{2} \geqslant \frac{\varepsilon}{2}, \frac{\pi}{2} \geqslant \frac{\beta_0 - \bar{\beta}}{2} \geqslant \frac{\varepsilon}{2}$. Hence we conclude that

$$\left(\frac{\partial\alpha}{\partial x}(x_0, y)\right)_{l.l.} \geqslant \frac{2\sin^2\frac{\varepsilon}{2}}{x_0}. \tag{6.21}$$

By the definition of the left lower derivative,

$$\left(\frac{\partial\alpha}{\partial x}(x_0, y)\right)_{l.l.} = \lim_{x\to x_0-0}\frac{\alpha(x_0, y) - \alpha(x, y)}{x_0 - x}.$$

We have

$$\frac{1}{x_0} = \lim_{x\to x_0-0}\frac{\ln x_0 - \ln x}{x_0 - x}. \tag{6.22}$$

From (6.21) and (6.22) it follows that there is an $x_1 \in (0, x_0)$ such that

$$\alpha(x_0, y_0) - \alpha(x_1, y_0) > \sin^2\frac{\varepsilon}{2}(\ln x_0 - \ln x_1),$$

and we see that (6.20) is satisfied if we put $y_1 = y_0$. By construction $x_1 \leqslant x_0$, $y_1 \leqslant y_0$, $x_1 y_1 < x_0 y_0$.

Similarly we conclude that if $\gamma_0 - \bar{\gamma} \geqslant \varepsilon$ there is a $y_1 \in (0, y_0)$ such that

$$\alpha_0(x_0, y_0) - \alpha_0(x_0, y_1) > M\ln\frac{y_0}{y_1}.$$

Putting $x_1 = x_0$, we obtain a pair (x_1, y_1) for which (6.20) is satisfied, and $x_1 \leqslant x_0$, $y_1 \leqslant y_0$, $x_1 y_1 < x_0 y_0$. Let $t \geqslant 0$ be the greatest lower bound of $x_1 y_1$ on the set of all pairs (x_1, y_1) for which (6.20) is satisfied, and $0 < x_1 \leqslant x_0$, $0 < y_1 \leqslant y_0$, $x_1 y_1 < x_0 y_0$. We find a sequence of values $x = x_n$, $y = y_n$, $n = 1, 2, \ldots$, for which (6.20) is satisfied such that $x_n \to \bar{x}$, $y_n \to \bar{y}$ as $n \to \infty$, where $0 \leqslant \bar{x} \leqslant x_0$, $0 \leqslant \bar{y} \leqslant y_0$, $x_n y_n \to t$. For each n we have

$$\alpha_0(x_0, y_0) - \alpha_0(x_n, y_n) > M\ln\frac{x_0 y_0}{x_n y_n}. \tag{6.23}$$

Hence it follows that $\bar{x} > 0$, $\bar{y} > 0$, since otherwise the right-hand side of (6.23) tends to ∞ as $n \to \infty$, which is impossible, since its left-hand side does not exceed π. Proceeding to the limit, we obtain

$$\alpha_0(x_0, y_0) - \alpha_0(\bar{x}, \bar{y}) \geqslant M\ln\frac{x_0 y_0}{t}. \tag{6.24}$$

Since $t < x_0 y_0$, the right-hand side of (6.24) is positive. Consider a triangle $AB'C'$ with vertex B' on the shortest curve AB and vertex C' on the shortest curve BC such that $\rho(A, B') = \bar{x}$, $\rho(A, C') = \bar{y}$. The quantity $\bar{v}_A^+$, calculated from the triangle $AB'C'$, obviously does not exceed $\bar{v}_A$ for the triangle ABC, $\alpha_0(\bar{x}, \bar{y}) \leqslant \alpha_0(x_0, y_0)$, hence it follows that (6.18) is preserved for the triangle $AB'C'$. Similar arguments applied to $AB'C'$ enable us to conclude that there are numbers $x' > 0$, $y' > 0$ for which $x' \leqslant \bar{x}$, $y' \leqslant \bar{y}$, where at least one of the inequalities is strict and

$$\alpha_0(\bar{x}, \bar{y}) - \alpha(x', y') > M \ln \frac{\bar{x}\,\bar{y}}{x'y'}. \tag{6.25}$$

Adding (6.24) and (6.25), we obtain

$$\alpha_0(x_0, y_0) - \alpha(x', y') > M \ln \frac{x_0 y_0}{x'y'}.$$

At the same time, $x'y' < \bar{x}\,\bar{y} = t$. This obviously contradicts the definition of t. The resulting contradiction proves the lemma.

6.5. Proof of the Second Theorem on Approximation. The proof of the second theorem on approximation (Theorem 6.3.2), which we give here, follows the paper Reshetnyak (1962). It relies on an analogue of Theorem 5.6.1 for the case of manifolds of bounded curvature and is simpler than the proof given in Aleksandrov and Zalgaller (1962).

We shall say that a domain G in a metric space (R, ρ) with intrinsic metric is convex if G is metrically connected and $\rho_G(X, Y) = \rho(X, Y)$ for any $X, Y \in G$. We shall call a two-dimensional manifold Q with intrinsic metric that is homeomorphic to a closed disc a *convex cone* if Q is isometric to a convex domain on the cone $Q(\theta)$, where $\theta < 2\pi$.

Let (R, ρ) be a two-dimensional manifold with intrinsic metric, and $U \subset R$ an open domain in R. We assume that in U there is specified a sequence of metrics (ρ_n), $n = 1, 2, \ldots$, such that the following conditions are satisfied:

1) as $n \to \infty$ the metrics ρ_n converge to a metric ρ_U;

2) the space (U, ρ_n) is a polyhedron;

3) there is a constant $A < \infty$ such that $\omega_{\rho_n}^+(U) < A$ for all n;

4) as $n \to \infty$ the set functions $\omega_{\rho_n}^+(E)$ converge weakly to some set function $\omega_0(E)$.

Condition 4 implies the following: a) $\omega_0(A) \geqslant \overline{\lim\limits_{n\to\infty}}\, \omega_{\rho_n}^+(A)$ for any closed set $A \subset U$; b) $\omega_0(V) \leqslant \underline{\lim\limits_{n\to\infty}}\, \omega_{\rho_n}^+(V)$ for any open set $V \subset U$.

Lemma 6.5.1. *Suppose that all the conditions 1–4 listed above are satisfied. Let Γ be a simple closed curve contained in U, and $G \subset U$ the open set those boundary is the curve Γ. We assume that $\omega_0(G) < 2\pi$ and that the curve Γ is rectifiable with respect to the metric ρ. Then there is a convex cone Q such that $\omega(Q) \leqslant \omega_0(G)$, and there is a contraction φ of the cone Q onto the domain $G \cup \Gamma$ which maps the boundary of the cone one-to-one and preserves arc lengths.*

The lemma is obtained from Theorem 5.6.1 by a limiting process. We outline a proof of it that does not go into details of a technical character.

We first construct a sequence of simple closed curves (Γ_n), $n = 1, 2, \ldots$, that converges to the curve Γ and is such that Γ_n for each n is a polygonal line on a manifold with polyhedral metric (U, ρ_n), and $s_{\rho_n}(\Gamma_n) \to s_\rho(\Gamma)$ as $n \to \infty$. Let G_n be the domain bounded by the simple closed curve Γ_n. It is easy to verify that

$$\omega_0(G \cup \Gamma) \geqslant \varlimsup_{n\to\infty} \omega_{\rho_n}^+(G_n \cup \Gamma_n).$$

If $\omega_0(\Gamma) = 0$, it is not difficult to verify that, whatever the value of $\varepsilon > 0$, $\omega_{\rho_n}^+(G_n) \leqslant \omega_{\rho_n}^+(G_n \cup \Gamma_n) < \omega_0(G) + \varepsilon$ for sufficiently large n. In particular, $\omega_\rho^+(G_n) < 2\pi$ for sufficiently large n. Having constructed a convex cone Q_n and its contraction map φ_n onto the domain $G_n \cup \Gamma_n$ that satisfies the conditions of Theorem 5.6.1, by means of a limiting process we then obtain the required convex cone Q.

However, it may happen that $\omega_0(\Gamma) \neq 0$. In this case we cannot conclude that $\omega_{\rho_n}^+(G_n) < \omega_0(G) + \varepsilon$ for sufficiently large n. The quantity $\omega_0(\Gamma)$ will be non-zero if there occurs an accumulation of the positive part of the curvature of G_n close to its boundary. The difficulty that arises in this way can be overcome as follows. Suppose we are given $h > 0$. We denote by $\Gamma_n(h)$ the set of all points of G_n for which $\rho_n(X, \Gamma_n) < h$, and suppose that $G_n(h) = G_n \setminus K_n(h)$, $G(h) = \{X \in G | \rho_0(X, \Gamma) \geqslant h\}$. Then

$$\omega_0(G(h)) \geqslant \varlimsup_{n\to\infty} \omega_{\rho_n}^+(G_n(h))$$

and so, whatever the value of $\varepsilon > 0$, for sufficiently large n we have

$$\omega_{\rho_n}^+(G_n(h)) < \omega_0(G(h)) + \varepsilon \leqslant \omega_0(G) + \varepsilon.$$

This shows that the excess of the curvature $\omega_{\rho_n}^+$, because of which we cannot assert that $\omega_{\rho_n}^+(G)$ does not exceed $\omega(G) + \varepsilon$ for sufficiently large n, is concentrated in the domain $\Gamma_n(h)$. It is impossible to apply Theorem 5.6.1 directly to the domain $G_n(h)$, since its topological nature may be rather complicated, and what is more we cannot say anything about the length of the boundary of $G_n(h)$. We therefore proceed as follows. We construct a sequence of values $h_n \to 0$, $h_n > 0$ for all n, such that $\varlimsup_{n\to\infty} \omega_{\rho_n}^+(G_n(h_n)) \leqslant \omega_0(G)$. Such a sequence is obtained by the standard diagonal process. We now transform each of the polyhedra (G_n, ρ_n) as follows. We consider all its vertices whose curvature is positive and which lie in the band $\Gamma_n(h_n)$. We join each of these vertices to the curve Γ_n by a shortest curve. We cut G_n along these shortest curves and to each of the cuts we paste a circular sector whose radius is equal to the distance from the corresponding vertex to Γ_n, and the angle is equal to the curvature at this vertex. After this procedure the curvature of each of those vertices of the polyhedron (G_n, ρ_n) lying in the band $\Gamma_n(h_n)$ at which it is positive becomes equal to zero, and the length of the boundary increases by no more than $h_n \omega_{\rho_n}^+(G_n) \leqslant M h_n$. As a result we obtain a sequence of polyhedra (G_n', ρ_n') homeomorphic to a disc such that $\varlimsup_{n\to\infty} \omega_{\rho_n}^+(G_n') \leqslant$

$\omega_0(G)$, and the length of the boundary of G_n' has the limit $s_{\rho_0}(\Gamma)$. Having constructed a convex cone Q_n satisfying the conditions of Theorem 5.6.1, for each of the polyhedra G_n' by a suitable limiting process we obtain a convex cone satisfying all the conditions of the lemma.

By means of Lemma 6.5.1, Theorem 6.3.2 is proved as follows.

We assume that a two-dimensional manifold R with intrinsic metric ρ satisfies the conditions of the lemma. Let X be an arbitrary point of R, let U be a neighbourhood of this point homeomorphic to a disc, and let ρ_n be a sequence of polyhedral metrics defined in U that converges to ρ as $n \to \infty$ and is such that $\omega_{\rho_n}^+(U) \leqslant \Omega < \infty$ for all n. Without loss of generality we can assume that the set functions $\omega_{\rho_n}^+(E)$ converge weakly in U to some set function $\omega_0(E)$. This can always be achieved by replacing the sequence ρ_n by a suitable subsequence.

Let T be an arbitrary simple triangle contained in U, let A, B, C be its vertices, and $K = [BC]$, $L = [AC]$, $M = [AB]$ its sides. The sides of T form a simple closed curve Γ. Let G be the interior of the triangle, and $\bar{\alpha}$, $\bar{\beta}$, $\bar{\gamma}$ the upper angles of the triangle at its vertices A, B, C. Let us prove that

$$\bar{\delta}(T) = \bar{\alpha} + \bar{\beta} + \bar{\gamma} - \pi \leqslant \omega_0(G).$$

If $\omega_0(G) \geqslant 2\pi$ the inequality is obvious, since each of the numbers $\bar{\alpha}$, $\bar{\beta}$, $\bar{\gamma}$ does not exceed π. Suppose that $\omega_0(G) < 2\pi$. Then by the lemma there is a convex cone Q that has a contraction map onto the domain G such that the boundary of Q is mapped onto the boundary of G with arc lengths preserved, and $\omega(Q) \leqslant \omega_0(Q)$. Let A_0, B_0, C_0 be points that belong to the boundary of Q and are such that $\varphi(A_0) = A$, $\varphi(B_0) = B$, $\varphi(C_0) = C$. Each of the arcs $[A_0B_0]$, $[B_0C_0]$, $[C_0A_0]$ that form the boundary of Q is a shortest curve. In fact, we have $l([A_0B_0]) = l([AB])$ by virtue of the fact that φ maps ∂Q onto the boundary of G with arc lengths preserved. Hence $l([A_0B_0]) = \rho(A, B) \leqslant \rho(A_0, B_0)$; since $l([A_0B_0]) \geqslant \rho(A_0, B_0)$, we have $l([A_0B_0]) = \rho(A_0, B_0)$. Let α_0, β_0, γ_0 denote the bounding angles of the cone Q at the points A_0, B_0, C_0. Then by the Gauss-Bonnet formula (Theorem 5.3.2) $\alpha_0 + \beta_0 + \gamma_0 - \pi = \omega(Q) \leqslant \omega_0(G)$. Let us prove that

$$\alpha_0 \geqslant \bar{\alpha}, \quad \beta_0 \geqslant \bar{\beta}, \quad \gamma_0 \geqslant \bar{\gamma}. \tag{6.26}$$

Let $X \in [AB]$ and $Y \in [AC]$ be arbitrary points on the sides of the triangle, and X' and Y' the corresponding points of the boundary of Q, so that $\varphi(X') = X$ and $\varphi(Y') = Y$. We put $z_0 = \rho(X', Y')$, $z = \rho(X, Y)$. Then $z_0 \geqslant z$. On the plane we construct triangles with sides x, y, z and x, y, z_0 respectively. Let $\gamma(X, Y)$ and $\gamma_0(X, Y)$ be the angles of these triangles at the vertices opposite to the sides z and z_0 respectively. Since $z_0 \geqslant z$ we have $\gamma_0(X, Y) \geqslant \gamma(X, Y)$. It remains to observe that if the points X and Y are sufficiently close to A, then $\gamma_0(X, Y) = \alpha_0$. Hence it obviously follows that

$$\bar{\alpha} = \varlimsup_{\substack{X \to A \\ Y \to A}} \gamma(X, Y) \leqslant \alpha_0;$$

the other two inequalities in (6.26) are proved similarly.

From (6.26) it follows that in the given case

$$\bar{\delta}(T) = \bar{\alpha} + \bar{\beta} + \bar{\gamma} - \pi \leqslant \omega(Q) \leqslant \omega_0(G).$$

Let $T_1, T_2, \ldots, T_m$ be an arbitrary finite collection of pairwise non-overlapping triangles homeomorphic to a disc and lying in the domain U. Let G_i be the interior of the triangle T_i. By what we have proved, for each i we have

$$\delta(T_i) \leqslant \omega_0(G_i).$$

Hence

$$\sum_{i=1}^{m} \delta(T_i) \leqslant \sum_{i=1}^{m} \omega_0(G_i) = \omega\left(\bigcup_{i=1}^{m} G_i\right) \leqslant \Omega.$$

Here we have used the fact that the sets G_i are pairwise disjoint. Thus, for any system of pairwise non-overlapping triangles homeomorphic to a disc and lying in U the sum of their excesses does not exceed Ω. We have thus established that for the two-dimensional manifold under consideration the axiom of boundedness of curvature is satisfied (even in a form somewhat stronger than that given in 6.2, we do not require the bounding convexity of the triangles). Thus Theorem 6.3.2 is proved.

§ 7. Analytic Characterization of Two-Dimensional Manifolds of Bounded Curvature

7.1. Theorems on Isothermal Coordinates in a Two-Dimensional Manifold of Bounded Curvature. As we mentioned above, in a neighbourhood of each point of a two-dimensional Riemannian manifold we can introduce a local coordinate system in which the metric quadratic form of the manifold is

$$ds^2 = \lambda(x, y)(dx^2 + dy^2). \tag{7.1}$$

Such a coordinate system is called *isothermal*. Finding an isothermal coordinate system reduces to the construction of a conformal map of a domain of a Riemannian manifold into the plane $\mathbb{R}^2$. We note that the size of the domain in a Riemannian manifold in which we can introduce a coordinate system of the stated type depends only on the topological structure of the Riemannian manifold. In particular, in any domain of a Riemannian manifold that is homeomorphic to an open disc on the plane we can introduce an isothermal coordinate system (the truth of this follows from some general results of the theory of functions of a complex variable, namely from the principle of uniformization, whose formulation will be given later).

Considering isothermal coordinate systems, it is advisable to identify the plane $\mathbb{R}^2$ with the set of complex numbers $\mathbb{C}$, and a point $(x, y) \in \mathbb{R}^2$ with the complex number $z = x + iy$. In this notation the differential quadratic form (7.1) is $ds^2 = \lambda(z)|dz|^2$.

The function $\lambda(z) \equiv \lambda(x, y)$ on the right-hand side of (7.1) is expressed in terms of the integral curvature of the Riemannian manifold by the formula

$$\lambda(z) = \exp\left\{\frac{1}{\pi}\int_G \ln\frac{1}{|z-\zeta|}\,\omega(d\zeta) + h(z)\right\}, \tag{7.2}$$

where $G \subset \mathbb{C}$ is the range of values for the given coordinate system and $h(z)$ is a harmonic function. As we shall show later, the concept of integral curvature can be defined for an arbitrary two-dimensional manifold of bounded curvature. For the latter it is a completely additive set function. In this connection there naturally arises the conjecture that in a neighbourhood of each point of a two-dimensional manifold of bounded curvature we can introduce a local coordinate system in which the metric of the manifold is defined by a line element of the form (7.1), and the function $\lambda(z)$ is expressed by a formula of the form (7.2) with the difference that in contrast to a Riemannian manifold ω is an additive set function of sufficiently general form. In particular, it can be not absolutely continuous with respect to area, as happens in the case of a Riemannian manifold. In favour of such a conjecture, in particular, is the circumstance that the possibility of introducing an isothermal coordinate system in a domain of a Riemannian manifold depends only on its topological structure. Let us explain this idea. To obtain a coordinate system of some special type in a neighbourhood of a point of a two-dimensional manifold of bounded curvature it is natural to consider first a sequence of Riemannian metrics ρ_n defined in a neighbourhood of the given point and converging to the metric of the manifold in this neighbourhood. For each of the Riemannian metrics we construct in a neighbourhood of the given point a coordinate system of the form of interest to us. After this we shall attempt to obtain the original coordinate system in the manifold under consideration by a limiting process. This attempt may turn out to be untenable, in particular because the size of the domain where the coordinate system in question is defined for the metric ρ_n tends to zero as $n \to \infty$. For an isothermal coordinate system of this kind the "vanishing" of the domain of definition cannot happen.

Here we give statements of theorems on the representation of a two-dimensional manifold of bounded curvature by an isothermal line element. The later parts of the present section will be devoted to a proof of them. We note that the different features of the proof have independent significance as a means of investigating two-dimensional manifolds of bounded curvature.

We first give some definitions. Let A be a Borel subset of the complex plane $\mathbb{C}$. We shall call an arbitrary totally additive set function defined on the totality of all Borel subsets A simply a measure in the set A (thus, it is not assumed that the measure is non-negative). For any measure ω there are defined non-negative measures ω^+ and ω^-, specified in A and such that $\omega = \omega^+ - \omega^-$. The measure ω^+ is called the *positive part* or *upper variation* of ω and is defined by the relation

$$\omega^+(E) = \sup_{H \subset E} \omega(H),$$

where the supremum is taken over the totality of all Borel sets $H \subset E$. Correspondingly, ω^- is called the *negative part* or *lower variation* of the measure ω. The sum $\omega^+ + \omega^-$ is denoted by $|\omega|$ and called the *total variation* of the measure ω.

Let G be a domain in $\mathbb{C}$. We assume that in G there are specified a measure ω and a harmonic function h. We put

$$\lambda(z; \omega, h) = \exp\left\{\frac{1}{\pi}\int_G \ln\frac{1}{|z-\zeta|}\omega(d\zeta) + h(z)\right\}. \tag{7.3}$$

Here the integral is understood in the sense of Lebesgue and Stieltjes. The function $\lambda(z; \omega, h)$ is defined and is finite for almost all $z \in G$.

Functions that have a representation of the form (7.3) are closely connected with a class of functions that plays an important role in the theory of partial differential equations, namely, with the class of subharmonic functions. Referring the reader to the corresponding literature (Privalov (1937), Hayman and Kennedy (1978)) for the details, we observe that a function $u(z)$ defined in a bounded domain G on the plane $\mathbb{C}$ is *subharmonic* if and only if it has the representation

$$u(z) \equiv u(z; \mu, h) = \frac{1}{\pi}\int_G \ln|z-\zeta|\mu(d\zeta) + h(z),$$

where μ is a non-negative measure in the domain G, and $h(z)$ is a harmonic function. From (7.2) we obtain

$$\ln\lambda(z) = \ln\lambda(z; \omega, h) = u_1(z) - u_2(z),$$

where $u_1(z) = u(z; \omega^-, h)$, $u_2(z) = u(z; \omega^+, 0)$. Thus, $\ln\lambda(z)$ is the difference of two subharmonic functions or, as we say, $\ln\lambda(z)$ is a δ-subharmonic function.

Let G be a bounded domain on the plane and $\lambda(z) \equiv \lambda(z; \omega, h)$ a function defined in G. Let L be an arbitrary rectifiable simple arc lying in the domain G, and $z(s)$, $0 \leqslant s \leqslant l$, a parametrization of it, where the parameter s is arc length. We consider the function $\lambda[z(s)]$. We observe that the function $\lambda(z)$ is only defined almost everywhere in the domain G. A rectifiable simple arc is a set of measure zero. Therefore a priori it is not clear whether the quantity $\lambda[z(s)]$ is defined for a set of values of s of positive measure. However, we can show that $\lambda[z(s)]$ is defined for almost all $s \in [0, l]$. This follows from the next lemma.

Lemma 7.1.1 (Reshetnyak (1960)). *Let $u(z)$ be a subharmonic function in a domain G. Then $u(z)$ is defined for all $z \in G$, and $-\infty \leqslant u(z) < \infty$. Let E be the set of those z for which $u(z) = -\infty$. Then for any $\varepsilon > 0$ and any $\alpha > 0$ there is a sequence of open discs ($B_m = B(c_m, r_m)$), $m = 1, 2, \ldots$, such that*

$$E \subset \bigcup_{m=1}^{\infty} B_m$$

and

$$\sum_{m=1}^{\infty} r_m^{\alpha} < \varepsilon.$$

The result of the lemma (which was known long before the cited work of Reshetnyak) can be restated as follows. For any $\alpha > 0$ the α-dimensional Hausdorff measure of the set of those $z \in G$ for which $u(z) = -\infty$ is equal to zero.

From Lemma 7.1.1 it follows, in particular, that if the function $\lambda(z)$ has a representation of the form (7.3), then the set of those z for which $\lambda(z)$ is not defined (they can be points for which the functions $u_1(z)$ and $u_2(z)$ in (7.3) become $-\infty$ simultaneously) can be covered by a sequence of discs for which the sum of the radii is less than ε, whatever $\varepsilon > 0$ is.

Thus, for almost all $s \in [0, l]$ the quantity $\lambda[z(s)]$ is defined and is finite. The function $\lambda[z(s)]$ is non-negative and measurable, and so the integral

$$s_\lambda(L) = \int_L \sqrt{\lambda(z)}\,|dz| \equiv \int_0^l \sqrt{\lambda[z(s)]}\,ds$$

is defined. We shall call $s_\lambda(L)$ the *length of the simple arc L with respect to the line element $ds^2 = \lambda(x, y)(dx^2 + dy^2) = \lambda(z)|dz|^2$*, where $\lambda(z) = \lambda(z; \omega, h)$. We note that $s_\lambda(L)$ may be equal to ∞.

Let z_1 and z_2 be two arbitrary points in the domain G, where $z_1 \neq z_2$. We denote the greatest lower bound of $s_\lambda(L)$ on the set of rectifiable simple arcs joining z_1 and z_2 by $\rho_\lambda(z_1, z_2)$. We put $\rho_\lambda(z, z) = 0$. It can happen that $\rho_\lambda(z_0, z) = \infty$ for some points $z_0 \in G$, whatever the point $z \in G$ is, where $z \neq z_0$. If z_0 has this property, then we shall call it a *point at infinity of G with respect to the line element $\lambda(z)|dz|^2$*. If z_0 is a point at infinity, then $\omega(\{z_0\}) \geqslant 2\pi$. Conversely, if $\omega(\{z_0\}) > 2\pi$, then z_0 is a point at infinity with respect to the line element $\lambda(z)|dz|^2$. If $\omega(\{z_0\}) = 2\pi$, the point z_0 may be at infinity or may not be. The role of the number 2π becomes clear if we observe that by representing the integral on the right-hand side of (7.2) in the form

$$\frac{1}{\pi}\int_G \ln\frac{1}{|z-\zeta|}\,\omega(d\zeta) = \frac{\omega(\{z_0\})}{2\pi}\ln\frac{1}{|z-z_0} + \frac{1}{\pi}\int_{G\setminus\{z_0\}} \ln\frac{1}{|z-\zeta|}\,\omega(d\zeta),$$

we deduce that $\sqrt{\lambda(z)}$ can be represented in the form

$$\sqrt{\lambda(z)} = |z - z_0|^{-\alpha}\lambda_0(z), \quad \alpha = \frac{\omega(\{z_0\})}{2\pi}.$$

The factor $\lambda_0(z)$ is such that the main contribution to the behaviour of the function $\sqrt{\lambda(z)}$ as $z \to z_0$ is given by the factor $|z - z_0|^{-\alpha}$. If $\omega(\{z_0\}) > 2\pi$, then $\alpha > 1$, and in this case the integral of $\sqrt{\lambda(z)}$ along any rectifiable curve passing through z_0 is equal to ∞. In the case $\omega(\{z_0\}) = 2\pi$ the convergence of the integral depends on the behaviour of $\lambda_0(z)$ as $z \to z_0$. By the properties of totally additive set functions, all the points at infinity with respect to the line element $\lambda(z)|dz|^2$ are isolated.

Theorem 7.1.1 (Reshetnyak (1960)). *Suppose that a function $\lambda(z) \equiv \lambda(z; \omega, h)$ is specified in a domain G on the plane $\mathbb{C}$. Let $\tilde{G}$ be the domain obtained from G by excluding points at infinity with respect to the given line element $\lambda(z)|dz|^2$. Then* *the function $\rho_\lambda(z_1, z_2)$ is the intrinsic metric in $\tilde{G}$ compatible with the topology of $\tilde{G}$ as a subset of $\mathbb{C}$, and the metric space $(\tilde{G}, \rho_\lambda)$ is a two-dimensional manifold of bounded curvature.*

Let G be a domain on the plane $\mathbb{C}$, and $\rho = \rho_\lambda$ a metric in this domain that can be defined from some function $\lambda(z)$ that has a representation of the form (7.3). For brevity we shall say that ρ is a *subharmonic metric* in the domain G.

Theorem 7.1.2. *Let M be a two-dimensional manifold of bounded curvature. Any point $p \in M$ has a neighbourhood U which in the induced metric ρ_U is isometric to some flat domain G with subharmonic metric ρ_λ.*

We make two remarks about the theorems we have stated.

1. In Reshetnyak (1960), where we present a proof of Theorems 7.1.1 and 7.1.2, the metric $\rho_\lambda(z_1, z_2)$ is defined somewhat differently from the way we have done here, namely as the greatest lower bound of $s_\lambda(L)$ on the set of all simple arcs that satisfy a stronger condition then rectifiability, namely the condition that the variation of turn is bounded (the exact statement of this condition is given in 7.2). However, by virtue of the results in Reshetnyak (1963a) such a definition of $\rho_\lambda(z_1, z_2)$ is equivalent to the one given above.

2. Let U be a domain in a two-dimensional manifold of bounded curvature, isometric to a flat domain G with some subharmonic metric ρ_λ. Any isometric map of (U, ρ_U) onto (G, ρ_λ) is called an *isothermal coordinate system* in U.

Suppose we are given an open set $U \subset \mathbb{C}$ and a map $f: U \to \mathbb{C}$, $f(z) = u(z) + iv(z)$ for any $z \in U$, where u and v are real functions. We shall say that *the map f is conformal* if one of the functions $f(z)$ and $\bar{f}(z) = u(z) - iv(z)$ is an analytic function of z. This means that the functions $u = \operatorname{Re} f$ and $v = \operatorname{Im} f$ satisfy either the system of equations $u_x = v_y$, $u_y = -v_x$ or the system of equations $u_x = -v_y$, $u_y = v_x$. In both cases we obtain

$$|f'(z)|^2 = 1/2([u_x(z)]^2 + [u_y(z)]^2 + [v_x(z)]^2 + [v_y(z)]^2).$$

The concept of a conformal map has a single geometrical meaning, which we do not dwell on, to save space, referring the reader to textbooks where the beginnings of the theory of functions of a complex variable are presented.

Theorem 7.1.3 (Huber (1960)). *Let $\varphi: U \to \mathbb{C}$, $\psi: V \to \mathbb{C}$ be isothermal coordinate systems in a two-dimensional manifold of bounded curvature. We assume that the set $U \cap V$ is not empty, and let $G = \varphi(U \cap V)$, $H = \psi(U \cap V)$. Then G and H are open sets in $\mathbb{C}$ and the function $\theta(z) = \psi[\varphi^{-1}(z)]$ is a conformal map of G onto H. If $ds^2 = \lambda(z)|dz|^2$ is the line element corresponding to the coordinate system φ, and $d\sigma^2 = \mu(w)|dw|^2$ is the line element corresponding to the coordinate system ψ, then for all $z \in G$ we have*

$$\lambda(z) = \mu[\theta(z)]|\theta'(z)|^2. \tag{7.4}$$

The assertion that the map $\theta = \psi \circ \varphi^{-1}$ is conformal is established in Huber (1960) and Reshetnyak (1963b). The rule for transforming a line element on going over from one isothermal coordinate system to another, which can be defined by (7.4), is proved in Huber (1960).

As a corollary of Theorem 7.1.3 we obtain an analytic characterization of an arbitrary two-dimensional manifold of bounded curvature. For this purpose we require the concept of a Riemann surface. Suppose we are given a two-dimensional manifold M. Then we shall say that M is a *Riemann surface* if there is specified in M a set $\mathfrak{A}$ of local coordinate systems whose domains of definition cover M, and for any two charts $\varphi: U \to \mathbb{C}$ and $\psi: V \to \mathbb{C}$ that belong to $\mathfrak{A}$ and are such that $U \cap V$ is non-empty the maps $\varphi \circ \psi^{-1}$ and $\psi \circ \varphi^{-1}$ are conformal. Charts belonging to the set $\mathfrak{A}$ are called *basic charts of the Riemann surface M*. If for any overlapping basic charts $\varphi: U \to \mathbb{C}$, $\psi: V \to \mathbb{C}$ of a Riemann surface M the maps $\varphi \circ \psi^{-1}$ and $\psi \circ \varphi^{-1}$ are holomorphic functions, then M is said to be *oriented*.

Any domain on the plane $\mathbb{C}$ naturally represents a Riemann surface if as a unique basic chart we take the identity map of this domain into the plane.

Any two-dimensional Riemannian manifold turns into a Riemann surface if for the basic charts we take all possible isothermal coordinate systems of this manifold. In fact, from the formulae for transforming the components of the metric tensor of a Riemannian manifold on going over from one local coordinate system to another it follows easily that when these coordinate systems are isothermal the components of the transition function (u, v) satisfy either the Cauchy-Riemann system of equations $u_x = v_y$, $u_y = -v_x$ or the system $u_x = -v_y$, $u_y = v_x$.

In particular, a sphere and more generally any surface in three-dimensional Euclidean space that satisfies the regularity conditions adopted in differential geometry is a Riemann surface.

Let M_1 and M_2 be arbitrary Riemann surfaces, and $f: M_1 \to M_2$ a continuous map. Then we shall say that the map f is conformal if for any basic chart $\varphi: U \to \mathbb{C}$ of M_1 and any basic chart $\psi: V \to \mathbb{C}$ of M_2 the map $\psi \circ f \circ \varphi^{-1}$ is conformal. We shall call $\psi \circ f \circ \varphi^{-1}$ a *representation* of the map f by means of the charts φ and ψ. The Riemann surfaces M_1 and M_2 are said to be *equivalent* if there is a topological map of M_1 onto M_2 that is conformal.

The *principle of uniformization*, of which we spoke earlier, is expressed by the following statements. Any Riemann surface homeomorphic to an open disc in $\mathbb{C}$ is conformally equivalent to either a disc or the plane $\mathbb{C}$. A Riemann surface homeomorphic to a sphere is conformally equivalent to a sphere.

Let M be a two-dimensional Riemann surface. Then we shall say that on M there is determined a *quadratic differential* σ if for any basic chart $\varphi: U \to \mathbb{C}$ there is defined a differential quadratic form $ds^2 = \lambda(z)|dz|^2$, specified on the set $G = \varphi(U)$ and called a representation of the quadratic differential σ in the local coordinate system φ, and the differential quadratic forms $\lambda(z)|dz|^2$ and $\mu(w)|dw|^2$ corresponding to two arbitrary overlapping basic charts φ and ψ are connected by (7.4).

From Theorem 7.1.3 it obviously follows that if M is a two-dimensional manifold of bounded curvature, then M is a Riemann surface for which the isothermal local coordinate systems form a set of basic charts. To any isothermal coordinate system $\varphi: U \to \mathbb{C}$ of the manifold M there corresponds a differential quadratic form $ds^2 = \lambda(z)|dz|^2$, where the function $\lambda(z)$ is defined on the set $G = \varphi(U)$ and has a representation of the form (7.2). If $\lambda(z)|dz|^2$ and $\mu(w)|dw|^2$ are the differential quadratic forms that correspond to two isothermal coordinate systems $\gamma: U \to \mathbb{C}$ and $\psi: V \to \mathbb{C}$, then the functions λ and μ are connected by (7.4). Thus, on the Riemann surface M there is defined a quadratic differential σ.

The converse assertion is true. Namely, we have the following theorem.

Theorem 7.1.4 (Huber (1960)). *Let M be a two-dimensional Riemann surface. We assume that on M there is specified a quadratic differential σ such that for any basic local coordinate system of the surface the coefficient $\lambda(z)$ in the representation of the differential σ in this coordinate system has a representation of the form* (7.3). *If M is connected, then on it we can define an intrinsic metric ρ_σ such that for any basic coordinate system $\varphi: U \to \mathbb{C}$ any point $p \in U$ has a neighbourhood $V \subset U$ having the property that for any $p_1, p_2 \in V$ we have*

$$\rho_\sigma(p_1, p_2) = \rho_\lambda[\varphi(p_1), \varphi(p_2)],$$

where ρ_λ is the metric defined in the domain $G = \varphi(U)$ by the line element $ds^2 = \lambda(z)|dz|^2$, the representation of σ in the coordinate system φ, as described above. The metric ρ_σ satisfying the given condition is unique and the metric space (M, ρ_σ) is a two-dimensional manifold of bounded curvature.

Theorem 7.1.4 enables us to establish a connection between the theory of manifolds of bounded curvature and the theory of functions of a complex variable. This gives a way of studying the structure in the large of two-dimensional manifolds of bounded curvature by the methods of the theory of functions. Some examples on the application of this theorem to the study of two-dimensional Riemannian manifolds of bounded curvature can be found in Huber (1954), for example.

If the function ω in (7.2) is an indefinite integral of some function $\mathscr{K}(z)$, continuous and satisfying a Hölder condition with exponent α, $0 < \alpha \leqslant 1$, inside G, then the function $\lambda(z)$ has continuous partial derivatives of the first and second order[2], and these derivatives also satisfy a Hölder condition with exponent[3] α. In this case the metric ρ_λ is Riemannian and for the point $z = x + iy$ the number $\mathscr{K}(z)$ is the Gaussian curvature of the metric at this point. Thus, from Theorem 7.1.1 it follows that a two-dimensional Riemannian manifold is a manifold of bounded curvature. It is true that we can still only prove

[2] We say that a function $\mathscr{K}: G \to \mathbb{R}$ satisfies a *Hölder condition* with exponent α, where $0 < \alpha \leqslant 1$, if there is a constant L such that $0 \leqslant L < \infty$ and $|\mathscr{K}(z_1) - \mathscr{K}(z_2)| \leqslant L|z_1 - z_2|^\alpha$ for any $z_1, z_2 \in G$. We shall say that $\mathscr{K}$ satisfies a Hölder condition with exponent α, $0 < \alpha \leqslant 1$, inside G if for any compact $A \subset G$ there is an $L(A) < \infty$ such that $|\mathscr{K}(z_1) - \mathscr{K}(z_2)| \leqslant L(A)|z_1 - z_2|^\alpha$ for any z_1, z_2 of A.

[3] Continuity of the function $K(z)$ is insufficient to guarantee the existence of continuous second derivatives of $\lambda(z)$, since, as we know, the solution of the Poisson equation $\Delta u(z) = f(z)$, where $f(z)$ is an arbitrary continuous function, may be a function that does not have continuous second derivatives (in this case the equation itself must be understood in a generalized sense, Petrovskij (1961)).

this under certain assumpions about a Riemannian metric. Here it is relevant to mention the following result. If in a domain U of a Riemannian manifold we introduce a local coordinate system such that the coefficients of the metric tensor of the manifold in this coordinate system are functions that have continuous second derivatives satisfying a Hölder condition with exponent α, then in this domain we can introduce an isothermal coordinate system, and the coefficient $\lambda(z)$ of the metric tensor in this coordinate system will have continuous second derivatives satisfying a Hölder condition.

As an example of the application of Theorem 7.1.4 we mention a theorem about isothermal coordinates on polyhedra.

Let M be a two-dimensional manifold with intrinsic metric ρ. We shall say that M is a *manifold of type* Σ if M is complete and homeomorphic to a sphere with finitely many points deleted.

We assume that M is a manifold of type Σ, M' is a set homeomorphic to M on the sphere $S(0, 1) \subset \mathbb{R}^3$ obtained by excluding certain points $a_1, a_2, \ldots, a_n$, and φ is a topological map of M' onto M. We take a point $y_0 \in M$ arbitrarily and let $\bar{B}(y_0, r)$ be the disc in M with centre y_0 and radius r. Since M is a complete two-dimensional manifold with intrinsic metric, the set $\bar{B}(y_0, r)$ is compact, whatever the value of $r > 0$. Hence it follows that the set $\varphi^{-1}[\bar{B}(y_0, r)] \subset M'$ is compact, and so none of the points $a_1, a_2, \ldots, a_n$ is a limit point. This enables us to conclude that for any sequence (x_m), $m = 1, 2, \ldots$, of points of the sphere $S(0, 1)$ that converge to one of the points a_i, for the points $y_m = \varphi(x_m)$ we have $\rho(y_0, y_m) \to \infty$ whatever the point $y_0 \in M$. In this connection we shall say later that to the point a_i there corresponds a point at infinity of the manifold M.

We now consider the case when M is a two-dimensional manifold of type Σ with a polyhedral metric, and the set of its vertices is finite. It is convenient to regard points at infinity of the manifold as vertices of it.

Theorem 7.1.5. *Let M be a polyhedron of type Σ satisfying all the conditions listed above. Then in M there is an isothermal coordinate system whose domain of definition is the whole manifold M except possibly for one point. For any such coordinate system the range of values is the whole plane except for a finite set of points, and the function $\lambda(z)$ in the representation $ds^2 = \lambda(z)|dz|^2$ of the line element of the polyhedron in this coordinate system has the representation*

$$\ln \lambda(z) = \sum_{j=1}^{n} \frac{\omega_j}{\pi} \ln \frac{1}{|z - z_j|} + C, \tag{7.5}$$

where $z_1, z_2, \ldots, z_n$ are the points corresponding to the vertices of the polyhedron, and C is a constant (we recall that to the number of vertices we also assign points at infinity). If z_j is an ordinary vertex of M, then ω_j is the curvature of this vertex. We have

$$\omega_1 + \omega_2 + \cdots + \omega_n \leqslant 4\pi. \tag{7.6}$$

The equality (7.5) can be rewritten in the form

$$\ln \lambda(z) = \frac{1}{\pi} \int_C \ln \frac{1}{|z - \zeta|} \omega(d\zeta) + C,$$

where ω is the measure, concentrated at the points $z_1, z_2, \ldots, z_n$, and $\omega(\{z_j\}) = \omega_j$.

We observe that if in the isothermal coordinate system an ordinary point of M corresponds to the point ∞ in the plane $\mathbb{C}$, then the equality sign holds in (7.6).

We mention the special case when the complete polyhedral manifold M is homeomorphic to a plane and has a unique finite vertex. In this case M is a cone. Choosing an isothermal coordinate in M so that to a point at infinity there corresponds the point ∞ of the extended complex plane $\overline{\mathbb{C}}$, we deduce that the matrix of M is defined by the line element

$$ds^2 = C\left(\frac{1}{|z|}\right)^{\omega/\pi} |dz|^2, \tag{7.7}$$

where $\omega < 2\pi$ is the curvature of the cone, and C is a constant. By changing the variables it is easy to arrange that the constant C in (7.7) is equal to 1.

The plane $\mathbb{C}$ with line element

$$ds^2 = C\frac{1}{|z|^2} \cdot |dz|^2 \tag{7.8}$$

is isometric to the surface of an infinite right circular cylinder. In this case the constant C has the following geometrical meaning: the quantity $2\pi\sqrt{C}$ is equal to the length of the section of a cylinder by a plane perpendicular to its generators.

7.2. Some Information about Curves on a Plane and in a Riemannian Manifold. Here we describe some auxiliary concepts that can be used in the proof of theorems stated in 7.1 on the representation of a two-dimensional manifold of bounded curvature by means of an isothermal line element.

Let (p, q) be an arbitrary pair of non-zero vectors on the plane $\mathbb{C}$. We assume that the vectors p and q do not lie on one line. The symbol $(\widehat{p, q})$ denotes the angle between them, taken with the + sign if the pair of vectors (p, q) is right, and with the − sign if the pair is left[4]. If the vectors p and q lie on one line, we put $(\widehat{p, q}) = 0$ if p and q have the same direction, and $(\widehat{p, q}) = \pi$ if they have opposite directions.

[4] Let $p = (p_1, p_2)$, $q = (q_1, q_2)$, $[p, q] = p_1 q_2 - p_2 q_1$. The pair (p, q) is right if $[p, q] > 0$, and left if $[p, q] < 0$. The angle $(\widehat{p, q})$ is formally defined as the number φ such that $-\pi < \varphi \leqslant \pi$ and $[p, q] = |p| \cdot |q| \cdot \sin \varphi$.

A *polygonal line on the plane* $\mathbb{C}$ is any finite sequence L of points $X_0, X_1, \ldots, X_m$ and vectors $p_i = \overline{X_{i-1}X_i}$ such that $X_{i-1} \neq X_i$ for any $i = 1, 2, \ldots, m$. We call the points X_i, $i = 0, 1, \ldots, m$, the *vertices* of the polygonal line L, and the vectors $p_i = \overline{X_{i-1}X_i}$ its *links*. We put

$$s(L) = \sum_{i=1}^{m} |p_i|, \quad \kappa(L) = \sum_{i=1}^{m-1} (\widehat{p_i, p_{i+1}}),$$

$$|\kappa|(L) = \sum_{i=1}^{m-1} |(\widehat{p_i, p_{i+1}})|.$$

The quantity $s(L)$ is the length of L, and we shall call $x(L)$ the *rotation* of L, and $|x|(L)$ the *absolute rotation* of L.

One of the main concepts of the theory presented here is that of the turn of a simple arc in a two-dimensional manifold of bounded curvature. If a simple arc K in such a manifold has a definite direction at each of its end-points, then we can define for it two numbers $\kappa_l(K)$ and $\kappa_r(K)$, called the left and right turns of the arc K. In the case when the manifold is the Euclidean plane with its usual metric, and the simpler arc K is a polygonal line, $\kappa_l(K) = \kappa(K)$, and $\kappa_r(K) = -\kappa(K)$. In the given case the terms "left turn" and "right turn" can be justified as follows. Let α_i and β_i be the angles between the rays X_iX_{i-1} and X_iX_{i+1} on the left and right of L respectively (see Fig. 36). Then $\alpha_i + \beta_i = 2\pi$ and $\kappa_l(L) = \sum_{i=1}^{m} (\pi - \alpha_i)$, $\kappa_r(L) = -\kappa(L) = \sum_{i=1}^{m} (\pi - \beta_i)$.

Let Γ be a simple closed polygonal line. It bounds a domain D homeomorphic to a disc. Let us orient Γ positively. Let $X_1, X_2, \ldots, X_m$ be successive vertices of it. We put $X_mX_1 = p_1, X_1X_2 = p_2, \ldots, X_{m-1}X_m = p_m$. Let α_i be the angle of the domain D at the point X_i. By the Gauss-Bonnet theorem for polyhedra we have

$$\kappa(\Gamma) = \sum_{i=1}^{m} (\pi - \alpha_i) = \sum_{i=1}^{m} (\widehat{p_i, p_{i+1}}) = 2\pi \tag{7.9}$$

(in the second sum we put $p_{m+1} = p_1$).

Let K and L be two oriented simple arcs with common beginning A and common end B, having no points in common other than A and B. We assume that K and L are polygonal lines. The simple arcs K and L bound some domain D. Let α and β be the angles of this domain at the points A and B. We assume that L lies to the left of K. The polygonal lines K and L form a simple closed

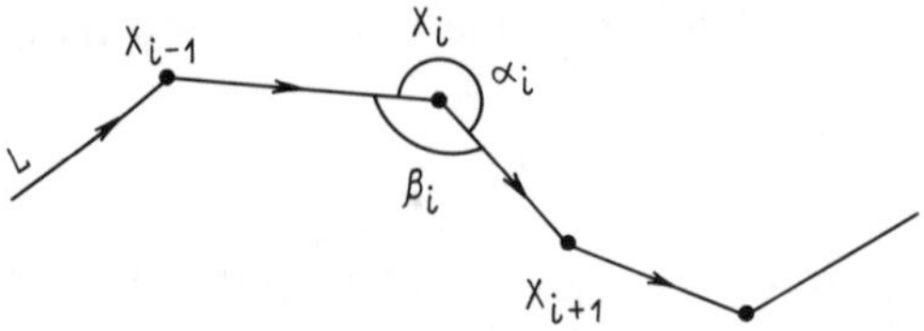

Fig. 36

polygonal line Γ, which we orient positively. Then by (7.9) we have $\kappa(\Gamma) = \kappa(K) + (\pi - \alpha) + (\pi - \beta) - \kappa(L) = 2\pi$. Hence $\kappa(K) = \alpha + \beta + \kappa(L)$. This equality can also be written as follows:

$$\kappa_l(K) = \alpha + \beta + \kappa_l(L). \tag{7.10}$$

We have $\kappa_l(K) = -\kappa_r(K)$, $\kappa_l(L) = -\kappa_r(L)$. Hence $\kappa_r(L) = \alpha + \beta + \kappa_r(K)$. Changing the notation, we deduce that if L lies to the left of K, then (7.10) is satisfied, and if L lies to the right of K, then

$$\kappa_r(K) = \alpha + \beta + \kappa_r(L). \tag{7.11}$$

Let L be a simple arc on the plane. Then we shall say that L is a *curve with bounded rotation* if the least upper bound of the absolute rotations of polygonal lines inscribed in L is finite. This least upper bound will be denoted by $|\kappa|(L)$ and called the *absolute rotation* of L.

Lemma 7.2.1. *Any simple arc L on a plane such that $|\kappa|(L) < \infty$ is rectifiable, and*

$$s(L) \leqslant \frac{d(L)}{2}[|\kappa|(L) + \pi], \tag{7.12}$$

where $d(L)$ is the diameter of L.

If a sequence of simple arcs (L_m), $m = 1, 2, \ldots$, converges to the simple arc L as $m \to \infty$, then

$$|\kappa|(L) \leqslant \varliminf_{m\to\infty} |\kappa|(L_m).$$

If the absolute rotations of the curves L_m, $m = 1, 2, \ldots$, are bounded in aggregate, that is, $|\kappa|(L_m) \leqslant H = \text{const} < \infty$ for all m, then the limiting curve L is rectifiable, and as $m \to \infty$ the lengths of L_m converge to the length of L.

Let L be a rectifiable simple arc on the plane. We shall say that L is *one-sidedly smooth* if it has a parametrization $z(t)$, $a \leqslant t \leqslant b$, such that for any $t \in (a, b]$ there is a *left derivative* $z_l'(t)$ and for any $t \in [a, b)$ there is a *right derivative* $z_r'(t)$, and $|z_l'(t)| \neq 0$, $|z_r'(t)| \neq 0$ at each point at which the corresponding derivative is defined, and the following conditions are also satisfied: if $a < t \leqslant b$, then

$$z_l'(t) = \lim_{u\to t-0} z_l'(u) = \lim_{u\to t-0} z_r'(u),$$

and if $a \leqslant t < b$, then

$$z_r'(t) = \lim_{u\to t+0} z_l'(u) = \lim_{u\to t+0} z_r'(u).$$

Any unilaterally smooth simple arc is rectifiable, and if $\zeta(s)$, $0 \leqslant s \leqslant l$, is its parametrization, where the parameter s is arc length, then the function $\zeta(s)$ also satisfies all the conditions stated in the definition of a unilaterally smooth simple arc. We have $|\xi_l'(s)| = 1$, $|\zeta_r'(s)| = 1$ at each point at which the corresponding derivatives are defined. We shall call the vectors $t_l(s) = \zeta_l'(s)$ and $t_r(s) = \zeta_r'(s)$ respectively the left and right tangent vectors at the point $\zeta(s)$ of the simple arc L.

Lemma 7.2.2. *Any simple arc with bounded rotation is unilaterally smooth. If L is a unilaterally smooth curve and $\zeta(s)$, $0 \leqslant s \leqslant l$, is its parametrization, where the parameter s is arc length, then in order that L should be a curve with bounded rotation it is necessary and sufficient that at least one of the functions $\zeta_l'(s)$ and $\zeta_r'(s)$ (we put $\zeta_l'(0) = \zeta_r'(0)$, $\zeta_r'(l) = \zeta_l'(l)$) should be a function of bounded variation.*

We observe that if one of the derivatives mentioned in the statement of the lemma is a function of bounded variation, then the other is automatically a function of bounded variation.

Let L be an oriented simple arc on the plane, p a point not belonging ro L, and $z(t)$, $a \leqslant t \leqslant b$, an arbitrary right parametrization of L. Then we can define a continuous function $\theta(t)$, $a \leqslant t \leqslant b$, such that $\theta(t)$ is the polar angle of the vector $z(t) - p$ for each t. We put $\varphi(t, p, z) = \theta(t) - \theta(a)$. We shall call $\theta(t)$ the angular function of the simple arc L with respect to the point p and the parametrization z. We put $\varphi(b, p, z) = \varphi(p, L)$. The quantity $\varphi(p, L)$ does not depend on the choice of right parametrization z of the simple arc L, and we shall call it the angle at which L is seen from the point p.

Lemma 7.2.3. *Let L be a simple arc on the plane, p a point not belonging to L, and $\varphi(t) = \varphi(t, p, z)$, $a \leqslant t \leqslant b$, the angular function of L with respect to the point p and the parametrizaion z of L. Then we have*

$$\operatorname*{Var}_a^b \varphi(t) \leqslant |\kappa|(L) + \pi. \tag{7.13}$$

Proofs of Lemmas 7.2.2 and 7.2.2 can be found in Reshetnyak (1960) and Reshetnyak (1963b), for example. Lemma 7.2.3 was obtained by Radon in his research into potential theory. Its proof reduces to a consideration of the case when the simple arc L is a polygonal line. For this case the proof of the unequality (7.13) is a problem in elementary geometry.

Let K be a simple arc on the plane. Let us orient K, and let A and B the its beginning and end respectively. We assume that K has tangents at A and B. Then with the arc K we can associate a number $\kappa(K)$, which we call the rotation of K. Let us construct arbitrarily a simple polygonal line L with ends A and B lying to the left of K and having no points in common with K other than A and B. Such a polygonal line exists, since K has tangents at A and B. The polygonal line L and the curve K bound a domain D homeomorphic to a disc. Let α and β be the angles of this domain at A and B. We put

$$\kappa(K) = \kappa_l(K) = \alpha + \beta + \kappa_l(L). \tag{7.14}$$

By means of (7.8) it is easy to show that the right-hand side of (7.14) does not depend on the choice of the polygonal line L, so it follows that the given definition is reasonable. We put $\kappa_r(K) = -\kappa(K)$. It is easy to verify (7.10) and (7.11) if K and L are not polygonal lines, but arbitrary simple arcs having tangents at A and B. The quantity $\kappa(K)$ will be called the rotation of the simple arc K, and the quantities $\kappa_l(K) = \kappa(K)$ and $\kappa_r(K) = -\kappa(K)$ the left and right turns of K respectively.

Let K be a one-sidedly smooth simple arc on the plane, and $x(t)$, $a \leqslant t \leqslant b$, a right parametrization of it. We denote the rotation of the arc $[x(a), x(t)]$ of the curve K by $\kappa(t)$. Then we have

$$|\kappa|(K) = \operatorname*{Var}_{a}^{b} \kappa(t),$$

which establishes a connection between the rotation and absolute rotation of a simple arc.

Let us consider the case when the simple arc L is of class C^2 and $z(s)$, $0 \leqslant s \leqslant l$, is a parametrization of it, where s is arc length. Then $z'(s) \neq 0$ for all $s \in [0, l]$, and there is defined a number $k(s)$, the curvature of L at the point $z(s)$. We have $z''(s) = k(s)v(s)$, where $v(s)$ is the unit normal vector at the point $z(s)$, chosen so that the pair of vectors $(z'(s), v(s))$ on the plane is a right pair. In this case the turn of the simple arc L is equal to

$$\kappa(L) = \int_0^l k(s)\, ds.$$

In the given case the quantity $|\kappa|(L)$ is equal to

$$\int_0^l |k(s)|\, ds.$$

The concepts and results concerning plane curves mentioned here form part of a general theory constructed in the monograph of Aleksandrov and Reshetnyak (1989).

Representation of the turn of a curve in a Riemannian manifold in an isothermal coordinate system. Henceforth we shall require an explicit expression for the geodesic curvature and the turn of a curve in a Riemannian manifold. For our purposes it is sufficient to consider the case when the coordinate system specified in the Riemannian manifold is isothermal.

We assume that in a flat domain G there is specified a Riemannian metric with line element

$$ds^2 = \lambda(x, y)(dx^2 + dy^2). \tag{7.15}$$

We shall assume that the function $\lambda(x, y)$ has all the partial derivatives of the first and second orders, and that these derivatives are continuous. Then, as we mentioned above, the function $\lambda(x, y) = \lambda(z)$ has the representation

$$\ln \lambda(z) = \frac{1}{\pi} \iint_G \ln \frac{1}{|z - \zeta|}\, d\omega(\zeta) + h(z), \tag{7.16}$$

where the set function $\omega(E)$ is the integral curvature and $h(z)$ is a harmonic function.

Let L be a smooth (of class C^2) oriented simple arc contained in the domain G, and $z(s) = x(s) + iy(s)$, $0 \leqslant s \leqslant l$, a parametrization of it, where the parameter

s is arc length in the sense of the geometry of the plane $\mathbb{R}^2$. The function $z(s)$ has a continuous second derivative and $|z'(s)| = 1$ for all $s \in [0, l]$. We denote by $k(s)$ the curvature in the sense of the geometry of $\mathbb{R}^2$ at the point $z(s)$ of the arc L. The geodesic curvature with respect to the line element (7.15) at the point $z(s)$ of L is then expressed by the formula

$$k_g(s) = \frac{k(s)}{\sqrt{\lambda}} + \frac{1}{2\sqrt{\lambda}}[-(\ln \lambda)_y x'(s) + (\ln \lambda)_x y'(s)].$$

Multiplying both sides of this equality by $\sqrt{\lambda}$ and integrating with respect to s, we deduce that the turn of the arc L with respect to the Riemannian metric with line element (7.15) is equal to

$$\kappa_g(L) = \int_0^l k(s)\,ds + \frac{1}{2}\int_0^l [-(\ln \lambda)_y\,dx + (\ln \lambda)_x\,dy].$$

Let $v(s)$ be the unit normal vector at the point $z(s)$ of the arc L, directed so that the pair of vectors $(t(s), v(s))$ is a right pair (that is, such that $[t(s), v(s)] > 0$). We have $t(s) = z'(s) = (x'(s), y'(s))$, $v(s) = (-y'(s), x'(s))$, from which it follows that

$$-(\ln \lambda)_y\,dx + (\ln \lambda)_x\,dy = -\frac{\partial}{\partial v}\{\ln \lambda\}\,ds.$$

Hence we obtain

$$\kappa_g(L) = \kappa(L) - \frac{1}{2}\int_0^l \frac{\partial}{\partial v}\{\ln \lambda(z)\}\,ds.$$

Substituting into the integrand the value of $\ln \lambda(z)$ from (7.16), we have

$$\kappa_g(L) = \kappa(L) - \frac{1}{2\pi}\iint\limits_G \left\{\int_0^l \frac{\partial}{\partial v}\left(\ln\frac{1}{|z(s) - \xi|}\right) ds\right\} \omega(d\zeta) - \frac{1}{2}\int_0^l \frac{\partial h(z)}{\partial v}\,ds.$$

We now observe that for the function

$$\lambda(z) = \ln\frac{1}{|z - \zeta|} = -\ln\sqrt{(x - \xi)^2 - (y - \eta)^2} \tag{7.17}$$

(we assume, as usual, that $z = x + iy$, $\zeta = \xi + i\eta$) we have

$$\begin{aligned}(\ln \lambda)_y\,dx - (\ln \lambda)_x\,dy &= -\frac{(y - \eta)\,dx - (x - \xi)\,dy}{(x - \xi)^2 + (y - \eta)^2} \\ &= d\left(\arctan\frac{y - \eta}{x - \xi}\right) = d\left(\arctan\frac{x - \xi}{y - \eta}\right).\end{aligned}$$

Hence it is clear that $(\ln \lambda)_y\,dx - (\ln \lambda)_x\,dy$ for the function $\lambda(z)$ defined by (7.17) is the angle at which the infinitesimal element (dx, dy) is seen from the point $\zeta = (\xi, \eta)$, and so the integral

$$\int_0^l \left(\frac{\partial}{\partial v}\ln\frac{1}{|z(s) - \zeta|}\right) ds$$

is equal to $\varphi(\zeta, L)$, the angle at which L is seen from the point ζ. Finally we obtain

$$\kappa_g(L) = \kappa(L) - \frac{1}{2\pi} \iint_G \varphi(\zeta, L)\, d\omega(\zeta) - \frac{1}{2} \int_L \frac{\partial h}{\partial \nu}(z)\, ds. \tag{7.18}$$

This formula is fundamental for the investigation of the differential properties of shortest curves on a manifold of bounded curvature in isothermal coordinates.

The equality (7.18) also enables us to get rid of the function $h(z)$ in the representation of the function $\lambda(z)$, which is given by (7.16), as the following proposition shows.

Lemma 7.2.4. *Let us assume that in the disc $\bar{B}(0, 1)$ on the plane $\mathbb{R}^2$ there is specified a Riemannian metric ρ defined by the line element $ds^2 = \lambda(z)(dx^2 + dy^2)$, and that the following conditions are satisfied:*

1) *the boundary of the disc $\bar{B}(0, 1)$ is a smooth curve of class C^2 with respect to the metric ρ, and the absolute turn of the circle $\Gamma(0, 1)$ in this metric does not exceed some constant $M_1 < \infty$;*

2) *there is a constant $M_2 < \infty$ such that $|\omega|(\bar{B}(0, 1)) \leqslant M_2$.*

Then the function $\lambda(z)$ has the representation

$$\lambda(z) = \exp\left\{\frac{1}{\pi} \int_{\bar{B}(0,1)} \ln \frac{1}{|z - \zeta|}\, d\tilde{\omega}(\zeta) + C\right\},$$

where C is a constant, and $\tilde{\omega}$ is a measure in the closed disc $\bar{B}(0, 1)$ such that

$$|\tilde{\omega}|(\bar{B}(0, 1)) \leqslant (M_1 + M_2) + 2\pi$$

and $\tilde{\omega}(E) = \omega(E)$ for any set E lying in the open disc $B(0, 1)$.

The measure $\tilde{\omega}$ is thus obtained from the measure ω by adding the measure concentrated on the circle $\Gamma(0, 1)$.

The proof of the lemma is based on the use of the following well-known representation of a harmonic function in terms of the value of its normal derivative on the boundary of a circle:

$$h(z) = h(0) + \frac{1}{\pi} \int_{\Gamma(0,1)} \frac{\partial h}{\partial \nu}(\zeta) \ln \frac{1}{|z - \zeta|}\, |d\zeta|. \tag{7.19}$$

On the plane we introduce polar coordinates (r, φ) and let Γ_θ be the arc of the circle $\Gamma(0, 1)$ corresponding to values $\varphi \in [0, \theta]$. We put

$$u(\theta) = \int_0^\theta \frac{\partial h}{\partial \nu}(e^{i\varphi})\, d\varphi = \int_{\Gamma_\theta} \frac{\partial h}{\partial \nu}(\zeta)\, |d\zeta|.$$

The equality (7.15) enables us to estimate the variation of the function $u(\theta)$ in terms of the variations of the functions

$$\kappa_g(\Gamma_\theta) = \int_0^\theta k_g(e^{i\varphi})\, d\varphi, \quad \kappa(\Gamma_\theta) = \theta$$

and

$$v(\theta) = \frac{1}{2\pi} \int_{\bar{B}(0,1)} \varphi(\zeta, \Gamma_\theta)\, d\omega(\zeta).$$

To estimate the variation of the function $v(\theta)$ we must take into account that $\varphi(\zeta, \Gamma_\theta)$ as a function of the variable θ is monotonic when $\zeta \in \bar{B}(0, 1)$, and $0 \leqslant \varphi(\zeta, \Gamma_\theta) \leqslant 2\pi$.

7.3. Proofs of Theorems 7.1.1, 7.1.2 and 7.1.3. The proofs of these theorems are based on a proposition which we call a theorem on the convergence of metrics (Reshetnyak (1960)).

Let G be a closed domain on the plane, bounded by finitely many simple closed curves of bounded rotation. Further, we consider various measures ω defined in G and satisfying the following condition: the integral

$$\ln \lambda(z; \omega) \equiv \frac{1}{\pi} \int_G \ln \frac{1}{|z - \zeta|}\, \omega(d\zeta)$$

is defined and finite for almost all $z \in D$.

Lemma 7.3.1. *Let ω be a measure defined in a domain $G \subset \mathbb{C}$ and let $\lambda(z) = \lambda(z; \omega)$. Then for any $z_1, z_2 \in G$ the quantity $\rho_{\lambda, G}(z_1, z_2)$ is equal to the greatest lower bound of the integral*

$$\int_K \lambda(z)\, |dz| \tag{7.20}$$

on the set of all curves K with bounded rotation contained in G and joining the points z_1 and z_2. On the set $\tilde{G}$, obtained from G by excluding points at infinity in the sense of the metric ρ_λ, the metric ρ_λ is intrinsic and compatible with the natural topology of G.

We observe that by definition $\rho_\lambda(z_1, z_2)$ is the greatest lower bound of the integral (7.20) on the set of all rectifiable curves joining the points z_1 and z_2 and contained in G.

Theorem 7.3.1. *Let (ω_n^1) and (ω_n^2), $n = 1, 2, \ldots$, be sequences of non-negative measures defined in the domain G and weakly converging to the measures ω_1 and ω_2 respectively as $n \to \infty$. Let $\omega_n = \omega_n^1 - \omega_n^2$, $\omega = \omega_1 - \omega_2$, $\lambda_n(z) = \lambda[z, \omega_n]$, $\lambda(z) = \lambda(z, \omega)$. Then as $n \to \infty$ the functions $\bar{\rho}_{\lambda_n}(z, \zeta)$ converge to the function $\bar{\rho}_\lambda(z, \zeta)$ uniformly on any closed set $A \subset G$ not containing points z such that $\omega_1(z) \geqslant 2\pi$.*

Proof of Theorem 7.1.1. Let G be a domain on the plane in which there is specified a function λ defined by (7.2), and suppose that the metric $\rho_\lambda(z_1, z_2)$ is defined with respect to λ as stated above. It is required to prove that the domain G, endowed with the metric ρ_λ, is a two-dimensional manifold of bounded curvature. The required proof is obtained by applying Theorem 6.2.2 and consists roughly in the following. We first establish that in a neighbourhood U of an arbitrary point $z_0 \in G$ the function $\lambda(z; \omega, h)$ can be represented in the form

$$\lambda(z; \omega, h) = C \exp \frac{1}{\pi} \int_G \ln \frac{1}{|z - \zeta|} \tilde{\omega}(d\zeta), \tag{7.21}$$

where $C > 0$ is a constant (it is necessary to distinguish the cases $\omega(\{z_0\}) < 2\pi$ and $\omega(\{z_0\}) = 2\pi$; see below). We then construct sequences of non-negative measures (ω_n^1) and (ω_n^2), each of which is concentrated on a finite set, and as $m \to \infty$ they converge weakly to $\tilde{\omega}^+$ and $\tilde{\omega}^-$ respectively. Let $\omega_n = \omega_n^1 - \omega_n^2$ and

$$\lambda_n(z) = C \exp \left\{ \frac{1}{\pi} \int_G \ln \frac{1}{|z - \zeta|} \omega_n(d\zeta) \right\}.$$

Since the measure ω_n is concentrated on a finite set, the metric ρ_{λ_n} is polyhedral. By Theorem 7.3.1, as $n \to \infty$ the metrics ρ_{λ_n} converge to the metric ρ_λ in a neighbourhood of the point z_0. The absolute curvature of the polyhedron (U, ρ_{λ_n}) is equal to $\omega_n^1(U) + \omega_n^2(U) \leqslant M = \text{const} < \infty$. Application of Theorem 6.2.2 now enables us to conclude that (G, ρ_λ) is a manifold of bounded curvature.

We now show how to realize the given programme in detail. First of all we show how to represent the function $\lambda(z; \omega, h)$ in the form (7.21).

We take a point $z_0 \in G$ arbitrarily. Let $\delta > 0$ be such that the closed ball $B_0 = \bar{B}(z_0, \delta) \subset G$. We have

$$\ln \lambda(z; \omega, h) = \frac{1}{\pi} \int_{B_0} \ln \frac{1}{|z - \zeta|} \omega(d\zeta) + \int_{G \setminus B_0} \ln \frac{1}{|z - \zeta|} \omega(d\zeta) + h(z).$$

We denote the second integral on the right by $h_1(z)$. The function $h_1(z)$ is harmonic in the disc B_0. Using the well-known representation of a harmonic function in a disc in terms of the values of its normal derivative along a circle, for $z \in B(z_0, \delta)$ we obtain

$$h(z) + h_1(z) = \ln C + \frac{1}{\pi} \int_0^{2\pi} \ln \frac{1}{|z - z_0 - e^{i\varphi}|} dv_0(\varphi), \tag{7.22}$$

where $v_0(\theta) = v(\theta) + v_1(\theta)$, $C = \text{const} > 0$,

$$v(\theta) = \delta \int_0^\theta \frac{\partial h}{\partial \nu}(z_0 + \delta e^{i\varphi}) \, d\varphi, \quad v_1(\theta) = \delta \int_0^\theta \frac{\partial h_1}{\partial \nu}(z_0 + \delta e^{i\varphi}) \, d\varphi.$$

The quantity $v_1(\theta)$ has the representation

$$v_1(\theta) = \int_{G \setminus B_0} \varphi(\zeta, \Gamma_\theta) \omega(d\zeta),$$

where Γ_θ is the arc $\{z | z = z_0 + \delta e^{i\varphi}, 0 \leqslant \varphi \leqslant \theta\}$ of the circle $\{z | |z - z_0| = \delta\}$, and $\varphi(\zeta, \Gamma_\theta)$ is its angular function. Obviously $v(\theta)$ and $v_1(\theta)$ are functions of bounded variation. Because of this the representation for the function $\lambda(z)$ in the disc B_0 can be written as follows:

$$\ln \lambda(z) = \frac{1}{\pi}\int_{B_0} \ln\frac{1}{|z-\zeta|}\tilde{\omega}(d\zeta) + \ln C.$$

Here $\tilde{\omega}(E) = \omega(E) + \sigma(E)$, where σ is a measure concentrated on the circle $\{z \,||z - z_0| = \delta\}$.

We now construct a sequence of polyhedral metrics defined in the disc B_0 and convergent in a neighbourhood of z_0 to the metric ρ_λ. We first assume that $\omega(\{z_0\}) < 2\pi$. On the basis of Theorem 6.2.2 this will prove that the domain G with metric ρ_λ is a two-dimensional manifold of bounded curvature.

Let $\{A_1, A_2, \ldots, A_{r_n}\}$ be a partition of the disc B_0 into pairwise disjoint Borel sets, the diameter of each of which does not exceed $1/n$, where n is an arbitrarily specified natural number. For each $i = 1, 2, \ldots, r_n$ we choose a point $z_i \in A_i$ arbitrarily and let ω_n^1 and ω_n^2 be measures in the disc B_0 concentrated on a finite set $\{z_1, z_2, \ldots, z_{r_n}\}$, and let $\omega_n^1(\{z_i\}) = \tilde{\omega}^+(A_i)$, $\omega_n^2(\{z_i\}) = \tilde{\omega}^-(A_i)$. It is easy to show that as $n \to \infty$ the set functions ω_n^1 and ω_n^2 defined in this way converge weakly to the set functions $\tilde{\omega}^+$ and $\tilde{\omega}^-$ respectively. We put $\omega_n = \omega_n^1 - \omega_n^2$ and define $\lambda_n(z)$ by putting

$$\lambda_n(z) = C\lambda(z; \omega_n),$$

where C is the constant on the right-hand side of (7.22). The metric ρ_λ defined by the function $\lambda_n(z)$ on the plane is polyhedral. By Theorem 7.3.1, as $n \to \infty$ the functions ρ_{λ_n} converge uniformly to the metric ρ_{λ, B_0} on any closed set $A \subset B_0$ not containing points z with $\omega^+(\{z\}) \geqslant 2\pi$. Since the metric ρ_λ is intrinsic, there is an $\varepsilon \in (0, 1)$ such that $\rho_{\lambda, B_0}(z_1, z_2) = \rho_\lambda(z_1, z_2)$ for $z_1, z_2 \in B(z_0, \varepsilon)$. We thus have a sequence of polyhedral metrics ρ_{λ_n} converging to the metric ρ_λ in the neighbourhood $B(0, \varepsilon)$ of the point $z_0 = 0$. The absolute curvatures of the polyhedral metrics ρ_{λ_n} are bounded in aggregate. The sequence of metrics that we have constructed actually satisfies the conditions of Theorem 6.2.3.

It was assumed above that $\omega(\{z_0\}) < 2\pi$. We now consider the case when $\omega(\{z_0\}) = 2\pi$ and z_0 is not a point at infinity with respect to the metric ρ_λ. In the given case the approximating polyhedral metric is constructed in two steps. We first replace the set function $\omega(E)$ by the one obtained from it by a uniform distribution of the load of the set function ω at the point 0 over a small neighbourhood. The function λ corresponding to the modified function ω determines the metric ρ_λ. We then approximate the metric ρ_λ by a polyhedral metric, as indicated above. Formally the construction is carried out as follows. We have

$$\ln \lambda(z) = \frac{1}{\pi}\iint\limits_{B_0\setminus\{z_0\}} \ln\frac{1}{|z-\zeta|}\tilde{\omega}(d\zeta) + 2\ln\frac{1}{|z-z_0|} < 2\ln\frac{1}{|z-z_0|} + \ln\lambda_0(z)$$

and let

$$\gamma_h(z) = \frac{1}{\pi}\int_0^{2\pi} \ln\frac{1}{|z - z_0 - he^{i\varphi}|}\,d\varphi.$$

Then $\gamma_h(z) = 2 \ln \frac{1}{|z - z_0|}$ when $|z - z_0| > h$, and $\gamma_h(z) = 2 \ln \frac{1}{h}$ when $|z - z_0| \leqslant h$. We denote by $\delta_n(E)$ a measure uniformly distributed along the circle $\{z \,|\, |z - z_0| = h\}$, where its value on the whole circle is equal to 2π. Then

$$\gamma_h(z) = \frac{1}{\pi} \int \ln \frac{1}{|z - \zeta|} \delta_h(d\zeta).$$

We define $\lambda_h(z)$ by putting

$$\ln \lambda_h(z) = \frac{1}{\pi} \iint\limits_{B_0 \setminus \{0\}} \ln \frac{1}{|z - \zeta|} \tilde{\omega}(d\zeta) + \gamma_h(z) = \frac{1}{\pi} \iint\limits_{B_0} \ln \frac{1}{|z - \zeta|} \omega_h(d\zeta),$$

where the measure $\omega_h(E)$ is defined by

$$\omega_h(E) = \tilde{\omega}(E \setminus \{z_0\}) + \delta_h(E).$$

Obviously $\lambda_h(z) = \lambda(z)$ when $|z| \geqslant h$ and $\lambda_h(z) \leqslant \lambda(z)$ for all z. Hence we conclude that as $h \to 0$ the metric ρ_{λ_h} converges to ρ_λ uniformly in the disc B_0. For $\varepsilon > 0$ we choose $h \in (0, 1)$ such that $|\rho_{\lambda_h}(z_1, z_2) - \rho_\lambda(z_1, z_2)| < \varepsilon/2$ for all $z_1, z_2 \in B_0$. For the function ω_h we have $\omega_h(\{z_0\}) = 0$. From what we have proved there is a polyhedral metric ρ_μ such that $|\rho_{\lambda_h}(z_1, z_2) - \rho_\mu(z_1, z_2)| < \varepsilon/2$ for any $z_1, z_2 \in B_0$, and the absolute curvature of the metric ρ_μ does not exceed

$$|\omega_h|(B_0) \leqslant 2\pi + |\tilde{\omega}|(B_0 \setminus \{z_0\}) = \tilde{\omega}(B_0).$$

Thus, in a neighbourhood of each point of the domain G that is not a point at infinity in the sense of the metric ρ_λ there is a sequence of polyhedral metrics satisfying the conditions of Theorem 6.2.2. We have thus proved that (G, ρ_λ) is a two-dimensional manifold of bounded curvature.

In particular, we finally obtain a proof of the fact that any two-dimensional Riemannian manifold is a manifold of bounded curvature.

The proof of Theorem 7.1.2 relies on Theorem 6.2.1. Let X be a point of a two-dimensional manifold of bounded curvature. Then according to Theorem 6.2.1 the point X has a boundedly convex neighbourhood U, homeomorphic to a closed disc, in which we can specify a sequence of polyhedral metrics ρ_n, $n = 1, 2, \ldots$, that converges uniformly in U to the metric of the manifold and such that the sums of the absolute turns of the boundary and of the absolute curvatures of the polyhedra (U, ρ_n) do not exceed some constant $M < \infty$ that does not depend on n. In the statement of Theorem 6.2.1 that we gave above we spoke of the possibility of approximation by Riemannian metrics. However, up to now nothing has been said about how to obtain such an approximation. We now have at our disposal all the necessary methods for constructing an approximation of the metric by Riemannian metrics. This is obtained by smoothing the polyhedral metrics that converge to the metric of the manifold in the given domain U.

We first show how to smooth the metric of a circular cone and a circular sector. For this we use an isothermal coordinate system. As we mentioned above, the plane $\mathbb{C}$, endowed with the metric ρ_λ generated by the line element $\lambda(z, \alpha)|dz|^2$, where $\lambda(z, \alpha) = (1/|z|)^{\omega/\pi}$, $\omega < 2\pi$, is isometric to the cone $Q(2\pi - \omega)$. Let $\mathbb{C}^+$ be the half-plane $\{z | \operatorname{Re} z \geqslant 0\}$. The half-plane $\mathbb{C}^+$, endowed with the same metric, is isometric to the angular domain $A(\pi - (\omega/2))$. Smoothing the metric of the cone is brought about, roughly speaking, by spreading the curvature over some neighbourhood of the point 0. This is done as follows.

Let ψ be the function of the real variable t defined by the following condition: $\psi(t) = 0$ when $t \leqslant 0$, $\psi(t) = \exp(-1/t)$ when $t > 0$. The function ψ belongs to the class C^∞. We put $H(z) = H(x, y) = C\psi((1/4) - |z|^2)$, where $C > 0$ is a constant. The function H is non-negative and belongs to the class C^∞, $H(z) = 0$ when $|z| > 1/2$, and $H(0) > 0$. We define the constant C from the condition

$$\iint_{\mathbb{R}^2} H(x, y)\, dx\, dy = 1.$$

Let us specify a number $h > 0$ arbitrarily and put

$$\lambda_h(z; \omega) = \exp\left\{\frac{\omega}{\pi} \int_{\mathbb{C}} \ln \frac{1}{|z - \zeta|} \cdot \frac{1}{h^2} H\left(\frac{\zeta}{h}\right) d\xi\, d\eta\right\}.$$

The function $\lambda_h(z) \equiv \lambda_h(z; \omega)$ belongs to the class C^∞, $\lambda_h(z) = (1/|z|)^{\omega/\pi}$ when $|z| > h/2$. In $\mathbb{C}$ we consider the Riemannian metric ρ_{λ_h}. As $h \to 0$ this metric converges to the metric of the cone $Q(2\pi - \omega)$. The Gaussian curvature of the metric ρ_{λ_h} at the point z is equal to $\dfrac{\omega}{h^2} H(z/h)/\lambda_h(z)$. Hence it follows that the absolute curvature of the plane in the metric ρ_λ is equal to $|\omega|$. The same metric ρ_{λ_h}, considered in the half-plane $\mathbb{C}^+$ as $h \to 0$, converges uniformly to the metric of the angular domain $A(\pi - (\omega/2))$. In view of the fact that the function $H(z)$ depends only on $|z|$, the metric ρ_{λ_h} is invariant under rotations around the point 0. Any line passing through the point 0 is a geodesic. The absolute curvature of the half-plane $\mathbb{C}^+$ in the metric ρ_λ is equal to

$$\frac{|\omega|}{h^2} \int_{\mathbb{C}^+} H\left(\frac{z}{h}\right) dx\, dy = \frac{|\omega|}{2}.$$

The disc $B(0, r)$ on the plane $\mathbb{C}$ with metric ρ_λ is isometric to a finite circular cone. When $h < r$ the metric ρ_{λ_h} outside the disc $B(0, h/2)$ coincides with the metric ρ_λ and as $h \to 0$ these metrics converge uniformly to the metric of the cone $Q(2\pi - \omega, r)$.

Let P be an arbitrary polyhedron homeomorphic to a closed disc, and $\delta > 0$ the smallest of the distances between its vertices. Let X be an arbitrary vertex of P, and $U = B(X, r)$, where $r < \delta/2$, its circular neighbourhood on the polyhedron P. If X is an interior vertex of P, then U is isometric to the cone $Q(\theta, r)$, where $\theta = \theta(X)$, and if X is a boundary vertex, then U is isometric to the circular sector $A(\theta, r)$. Let us map U isometrically into a plane, endowed with the line element $(1/|z|)^{\omega/\pi}|dz|^2$ so that to the point X there corresponds the point 0. In the case when X is an interior vertex, the image of U is a disc, and if X is a boundary vertex the image of U is a half-disc. We shall assume that the diameter of this half-disc lies on the line $\operatorname{Re} z$. We now construct the approximating Riemannian metric ρ_{λ_h}, where h is sufficiently small, $h < h_0$. Transferring the metric ρ_{λ_h} back to the disc $B(X, r)$, we obtain a Riemannian metric on P that is specified in a neighbourhood of the point X. Carrying out the given construction for all the vertices of P, we obtain on it a family of Riemannian metrics ρ_h that depend on a parameter $h \in (0, h_0)$. As $h \to 0$ these metrics converge uniformly to the metric of the polyhedron P. It is not difficult to see that the absolute curvature of each of them is equal to $|\omega|(P) + |\kappa|(\partial P)$. The turn of the boundary in the metric ρ_h is equal to zero.

We now consider the original manifold of bounded curvature. We have a point X, its neighbourhood G homeomorphic to a closed disc, and a sequence of polyhedral metrics (ρ_n) defined in G and converging as $n \to \infty$ to the metric of G such that the quantities $|\omega_{\rho_n}|(G) + |\kappa_{\rho_n}|(G)$ are bounded in aggregate. Approximating the metrics ρ_n by a Riemannian metric, as we mentioned above, we obtain a sequence of Riemannian metrics (r_n), $n = 1, 2, \ldots$, converging in G to the metric of the manifold and such that the sequence of absolute curvatures $\{|\omega_{r_n}|(G))$, $n = 1, 2, \ldots$, is bounded, and the turn of the boundary in each of these metrics is equal to zero.

By the theorem on the existence of isothermal coordinates in a Riemannian manifold, for each n in the disc $\bar{B}_1 = \bar{B}(0, 1)$ we can introduce a Riemannian metric ρ_n defined by the line element $ds^2 = \lambda_n(z)|dz|^2$ such that the disc $\bar{B}_1$ in this metric is isometric to the domain G with metric r_n. We can assume that to the point 0 there corresponds by isometry a given point X. The function $\lambda_n(z)$ has the representation

$$\lambda_n(z) = \exp\left\{\frac{1}{\pi}\int_{B_1} \ln\frac{1}{|z-\zeta|}\,\omega_n(d\zeta) + h_n(z)\right\},$$

where h_n is a harmonic function and the sequence $(|\omega_n|(B_1))$, $n = 1, 2, \ldots$, is bounded. The turn of the circle $\Gamma(0, 1)$ in each of the Riemannian metrics ρ_{λ_n} is equal to zero. Because of this the function λ_n has the representation

$$\lambda_n(z) = \gamma_n\tilde{\lambda}_n(z) = \gamma_n \exp\left\{\frac{1}{\pi}\int_{B_1} \ln\frac{1}{|z-\zeta|}\,\tilde{\omega}_n(d\zeta)\right\},$$

where $\gamma_n > 0$ is a constant and the variations of the set functions $\tilde{\omega}_n(E)$ are bounded in aggregate.

From the sequence of set functions $(\tilde{\omega}_n)$ we choose a subsequence for which the functions $\tilde{\omega}_n^+$ and $\tilde{\omega}_n^-$ converge weakly to certain set functions $\tilde{\omega}^0$ and $\tilde{\omega}^1$ respectively. Let $\tilde{\omega}_0 = \tilde{\omega}^0 - \tilde{\omega}^1$. We have $\tilde{\lambda}_n(z) = \lambda(z, \tilde{\omega}_n)$. Let $\tilde{\lambda}_0(z) = \lambda(z, \tilde{\omega}_0)$. For simplicity we shall assume that the chosen sequence coincides with the original one. On the basis of Theorem 7.3.1 the Riemannian metrics $\rho_{\tilde{\lambda}_n}$ converge uniformly to the metric $\rho_{\tilde{\lambda}_0}$ on any closed subset of the disc $\bar{B}_1$ that does not contain points at infinity with respect to the metric $\rho_{\tilde{\lambda}_0}$. The metric ρ_{λ_n} differs from $\rho_{\tilde{\lambda}_n}$ by a constant factor γ_n. The disc $\bar{B}_1$ with metric ρ_{λ_n} is isometric to (G, r_n). Since the metrics ρ_n converge to some metric ρ_λ as $n \to \infty$, this enables us to conclude that the factors γ_n converge as $n \to \infty$ to some finite non-zero limit γ_0, and the metric $\rho_{\tilde{\lambda}_0}$ does not have points at infinity. Hence it follows that the metrics ρ_{λ_n} converge to the metric $\rho_{\lambda_0} = \gamma_0 \rho_{\tilde{\lambda}_0}$ as $n \to \infty$. Since the disc $\bar{B}_1$ with metric ρ_{λ_n} is isometric to (G, r_n), it follows that the disc B_1 with metric ρ_{λ_0} is isometric to the domain G endowed with the natural metric of the manifold. Thus Theorem 7.1.2 is proved.

The proof of Theorem 7.1.3 is based on an additional assumption, which is expressed by Lemma 7.3.2 stated below. As a preliminary we describe some constructions that are necessary for the statement of this lemma.

Let G be a domain in $\mathbb{C}$, and $\lambda(z) = \lambda(z; \omega, h)$ a function defined in G. We specify a point $z_0 \in G$ arbitrarily. For $r > 0$ we put $\tilde{\lambda}_r(z) = \lambda(z_0 + rz)$. The domain of definition of the function $\tilde{\lambda}_r(z)$ is the set (G, r) obtained from G by the transformation $z \to \dfrac{z - z_0}{r}$. Obviously there is an $r_0 > 0$ such that when $0 < r < r_0$ the domain G_r contains the disc $K_2 = \bar{B}(0, 2)$. We define the constant $\gamma_r > 0$ from the condition

$$\int_{|z|=1} \sqrt{\gamma_r \tilde{\lambda}_r(z)}\,|dz| = 2\pi$$

and let $\lambda_r(z) = \gamma_r \tilde{\lambda}_r(z)$.

For $z \in \mathbb{C}$ we put $\varphi_r(z) = z_0 + rz$. We have $\lambda(z) = \lambda(z; \omega, h)$. Let ω_r be the measure defined by the condition $\omega_r(E) = \omega[\varphi_r(E)]$. It is not difficult to see that $\lambda_r(z) = \lambda(z, \omega_r, C_r h_r)$, where $C_r = \ln \gamma_r$, $h_r(z) = h(z_0 + rz)$. We shall assume that $0 < r < r_0$. In this case, in the disc K_2 there is defined a metric ρ_{λ_r, K_2}. We put $\omega_0 = \omega(\{z_0\})$ and let $\lambda_0(z) = |z|^{-\omega_0/\pi}$. As $r \to 0$ the set functions $\omega_r(E)$ converge weakly to the measure δ defined by the condition $\omega_0\delta(E) = 0$ if $0 \notin E$ and $\omega_0\delta(E) = \omega_0$ if $0 \in E$. We have $\lambda \equiv \lambda(z; \omega, h)$. The theorem on convergence of metrics (Theorem 7.3.1) enables us to conclude that the following proposition is true.

Lemma 7.3.2. *If the point $z_0 \in G$ and the functions $\lambda(z) \equiv \lambda(z; \omega, h)$ are defined in G so that $\omega(\{z_0\}) < 2\pi$, then as $r \to 0$ the metric ρ_{λ_r, K_2} converges in the disc $K_2 = \bar{B}(0, 2)$ to the metric ρ_{λ_0, K_2}.*

Let z_0 be an arbitrary point on the plane $\mathbb{C}$, and (Γ_m), $m = 1, 2, \ldots$, a sequence of simple closed curves such that the point z_0 lies inside each of them. We put $r_m = \inf_{z \in \Gamma_m} |z - z_0|$, $R_m = \sup_{z \in \Gamma_m} |z - z_0|$. We shall say that the sequence of curves (Γ_m) converges regularly to the point z_0 if $R_m \to 0$ as $m \to \infty$ and the ratio r_m/R_m tends to 1 as $m \to \infty$.

Theorem 7.3.2 (Men'shov (1948)). *Let G and H be open subsets of the plane $\mathbb{C}$, and $\varphi\colon G \to H$ a topological map of G onto H. We assume that for all points $z \in G$, excluding points that generate a no more than countable set, the following condition is satisfied: for any sequence (Γ_m) of simple closed curves lying in the domain G that converge regularly to the point z their images $\varphi(\Gamma_m)$ converge regularly to the point $w = \varphi(z)$. Then the map φ is conformal.*

We assume that in G there is specified a metric ρ and let (Γ_m), $m = 1, 2, \ldots$, be a sequence of simple closed curves such that a given point z_0 of G lies inside each of them. We shall say that the sequence (Γ_m), $m = 1, 2, \ldots$, converges regularly to the point z_0 in the sense of the metric ρ if the ratio $\tilde{r}_m/\tilde{R}_m$, where $\tilde{r}_m = \inf_{z \in \Gamma_m} \rho(z, z_0)$, $\tilde{R}_m = \sup_{z \in \Gamma_m} \rho(z, z_0)$, tends to 1 as $m \to \infty$.

Let ρ be the metric ρ_λ, where $\lambda = \lambda(z; \omega, h)$. Lemma 7.3.2 enables us to conclude that the sequence of closed curves (Γ_m) converges regularly to a point $z_0 \in G$ such that $\omega(\{z_0\}) < 2\pi$ if and only if it converges to z_0 regularly in the sense of the metric ρ_λ.

Let G and H be domains on the plane $\mathbb{C}$, and let ρ_λ and ρ_μ be metrics defined in G and H respectively by line elements $\lambda(z)|dz|^2$ and $\mu(w)|dw|^2$, where $\lambda(z) = \lambda(z; \omega, h)$, $\mu(w) = \mu(w; \theta, g)$, ω and θ are measures defined in G and H respectively, and h and g are harmonic functions. We assume that the metric spaces (G, ρ_λ) and (H, ρ_μ) are isometric and that φ is an isometric map of (G, ρ_λ) onto (H, ρ_μ). The map φ is topological. Any sequence of closed curves converging to the point $z \in G$ regularly in the sense of the metric ρ_λ is transformed by the map φ into a sequence of closed curves converging to the point $w = \varphi(z)$ regularly in the sense of the metric ρ_μ. By the previous remark, a sequence of curves that converge to the point z regularly in the sense of the metric ρ_λ converges regularly to this point. An exception is formed by points z for which $\omega(\{z\}) \geqslant 2\pi$, $\theta(\{w\}) \geqslant 2\pi$, where $w = \varphi(z)$. Consequently, the map φ satisfies the conditions of Men'shov's theorem, and so it is conformal. This proves the assertion of Theorem 7.1.3 about the conformal property of transition functions of two distinct isothermal coordinate systems in a two-dimensional manifold of bounded curvature.

Formula (7.4) on the transformation of a line element on going over from one isothermal cordinate system to another is proved by arguments similar to those used above on deriving the formula for transforming the coefficients of the metric tensor of a Riemannian manifold on going over from one local coordinate system to another. The fact that in the given case $\lambda(z)$ and $\mu(w)$ may be discontinuous functions somewhat complicates the arguments. However, it is easy to overcome the difficulties of technical character that arise here.

Theorem 7.1.4, like Theorem 7.1.5, is a simple consequence of the preceding results contained in Theorems 7.1.1–7.1.3. For this reason we shall not dwell on its proof.

7.4. On the Proof of Theorem 7.3.1. Theorem 7.3.1, a *theorem on the convergence of metrics*, is an important instrument for the proof of theorems on the analytic representation of two-dimensional manifolds of bounded curvature by means of an isothermal line element. It is therefore necessary to say a few words about its proof.

Suppose that the conditions of Theorem 7.3.1 are satisfied, that is, we are given a domain G whose boundary is the union of finitely many simple closed curves of bounded rotation. We assume that in G there are specified sequences of non-negative measures (ω_n^1), (ω_n^2), $n = 1, 2, \ldots$, that converge weakly to certain measures ω^1 and ω^2 respectively. We put $\omega_n = \omega_n^1 - \omega_n^2$, $\omega = \omega^1 - \omega^2$, and let $\lambda_n(z) = \lambda(z, \omega_n)$, $\lambda(z) = \lambda(z, \omega)$. Metrics ρ_{λ_n} and ρ_λ are defined in G. We need to prove that on any compact set $A \subset G$ that does not contain points z such that $\omega^1(\{z\}) \geqslant 2\pi$, $\rho_{\lambda_n}(z_1, z_2)$ converges uniformly to $\rho_\lambda(z_1, z_2)$ as $n \to \infty$.

For the proof of this it is sufficient to show that if (z_{1n}), (z_{2n}), $n = 1, 2, \ldots$, are two arbitrary sequences of points of G that converge as $n \to \infty$ to points z_1 and z_2 such that $\omega^1(\{z_1\}) < 2\pi$ and $\omega^1(\{z_2\}) < 2\pi$, then $\rho_{\lambda_n}(z_{1n}, z_{2n}) \to \rho_\lambda(z_1, z_2)$ as $n \to \infty$. We shall prove this last assertion.

We recall that $\bar{\rho}_\lambda(z_1, z_2)$ is the greatest lower bound of the quantity

$$s_\lambda(K) = \int_K \sqrt{\lambda(z)}\, ds$$

on the set of all curves K with finite absolute rotation that join z_1 and z_2 and are contained in the domain G.

We shall say that a measure ω defined on the plane $\mathbb{C}$ is regular if it is the indefinite integral of a function of class C^∞. We assume that the measures ω_n^1 and ω_n^2 in the statement of Theorem 7.3.1 are regular (we suppose that measures defined in G can be extended to the plane $\mathbb{C}$ by means of the stipulation that $\omega(E) = 0$ if $E \cap G = \varnothing$). In this case the metrics ρ_λ are Riemannian. It is sufficient to establish the truth of Theorem 7.3.1 for this case. In fact, if this is done, then in the general case we can proceed as follows. For each n we approximate the measures ω_n^1 and ω_n^2 by regular measures $\tilde{\omega}_n^1$ and $\tilde{\omega}_n^2$. By hypothesis, for the case when the measures ω_n^i, $i = 1, 2, \ldots$, are regular Theorem 7.3.1 is true. Let us construct regular measures $\tilde{\omega}_n^1$ and $\tilde{\omega}_n^2$ sufficiently close in the sense of weak topology to the measures ω_n^1 and ω_n^2 respectively. We put $\tilde{\lambda}_n = \lambda(z, \tilde{\omega}_n^1 - \tilde{\omega}_n^2)$. For each n there is defined in the domain G a Riemannian metric $\rho_{\tilde{\lambda}_n}$. If $\tilde{\omega}_n^1$ and $\tilde{\omega}_n^2$ are sufficiently close to the measures ω_n^1 and ω_n^2, then as $n \to \infty$ the measures $\tilde{\omega}_n^1$ converge weakly to the measure ω_1, and $\tilde{\omega}_n^2$ to ω_2. Since Theorem 7.3.1 is assumed to be proved for the case when the measures ω_n^1 and ω_n^2 are regular, this enables us to conclude that $\rho_{\tilde{\lambda}_n}(z_{1n}, z_{2n}) \to \rho_\lambda(z_1, z_2)$ as $n \to \infty$ (here $\tilde{\lambda}_n = \lambda(z, \tilde{\omega}_n^1 - \tilde{\omega}_n^2)$). Using again the fact that Theorem 7.3.1 is assumed to be proved for the case of regularity of the measures ω_n^i, $i = 1, 2, \ldots$, we can assert that if the measures $\tilde{\omega}_n^i$, $i = 1, 2, \ldots$, are chosen sufficiently close to ω_n^i, then $|\rho_{\tilde{\lambda}_n}(z_{1n}, z_{2n}) - \rho_{\lambda_n}(z_{1n}, z_{2n})| \to 0$ as $n \to \infty$. Since $\rho_{\tilde{\lambda}_n}(z_{1n}, z_{2n}) \to \rho_\lambda(z_1, z_2)$, it follows that $\rho_{\lambda_n}(z_{1n}, z_{2n}) \to \rho_\lambda(z_1, z_2)$ as $n \to \infty$.

Thus, the general case reduces to the case when the measures ω_n^1 and ω_n^2, $n = 1, 2, \ldots$, are regular, so the metrics ρ_{λ_n} are Riemannian. Henceforth we shall assume that this condition is satisfied.

The subsequent arguments rely on the next two propositions.

Lemma 7.4.1. *Suppose that the sequences of measures (ω_n^1), (ω_n^2), $n = 1, 2, \ldots$, satisfy all the conditions listed above. We assume that we are given a sequence of curves (K_n) lying in G that converges to a curve K that also lies in G. If the absolute curvatures of the curves K_n are bounded in aggregate, $|\kappa|(K_n) \leqslant M =$*

const $< \infty$ *for all* n *and the curve* K *does not contain points* z *for which* $\omega^1(\{z\}) \geqslant 2\pi$, *then*

$$s_{\lambda_n}(K_n) \to s_\lambda(K)$$

as $n \to \infty$.

The proof of this lemma is based on arguments of purely technical character.

Lemma 7.4.2. *Let* ω_n^1, ω_n^2, $n = 1, 2, \ldots$, *be sequences of regular measures defined in a closed domain* G *and satisfying all the conditions stated above. Then for any point* $z \in G$ *such that* $\omega^1(\{z\}) + \omega^2(\{z\}) < 2\pi$ *there are numbers* $\delta > 0$ *and* $M < \infty$ *having the property that for any curve that lies in the set* $G \cap \bar{B}(z, \delta)$ *and is a shortest curve, in at least one of the Riemannian metrics* ρ_{λ_n} *the absolute rotation does not exceed* M.

In Lemma 7.4.2 shortest curves containing boundary points are admitted. The proof of the lemma is based on the use of formula (7.18) for the turn of a curve in an isothermal coordinate system. We assume that curve L is a shortest curve in the metric ρ_λ, where $\lambda = \lambda(z; \omega)$, and lies in the disc $B(z_0, r) \subset G$. We assume that the measure ω is regular. Let $z(t)$, $a \leqslant t \leqslant b$, be a parametrization of L, and let $\kappa(t)$ and $\varphi(t, \zeta)$ be the rotation and angular function of the arc of L corresponding to values of the parameter lying in the interval $[a, t]$. The geodesic turn of L is equal to zero, and so by (7.18) we have

$$\kappa(t) = \frac{1}{2\pi} \int_G \varphi(t, \zeta)\omega(dE_\zeta). \tag{7.23}$$

(We are considering the case when L does not contain boundary points of G.) Hence

$$|\kappa|(L) = \overset{b}{\underset{a}{V}} \kappa(t) \leqslant \frac{1}{2\pi} \int_G \overset{b}{\underset{a}{V}} \varphi(t, \zeta)|\omega|(dE_\zeta). \tag{7.24}$$

The quantity $\overset{b}{\underset{a}{V}} \varphi(t, \zeta)$ is estimated as follows. On the one hand, we have

$$\overset{b}{\underset{a}{V}} \varphi(t, \zeta) \leqslant \kappa(L) + \pi. \tag{7.25}$$

On the other hand, if the point ζ is at a distance h from L, then

$$\overset{b}{\underset{a}{V}} \varphi(t, \zeta) \leqslant \frac{s(L)}{h} \leqslant \frac{4r[\kappa(L) + \pi]}{\pi h}. \tag{7.26}$$

The result of Lemma 7.4.2 is obtained by a combination of these two estimates. We assume that L does not overlap the boundary of G. If this is not so, then on the right-hand side of (7.23) there appears a term that depends only on G and r; otherwise the arguments do not differ from the case when L does not contain boundary points of G.

The proof of Theorem 7.3.1 is now realized as follows. We specify $\varepsilon > 0$ arbitrarily, and specify a curve K of bounded curvature joining the points z_1 and z_2 and such that $s_\lambda(K) < \rho_\lambda(z_1, z_2) + \varepsilon$. On the curve K there may be points z for which $\omega^1(\{z\}) \geqslant 2\pi$. Replacing a small arc of K containing such a point z by an arc of a circle with centre z, we can arrange that K does not pass through such points. Let K_n be the curve obtained from K by joining the intervals $[z_1, z_{1n}]$ and $[z_2, z_{2n}]$. As $n \to \infty$ the curves K_n converge to the curve K and $|\kappa|(K_n) \leqslant |\kappa|(K) + 2\pi$ for each n, so the absolute rotations of the curves K_n are bounded in aggregate. By Lemma 7.4.1

$$s_{\lambda_n}(K_n) \to s_\lambda(K) < \rho_\lambda(z_1, z_2) + \varepsilon$$

as $n \to \infty$. For each n

$$\rho_{\lambda_n}(z_{1n}, z_{2n}) \leqslant s_{\lambda_n}(K)$$

and so

$$\varlimsup_{n\to\infty} \rho_{\lambda_n}(z_{1n}, z_{2n}) \leqslant \rho_\lambda(z_1, z_2) + \varepsilon,$$

hence, since $\varepsilon > 0$ is arbitrary, we conclude that

$$\varlimsup_{n\to\infty} \rho_{\lambda_n}(z_{1n}, z_{2n}) \leqslant \rho_\lambda(z_1, z_2). \tag{7.27}$$

We now show that

$$\varliminf_{n\to\infty} \rho_{\lambda_n}(z_{1n}, z_{2n}) \geqslant \rho_\lambda(z_1, z_2). \tag{7.28}$$

We single out all points z for which $\omega^1(\{z\}) + \omega^2(\{z\}) \geqslant 2\pi$ and which are not points at infinity in the sense of the metric ρ_λ. The set E of such points z is finite. We shall assume that $z_1 \notin E$ and $z_2 \notin E$. We specify $\varepsilon > 0$ arbitrarily and for each point $z \in E$ we construct a disc with centre z so that these discs do not overlap and the sum of the lengths of their circles in the metric ρ_λ is less than ε. Let K_n be the shortest curve in the metric ρ_{λ_n} joining the points z_{1n} and z_{2n}. We have

$$s_{\lambda_n}(K_n) = \rho_{\lambda_n}(z_{1n}, z_{2n}).$$

The curve K_n can be inside some of the discs constructed above. Replacing each arc of K_n contained in such a disc by an arc of the circle of the disc, we obtain a curve K_n' joining the points z_{1n} and z_{2n}. For sufficiently large n we shall have

$$s_{\lambda_n}(K_n') < s_{\lambda_n}(K_n) + \varepsilon.$$

Applying Borel's theorem on covering and using Lemma 7.4.2, it is not difficult to show that the absolute rotations of the curves K_n' are bounded in aggregate. Without loss of generality we can assume that as $n \to \infty$ the curves K_n' converge to some curve K joining the points z_1 and z_2. As $n \to \infty$ we have

$$\rho_{\lambda_n}(z_{1n}, z_{2n}) \geqslant s_{\lambda_n}(K_n') - 2\varepsilon \to s_\lambda(K) - 2\varepsilon \geqslant \rho_\lambda(z_1, z_2) - 2\varepsilon.$$

Hence

$$\varliminf_{n\to\infty} \rho_{\lambda_n}(z_{1n}, z_{2n}) \geqslant \rho_\lambda(z_1, z_2) - 2\varepsilon$$

and since $\varepsilon > 0$ is arbitrary the inequality (7.28) follows. The required relation follows from (7.27) and (7.28):

$$\rho_\lambda(z_1, z_2) = \lim_{n\to\infty} \rho_{\lambda_n}(z_{1n}, z_{2n}).$$

In the proof of (7.28) we have assumed that $\omega^1(\{z\}) + \omega^2(\{z\}) < 2\pi$ for points $z = z_1, z_2$. If this condition is not satisfied, then the required result can be deduced from what we have proved by a small shift of the points z_1 and z_2.

Chapter 3
Basic Facts of the Theory of Manifolds of Bounded Curvature

§ 8. Basic Results of the Theory of Two-Dimensional Manifolds of Bounded Curvature

8.1. A Turn of a Curve and the Integral Curvature of a Set. The concepts of a turn have been defined for curves in a Riemannian manifold and on a manifold with polyhedral metric. Next for such manifolds we also defined the concept of the integral curvature of a set. Our aim is to extend these concepts to the case of arbitrary two-dimensional manifolds of bounded curvature. Here, in contrast to the monograph of Aleksandrov and Zalgaller (1962), we shall rely on the analytic apparatus at our disposal – the representation of the metric of a manifold of bounded curvature by the isothermal line element $ds^2 = \lambda(z)|dz|^2$. We have $\lambda(z) = \lambda(z; \omega, h)$. We establish the geometrical meaning of the set function ω and then solve the problem of introducing the set function known as integral curvature. This is preceded by a study of the concept of the turn of a simple arc in a manifold whose metric is defined by an isothermal line element. We first define the turn formally by analogy with the case of curves in a Riemannian manifold. Having established the geometrical meaning of the quantities introduced formally by means of isothermal coordinates, we obtain an answer to the question of how to regard the turn of a curve in an arbitrary two-dimensional manifold of bounded curvature.

The total angle and curvature at a point of a two-dimensional manifold of bounded curvature. Let M be a two-dimensional manifold of bounded curvature, U a domain in M homeomorphic to a disc, and $\varphi\colon U \to \mathbb{C}$ an isothermal coordinate system in M. Let $G = \varphi(U)$. Then in G there is defined a function $\lambda(z) \equiv$

$\lambda(z; \omega, h)$ and a metric ρ_λ in it such that φ is an isometric map of (U, ρ_U) onto (G, ρ_λ).

Let K be a simple arc on a plane, a its beginning and b its end. We take a point $z_0 \in K$ arbitrarily. Let z be a point of K other than z_0. We denote by $l(z_0, z)$ the ray with beginning at z_0 and passing through z. If $l(z_0, z)$ converges to some limiting ray l_0 when z tends to z_0 from the right (left), we shall say that the curve K has *right* (*left*) *half-tangents* at the point z_0. We shall denote the left half-tangent of K at z_0 by $l_l(z_0, K)$ and the right half-tangent by $l_r(z_0, K)$.

Let K be a simple arc in a domain U of a manifold of bounded curvature. Then we shall say that K *belongs to the class* $\tilde{\Delta}$ if its image in the isothermal coordinate system has half-tangents at its end-points. The property that a curve has left and right half-tangents at any point of it is preserved by conformal maps. So clearly the property that a curve belongs to the class $\tilde{\Delta}$ does not depend on the choice of isothermal coordinate system in the domain U. As we shall show later, $\tilde{\Delta}$ coincides with the class of curves that have a definite direction at their end-points.

Let X be an arbitrary point of the domain U, and $z = \varphi(X)$ the corresponding point on the plane $\mathbb{C}$ in the given isothermal coordinate system. In the domain $G = \varphi(U)$ there is defined a set function ω. We show that $\omega(\{z\})$ does not depend on the choice of isothermal coordinate system.

Let $\Gamma(X, r)$ be a circle in a two-dimensional manifold of bounded curvature, $X \in U$ its centre, and r its radius.

Lemma 8.1.1. *Let $X \in U$, $z = \varphi(X)$. Then there is a number $\delta_1 > 0$ such that if $0 < r < \delta_1$, then $\Gamma(X, r)$ is a simple closed curve. Let $\sigma(X, r)$ be the length of the curve $\Gamma(X, r)$. Then*

$$\frac{\sigma(X, r)}{r} \to \theta(X) = 2\pi - \omega(\{z\}) \tag{8.1}$$

as $r \to 0$.

We shall denote the quantity $\omega(\{z\})$ (see the statement of the lemma) by $\omega(X)$ and call it the *curvature of the manifold M* at the point X. We shall call $\theta(X) = 2\pi - \omega(X)$ the *total angle* at the point X.

We shall call a point $X \in U$ a *peak point* if $\omega(X) = 2\pi$.

There arises the question of studying the structure of a circle of arbitrary radius in a two-dimensional manifold of bounded curvature. The papers of Zalgaller (1950b) and Burago and Stratilatova (1965) are devoted to this. If the radius of the circle is sufficiently large, then generally speaking the circle is not a simple closed curve. However, this may hold also in the case of Riemannian manifolds. Burago and Stratilatova gave a definition of the length of a circle in a two-dimensional manifold of bounded curvature for the case when the circle is not a simple closed curve. They also mentioned a formula that gives an expression for the length of a circle and the area of the disc bounded by it in terms of other characteristics of the manifold.

We note that in the case when a two-dimensional manifold of bounded curvature is a cone of curvature ω_0 with vertex X, then, as we showed above,

$$\sigma(X_0, r) = (2\pi - \omega_0)r,$$

so the definition given here agrees with the definition that we have in the case when the manifold is a cone, and hence also in the case when it is a polyhedron.

Let K_1 and K_2 be two simple arcs in a domain U that have a common starting point X. We assume that K_1 and K_2 have no common points other than X, and that the arcs $L_1 = \varphi(K_1)$, $L_2 = \varphi(K_2)$ have right half-tangents at the point $z = \varphi(X)$. The curves K_1 and K_2 split a neighbourhood of X into two sectors S_1 and S_2. Let Σ_1 and Σ_2 be the images of these sectors under the map $\varphi: U \to \mathbb{C}$. The curves L_1 and L_2 have right half-tangents at the point z. Let l_1 be the half-tangent of L_1 and l_2 the half-tangent of L_2. The rays l_1 and l_2 split the plane $\mathbb{C}$ into two angular domains V_1 and V_2. The value of one of them is naturally called the angle of the sector Σ_1, and the value of the other is called the angle of the sector Σ_2 (see Fig. 37). Let α' be the angle of the sector Σ_1, and α'' the angle of the sector Σ_2. Let $\omega_0 = \omega(X)$. We put

$$\alpha_1 = \left(1 - \frac{\omega_0}{2\pi}\right)\alpha', \quad \alpha_2 = \left(1 - \frac{\omega_0}{2\pi}\right)\alpha''. \tag{8.2}$$

We shall call α_1 and α_2 the angles of the sectors S_1 and S_2.

The next proposition establishes the geometrical meaning of the quantities α_1 and α_2.

Lemma 8.1.2. *We assume that the simple arcs K_1 and K_2 starting from a point $X \in U$ are such that the curves $L_1 = \varphi(K_1)$ and $L_2 = \varphi(K_2)$ have right half-tangents at the point $z = \varphi(X)$ and that K_1 and K_2 have no points in common other than X. Then there is a number $\delta_2 > 0$ such that when $0 < r < \delta_2$ the circle $\Gamma(X, r)$ in the manifold M is a simple closed curve and is split by K_1 and K_2 into two arcs, one of which lies in the sector S_1 and the other in the sector S_2. Let $\sigma_1(r)$ be the length of the first arc, and $\sigma_2(r)$ the length of the second. Then*

$$\lim_{r\to 0} \frac{\sigma_1(r)}{r} = \alpha_1, \quad \lim_{r\to 0} \frac{\sigma_2(r)}{r} = \alpha_2. \tag{8.3}$$

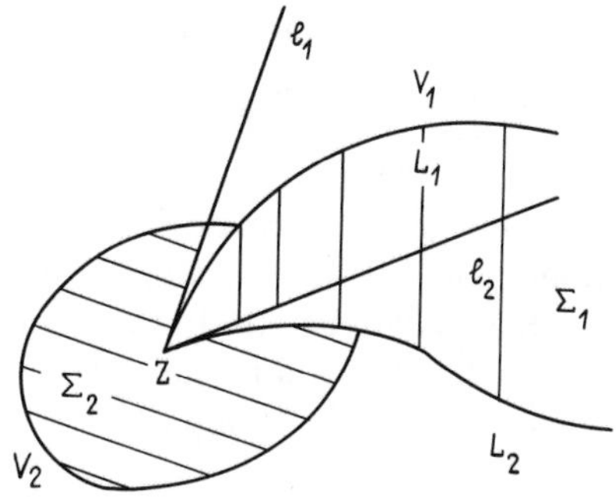

Fig. 37

Lemma 8.1.2 shows that the value of α_1 and α_2 is determined by the geometrical properties of a small neighbourhood of the point z_0 in a manifold of bounded curvature.

The formal turn of a curve in isothermal coordinates. As above, let U be a domain homeomorphic to a disc in a two-dimensional manifold of bounded curvature, $\varphi: U \to \mathbb{C}$ an isothermal coordinate system defined in U, $G = \varphi(U)$, $\lambda(z)|dz|^2$ the line element of the manifold M in this coordinate system, and $\lambda(z) \equiv \lambda(z; \omega, h)$.

For an oriented simple arc K on the plane $\mathbb{C}$ there is defined an angular function $z \mapsto \varphi(K, z)$. Up to now we have assumed that $\varphi(K, z)$ is defined only at points not belonging to K. Now we need to define it in the case $z \in K$ also. This definition is based on the following proposition.

Lemma 8.1.3. *Let K be an oriented simple arc on the plane, z_0 an arbitrary internal point of it, a its beginning and b its end. If the sequence (z_n), $n = 1, 2, \ldots$, converges to the point z_0, having kept on one side of K, then there is a finite limit*

$$\lim_{n\to\infty} \varphi(K, z_n). \tag{8.4}$$

We assume that K belongs to the class $\tilde{\Delta}$ and let (a_n), (b_n), $n = 1, 2, \ldots$, be sequences of points of the curve K such that $a < a_n < b_n < b$, and let K_n be the arc $[a_n b_n]$ of K. Then there are finite limits

$$\lim_{n\to\infty} \varphi(K_n, a), \quad \lim_{n\to\infty} \varphi(K_n, b). \tag{8.5}$$

If the sequence (z_n) converges to the point $z_0 \in K$ on the right with respect to K, then the limit (8.4) will be denoted by $\varphi_r(K, z_0)$, and if $z_n \to z_0$ on the left of K, then the limit (8.4) will be denoted by $\varphi_l(K, z_0)$. Clearly, the value of the limit (8.4) depends only on which side of K the sequence (z_n), $n = 1, 2, \ldots$, converges. The existence of the limits (8.5) is guaranteed by the fact that K has half-tangents at a and b. We put $\varphi(K, a)$ and $\varphi(K, b)$ equal to the limits (8.5) respectively.

If z is not an interior point of K, we put

$$\varphi_l(K, z) = \varphi_r(K, z) = \varphi(K, z).$$

We thus obtain functions $z \mapsto \varphi_l(K, z)$ and $z \mapsto \varphi_r(K, z)$ defined at all points of the plane $\mathbb{C}$. We note that if z is an interior point of K, then we have

$$\varphi_l(K, z) - \varphi_r(K, z) = 2\pi. \tag{8.6}$$

We assume that in the representation $\lambda(z) = \lambda(z; \omega, h)$ the measure ω is regular, that is, it is the indefinite integral of a function of class C^∞ and K is a smooth curve of class C^2 in the domain G. Let $z(s)$, $0 \leqslant s \leqslant l$, be a parametrization of the curve K, where the parameter s is arc length. In the given case the metric ρ_λ is Riemannian, and as we showed above (formula (7.18)) the turn $\kappa_\lambda(K)$ of the curve K in this metric is expressed by the formula

$$\kappa_\lambda(K) = \kappa(K) - \frac{1}{2\pi}\int_G \varphi(K, z)\omega(dz) - \frac{1}{2}\int_K \frac{\partial h}{\partial \nu}(z)|dz|.$$

Let K be a simple arc lying in the domain G, and $h(z)$ a harmonic function in G. We specify arbitrarily a neighbourhood V of the curve K homeomorphic to a disc and contained in G. Let g be a harmonic function defined in V and conjugate to h, that is, such that

$$\frac{\partial g}{\partial x}(z) = \frac{\partial h}{\partial y}(z), \quad \frac{\partial g}{\partial y}(z) = -\frac{\partial h}{\partial x}(z).$$

A function g satisfying these conditions exists since V is simply-connected. We assume that the simple arc K is smooth. Let a be its beginning and b its end, and let $z(s)$, $0 \leqslant s \leqslant l$, be a parametrization of K, where the parameter s is arc length. Then

$$\int_K \frac{\partial h}{\partial v}(z)|dz| = \int_K \frac{d}{ds}(g[z(s)])\, ds = g[z(l)] - g[z(0)] = g(b) - g(a).$$

We use the given relation to give a meaning to the integral

$$\int_K \frac{\partial h}{\partial v}(z)|dz|$$

in the case when the simple arc K is not smooth. Namely, in this case we put

$$\int_K \frac{\partial h}{\partial v}(z)|dz| = g(b) - g(a), \tag{8.7}$$

where g is a harmonic function defined in a neighbourhood of K and conjugate to h. It is easy to show that the difference $g(b) - g(a)$ does not depend on the choice of neighbourhood of K and the function g. (If the neighbourhood is chosen, then $g(z)$ is defined up to a constant term.)

We now assume that L is an arbitrary simple arc of class $\tilde{\Delta}$ in the domain U, and $K = \varphi(L)$. The left turn of the curve L in the coordinate system $\varphi: U \to \mathbb{C}$ is the quantity

$$\kappa_l(L) = \kappa(K) - \frac{1}{2\pi}\int_G \varphi_r(K, z)\omega(dz) - \frac{1}{2}\int_K \frac{\partial h}{\partial v}(z)|dz|. \tag{8.8}$$

The right turn of L is the quantity

$$\kappa_r(L) = -\kappa(K) + \frac{1}{2\pi}\int_G \varphi_l(K, z)\omega(dz) + \frac{1}{2}\int_K \frac{\partial h}{\partial v}(z)|dz|. \tag{8.9}$$

The second integral on the right-hand side of (8.8) and (8.9) is defined, as we described above, in accordance with (8.7).

We shall not indicate the coordinate system φ in the notation for the quantities $\kappa_l(L)$ and $\kappa_r(L)$ since, as we shall show later, they do not depend on the choice of this coordinate system.

Let K^0 be the totality of all interior points of the simple arc $K = \varphi(L)$. Adding (8.8) and (8.9) term by term and taking (8.7) into consideration, we obtain

$$\kappa_l(L) + \kappa_r(L) = \frac{1}{2\pi}\int_{K^0} 2\pi\omega(d\zeta) = \omega(K^0). \tag{8.10}$$

We now mention some simple properties of $\kappa_l(L)$ and $\kappa_r(L)$ that follow directly from the definition.

Theorem 8.1.1. *Let $L \subset U$ be a simple arc of class $\tilde{\Lambda}$, X an interior point of L, and L_1 and L_2 the arcs into which K is split by the point X. We assume that the arcs $K_1 = \varphi(L_1)$, $K_2 = \varphi(L_2)$ have half-tangents at the point $z = \varphi(X)$, and let $\theta_l(X, K)$ and $\theta_r(X, K)$ be the angles in the sense of the metric ρ_λ at the point Y between the arcs L_1 and L_2 on the left and right of L, respectively. Then*

$$\kappa_l(L) = \kappa_l(L_1) + \kappa_l(L_2) + (\pi - \theta_l(X, L)),$$

$$\kappa_r(L) = \kappa_r(L_1) + \kappa_r(L_2) + (\pi - \theta_r(X, L)).$$

The proof of the theorem follows directly from the representations of $\kappa_l(L)$ and $\kappa_r(L)$, which are given by (8.8) and (8.9).

Next we put

$$\pi - \theta_l(X, L) = \kappa_l(X, L), \quad \pi - \theta_r(X, L) = \kappa_r(X, L)$$

and the quantities $\kappa_l(X, L)$ and $\kappa_r(X, L)$ will be called the *left turn and right turn of the curve L at the point X*.

We assume that the domain U is oriented by the stipulation that the isothermal coordinate system $\varphi: U \to \mathbb{C}$ specified in it is right. For $E \subset U$ we put

$$\tilde{\omega}(E) = \omega[\varphi(E)].$$

Theorem 8.1.2. *Let L_1 and L_2 be two simple arcs of class $\tilde{\Lambda}$ with common beginning A and common end B, lying in the domain U. We assume that L_1 and L_2 have no points in common other than A and B, and that L_2 lies to the right of L_1. Let D be the domain included between the curves K and L, and let α and β be the angles of D at the points A and B. Then*

$$\kappa_l(L_2) = \kappa_l(L_1) + \alpha + \beta - \tilde{\omega}(D \cup K^0),$$

$$\kappa_r(L_2) = \kappa_r(L_1) - \alpha - \beta + \tilde{\omega}(D \cup K^0).$$

Corollary. *Let L and L_n, $n = 1, 2, \ldots$, be simple arcs of class $\tilde{\Lambda}$ joining the points $A, B \in U$, where $L \subset U$ and $L_n \subset U$ for all n. We assume that for each n the curves L and L_n have no points in common other than A and B. Let α_n and β_n be the angles of the domain D_n included between L and L_n at the points A and B respectively. If the curves L_n converge to the curve L on the right, then*

$$\kappa_r(L) = \lim_{n\to\infty} (\kappa_r(L_n) + \alpha_n + \beta_n).$$

Similarly, if $L_n \to L$ on the left, then

$$\kappa_l(L) = \lim_{n\to\infty} (\kappa_l(L_n) + \alpha_n + \beta_n).$$

Theorem 8.1.2 follows directly from the expressions for $\kappa_l(L)$ and $\kappa_r(L)$, which are given by (8.8) and (8.9).

For the proof of the corollary we observe that if L_n for each n lies to the right of L, then, according to the theorem,

$$\kappa_r(L_n) + \alpha_n + \beta_n = \kappa_r(L) + \tilde{\omega}(D_n \cup L_n^0).$$

By virtue of the property of complete additivity of a measure, $\omega(D_n \cap L_n^0) \to 0$ as $n \to \infty$, so it follows that $\kappa_r(L) = \lim_{n\to\infty} (\kappa_r(L_n) + \alpha_n + \beta_n)$. The assertion about the left turn is proved in exactly the same way.

Geometrical definition of the turn of a simple arc. Here we shall establish the geometrical meaning of the quantities $\kappa_l(L)$ and $\kappa_r(L)$ introduced above in a formal way by means of the isothermal coordinate system $\varphi: U \to \mathbb{C}$ defined in a domain U of a two-dimensional manifold of bounded curvature. The main instrument will be the following theorem, which also has independent significance.

Theorem 8.1.3. *Suppose we are given a metric $\rho_\lambda(z_1, z_2)$, where $\lambda(z) \equiv \lambda(z; \omega, h)$, in an open domain G on the plane $\mathbb{C}$. We assume that the curve K is a shortest curve in G in the sense of the metric ρ_λ. If K does not contain peak points of the metric ρ_λ (that is, such that $\omega(\{z\}) = 2\pi$), then K is a plane curve of bounded rotation. The left and right turns of K in the metric ρ_λ (defined by* (8.8) *and* (8.9)) *are non-positive.*

The proof of Theorem 8.1.3 is based on an approximation of the metric ρ_λ specified in the domain G by Riemannian metrics. Here we shall not dwell on the details, referring the reader to the article Reshetnyak (1963b). Let us make some remarks about Theorem 8.1.3.

First of all we observe that the limitation that consists in the absence of peak points of the metric ρ_λ on the curve is due to an important matter. In fact, it is easy to construct an example of a manifold of bounded curvature for which the two shortest curves K and L starting from the point A with $\omega(A) = 2\pi$ are arranged as in Fig. 38. The shortest curve K on approaching the point A makes an infinite set of coils. Under any topological map of an arbitrary neighbourhood of the point A into a plane the image of at least one of the curves K and L will not be a curve of bounded rotation.

In the case when the metric ρ_λ is Riemannian, by virtue of classical results of Riemannian geometry we can assert that the left and right turns of a shortest curve are equal to zero. In the general case it is impossible to assert this; it is easy to convince oneself in this just example of polyhedra.

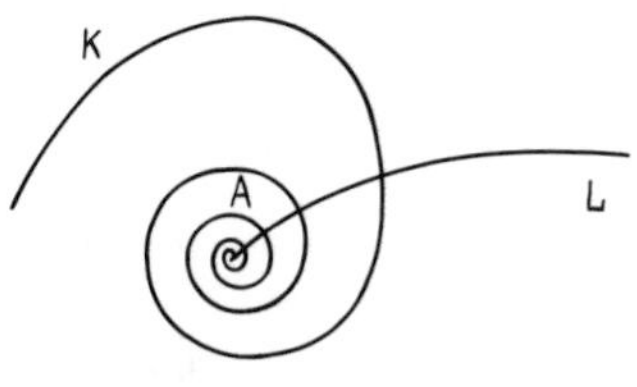

Fig. 38

Let K be a shortest curve in a two-dimensional manifold of bounded curvature that does not contain peak points. Any arc L of the curve K is also a shortest curve, and so by Theorem 8.1.3 its left and right turns are non-positive. Applying (8.10), we deduce that for any arc L of the curve K we have $\tilde{\omega}(L^0) \leqslant 0$. This enables us to conclude that for any set $E \subset K^0$ that is open with respect to K we have $\tilde{\omega}(E) \leqslant 0$, from which it follows that in general for any Borel set $E \subset K^0$ the quantity $\tilde{\omega}(E)$ is non-positive. Moreover, the relations $\kappa_l(L) \leqslant 0$, $\kappa_r(L) \leqslant 0$, $\kappa_l(L) + \kappa_r(L) = \tilde{\omega}(K^0)$ enable us to conclude that for any shortest curve L that does not contain peak points we have

$$\begin{aligned} |\kappa_l(L)| &\leqslant |\omega(L^0)| = |\omega|(L^0), \\ |\kappa_r(L)| &\leqslant |\omega(L^0)| = |\omega|(L^0). \end{aligned} \tag{8.11}$$

If the measure ω in the domain $G \subset \mathbb{C}$ is non-negative, then points z such that $\omega(\{z\}) = 2\pi$ are points at infinity with respect to the metric ρ_λ, where $\lambda(z) \equiv \lambda(z; \omega, h)$, so in this case the metric ρ_λ does not have peak points. For suppose that the point z_0 is such that $\omega(\{z_0\}) = 2\pi$. Then in the disc $B(z_0, r)$, where $0 < r < 1/2$, we have

$$\ln \lambda(z) = \frac{1}{\pi} \int_{B(z_0, r)} \ln \frac{1}{|z - \zeta|} \omega(dE_\zeta) + u(z),$$

where $u(z)$ is a harmonic function. When $z \in B(z_0, r)$, $\zeta \in B(z_0, r)$ we have $|z - \zeta| < 2r < 1$ and so $\ln \dfrac{1}{|z - \xi|} > 0$. Hence we conclude that in the disc $B(z_0, r/2)$

$$\ln \lambda(z) \geqslant 2 \ln \frac{1}{|z - z_0|} + \int_{B(z_0, r) \setminus \{z_0\}} \ln \frac{1}{|z - \xi|} \omega(dE_\zeta) + u(z) \geqslant 2 \ln \frac{1}{|z - z_0|} + C,$$

where C is a constant, and so

$$\lambda(z) \geqslant e^c \cdot |z - z_0|^{-2}$$

for any $z \in B(z_0, r/2)$. Hence it is clear that for any curve K starting from the point z_0 we have $s_\lambda(K) = \infty$, as we needed to prove.

If the set function ω is non-negative, then by what we have said any curve K that is a shortest curve in the metric ρ_λ, where $\lambda(z) \equiv \lambda(z; \omega, h)$, does not contain peak points of the metric ρ_λ, and so by Theorem 8.1.3 K is a curve with bounded rotation on the plane. We have $\kappa_l(K) \leqslant 0$, $\kappa_r(K) \leqslant 0$, $\omega(K^0) \geqslant 0$. The equality (8.10) enables us to conclude that in this case $\kappa_l(K) = \kappa_r(K) = 0$. Thus we obtain the following result.

Theorem 8.1.4. *If the measure ω in the domain G is non-negative, $\lambda(z) \equiv \lambda(z; \omega, h)$, then any curve K that is a shortest curve with respect to the metric ρ_λ is a curve with bounded rotation and its left and right turns in the metric ρ_λ are equal to zero.*

Let $L \subset U$ be a shortest curve in the domain U of a two-dimensional manifold of bounded curvature M. We assume that L does not contain peak points. Then it follows from Theorem 8.1.3 that the shortest curve L, like any arc of it, is a curve of class $\tilde{\Delta}$. We take an interior point X of the shortest curve L arbitrarily. Let (A_n), (B_n), $n = 1, 2, \ldots$, be sequences of points of L that converge to X on the left and right respectively. Let L_n and M_n be the arcs $[A_nX]$ and $[XB_n]$ of the curve L, and K_n the arc $[A_nB_n]$. Let $\kappa_l(X, L)$ and $\kappa_r(X, L)$ be the left and right turns of L at X. By Theorem 8.1.1 $\kappa_l(K_n) = \kappa_l(L_n) + \kappa_l(M_n) + \kappa_l(X, L)$. According to Theorem 8.1.3 $\kappa_l(K_n) \leqslant 0$. As $n \to \infty$, $\kappa_l(L_n) \to 0$ and $\kappa_l(M_n) \to 0$. This enables us to conclude that $\kappa_l(X, L) \leqslant 0$. Similarly we can show that $\kappa_r(X, L) \leqslant 0$. We have thus established the following proposition.

Theorem 8.1.5. *If a shortest curve L in a domain U of a manifold of bounded curvature does not contain peak points, then at any point of it the left and right turns are non-positive.*

We shall define a *geodesic polygonal line* in a domain U of a two-dimensional manifold of bounded curvature as any finite sequence L of oriented simple arcs $L_1, L_2, \ldots, L_m$, each of which is a shortest curve, where the end of the arc L_1 coincides with the beginning of the arc L_{i+1} when $i = 1, 2, \ldots, m-1$ and the union of the arcs $L_1, L_2, \ldots, L_m$ is a simple arc $\tilde{L}$. The arc $\tilde{L}$ is called the *carrier* of the polygonal line L, the arcs L_i, $i = 1, 2, \ldots, m$, are called its *links*, and their end-points are called the *vertices* of L. We note that by dividing any of the arcs L_i into two we obtain another geodesic polygonal line.

If the carrier $\tilde{L}$ of a geodesic polygonal line L does not contain peak points, then the image $\tilde{K}$ of $\tilde{L}$ in the isothermal coordinate system $\varphi\colon U \to \mathbb{C}$ is a curve with bounded rotation. We denote by $\tilde{\kappa}_r(L)$ the sum of the right turns of $\tilde{L}$ at interior points of it that are vertices of L. Similarly, let $\tilde{\kappa}_l(L)$ be the sum of the left turns of $\tilde{L}$ at these points. We shall call $\tilde{\kappa}_r(L)$ and $\tilde{\kappa}_l(L)$ the conditional right and left turns of L. Let X_{i-1} and X_i be the beginning and end of the shortest curve L_i. By Theorem 8.1.1 we have

$$\kappa_l(\tilde{L}) = \sum_{i=1}^{m} \kappa_l(L_i) + \sum_{i=1}^{m-1} \kappa_l(X_i, L_i) = \sum_{i=1}^{m} \kappa_l(L_i) + \tilde{\kappa}_l(L).$$

Since L_i is a shortest curve, by (8.11) it follows that

$$|\kappa_l(\tilde{L}) - \tilde{\kappa}_l(L)| \leqslant |\tilde{\omega}|(L_0). \tag{8.12}$$

Similarly,

$$|\kappa_r(\tilde{L}) - \tilde{\kappa}_r(L)| \leqslant |\tilde{\omega}|(L_0). \tag{8.13}$$

Theorem 8.1.6. *Let $L \subset U$ be a simple arc of class $\tilde{\Delta}$ with end-points A and B. Then there is a sequence of geodesic polygonal lines (L_n), $n = 1, 2, \ldots$, joining A and B and coverging to L on the right (left) (that is, such that the carriers $\tilde{L}_n$ of these polygonal lines converge to a curve K on the right, respectively left). Let (L_n), $n = 1, 2, \ldots$, be such a sequence, and α_n* and *β_n the angles of the domain D_n at the points A and B. If the polygonal lines L_n converge to the curve K on the right,* *then* $\tilde{\kappa}_r(L_n) - \alpha_n - \beta_n \to \kappa_r(L)$. *If* $L_n \to K$ *to the left, then* $\tilde{\kappa}_l(L_n) + \alpha_n + \beta_n \to \kappa_l(L)$ *as* $n \to \infty$.

Theorem 8.1.6 is proved by applying the corollary of Theorem 8.1.2 and the inequalities (8.12), (8.13).

Theorem 8.1.6 establishes the geometrical meaning of the quantities $\kappa_l(L)$ and $\kappa_r(L)$. In particular, it implies the equivalence of the definition of turn given in the monograph Aleksandrov and Zalgaller (1962) and turn in the sense of the definition of the present work.

Curvature as a set function in a manifold of bounded curvature. Let M be a two-dimensional manifold of bounded curvature, U a domain in M homeomorphic to a disc, and $\varphi: U \to \mathbb{C}$ an isothermal coordinate system in this domain. We shall say that a simple arc L in M belongs to the class Δ if any partial arc L_1 of it belongs to the class $\tilde{\Delta}$. This is equivalent to the fact that for any isothermal coordinate system defined in a neighbourhood of L the image of L is a simple arc on the plane that has left and right tangents at each of its points. Similarly we shall say that a simple closed curve Γ *belongs to the class* Δ if any simple arc contained in Γ belongs to the class $\tilde{\Delta}$.

As above, let $G = \varphi(U)$ and let $\lambda(z) \equiv \lambda(z; \omega, h)$ be a function defined in U such that the domain G with metric ρ_λ is isometric to the domain U with induced metric ρ_U and $\varphi: U \to G$ is an isometric map of (U, ρ_U) onto (G, ρ_λ).

Let Γ be a simple closed curve of class Δ lying in U and $\Lambda = \varphi(\Gamma)$ its image in a given isothermal coordinate system. We shall assume that U is oriented by stipulating that the coordinate system φ is right, and we orient the curve Γ positively. Let D be the domain bounded by Γ. We take points X and Y on Γ arbitrarily. They split Γ into two simple arcs Γ_1 and Γ_2, which we shall assume to be oriented so that when we go round Γ in the positive direction we also go round each of the arcs Γ_1 and Γ_2 in the positive direction. Let α and β be the angles of D at X and Y. We put

$$\kappa(\Gamma) = \kappa_l(\Gamma_1) + \kappa_l(\Gamma_2) + (\pi - \alpha) + (\pi - \beta).$$

It follows easily from Theorem 8.1.6 that $\kappa(\Gamma)$ does not depend on the choice of the points X and Y. Let $z_1 = \varphi(X)$, $z_2 = \varphi(Y)$, $\Lambda_1 = \varphi(\Gamma_1)$, $\Lambda_2 = \varphi(\Gamma_2)$. For the angular function $\varphi_l(\Lambda, z)$ we have $\varphi(\Lambda, z) = \varphi_l(\Lambda_1, z) + \varphi_l(\Lambda_2, z)$ for any point z other than z_1 and z_2. Let $\bar{\alpha}$ and $\bar{\beta}$ be the angles of the domain Q bounded by the curve Λ at the points z_1 and z_2. We have $\kappa(\Lambda_1) + \kappa(\Lambda_2) + (\pi - \bar{\alpha}) + (\pi - \bar{\beta}) = 2\pi$. At the point z_1 we have $\varphi(\Lambda, z) - \varphi_l(L_1, z) - \varphi_l(L_2, z) = 2\pi - \bar{\alpha}$ and similarly for the point z_2. Using this relation, after obvious transformations we obtain

$$\kappa(\Gamma) = 2\pi - \frac{1}{2\pi} \int_G \varphi(\Gamma, z)\omega(dz). \tag{8.14}$$

We now observe that $\varphi(\Gamma, z) = 2\pi$ if z belongs to Q and $\varphi(\Gamma, z) = 0$ otherwise. Hence we conclude that the integral on the right-hand side of (8.14) is equal to

$2\pi\omega[Q]$, and so

$$\kappa(\Gamma) + \omega(Q) = 2\pi.$$

For an arbitrary Borel set $E \subset U$ we put $\Omega(E) = \omega[\varphi(E)]$. Obviously Ω is a measure defined in a domain of the two-dimensional manifold of bounded curvature M.

For any positively oriented simple closed curve of class Δ contained in U we have

$$\kappa(\Gamma) + \Omega(D) = 2\pi, \tag{8.15}$$

where D is the domain bounded by the curve Γ. This property uniquely determines the set function Ω, and as a result we arrive at the following theorem.

Theorem 8.1.7. *On any two-dimensional manifold of bounded curvature M we can define a measure Ω such that for any simple closed curve Γ of class Δ lying in a domain U of M homeomorphic to an open disc and bounding a domain D the equality* (8.15) *is satisfied. The measure Ω satisfying this condition is unique. For any isothermal coordinate system $\varphi: U \to \mathbb{C}$ specified in a domain U we have $\Omega(E) = \omega[\varphi(E)]$, where $E \subset U$ is a Borel set and ω is the measure by means of which the function $\lambda(z) = \lambda(z; \omega, h)$ is defined in the representation $ds^2 = \lambda(z)|dz|^2$ of the metric tensor of the manifold in the given coordinate system.*

The measure Ω defined in Theorem 8.1.7 is called the curvature of a set of a given manifold of bounded curvature. The equality (8.15), by analogy with the case of a Riemannian manifold or a manifold with polygonal metric, is called the *Gauss-Bonnet formula.*

Curves with bounded variation of turn. Here we distinguish an important class of curves in a two-dimensional manifold of bounded curvature.

Let L be an arbitrary simple arc of class Δ in a two-dimensional manifold of bounded curvature M. We orient the arc L and specify a definite orientation of M. Let $X(t)$, $a \leqslant t \leqslant b$, be an arbitrary right parametrization of L, and L_t the arc of L corresponding in this parametrization to the interval $[a, t]$, $a < t$. Then there are defined the quantities

$$\kappa_l(t) = \kappa_l(L_t), \quad \kappa_r(t) = \kappa_r(L_t), \quad \Omega(t) = \Omega(L_t^0).$$

We also put $\kappa_l(a) = \kappa_r(a) = 0$. For all $t \in [a, b]$ we have $\kappa_l(t) + \kappa_r(t) = \Omega(t)$. Representing the measure Ω in the form $\Omega = \Omega^+ - \Omega^-$, where Ω^+ and Ω^- are its positive and negative parts, we obtain $\Omega(t) = \Omega^+(L_t^0) - \Omega^-(L_t^0)$. The functions $\Omega^+(L_t^0)$ and $\Omega^-(L_t^0)$ are obviously non-decreasing, and so $\Omega(t)$ is a function of bounded variation. We shall say that L is a curve of bounded variation of turn if one of the functions $\kappa_l(t)$ and $\kappa_r(t)$ is a function of bounded variation. In view of the equality $\kappa_l(t) + \kappa_r(t) = \Omega(t)$ it is then obvious that the other is also a function of bounded variation.

Theorem 8.1.8. *If a curve L in a two-dimensional manifold of bounded curvature does not contain peak points and is a curve of bounded variation of turn, then* *it is rectifiable. We assume that $\varphi \colon U \to \mathbb{C}$ is an isothermal coordinate system defined in a neighbourhood of L and let $K = \varphi(L)$. We assume that L does not contain peak points. Then for L to be a curve with bounded variation of turn it is necessary and sufficient that K should be a curve with bounded rotation.*

It was shown in Zalgeller (1965) that if a simple arc L with bounded variation of turn has peak points, then it may turn out to be non-rectifiable. In this case the image of a curve in an isothermal coordinate system may not be a curve of bounded rotation.

Area in a two-dimensional manifold of bounded curvature. Let M be a two-dimensional manifold of bounded curvature, and E a Borel set in M. We assume that E is contained in the domain of definition of an isothermal coordinate system $\varphi \colon U \to \mathbb{C}$ in M. Let $G = \varphi(U)$ and let $\lambda(z)|dz|^2$ be the line element in this cordinate system. We put

$$\sigma(E) = \iint\limits_{\varphi(E)} \lambda(z)\, dx\, dy.$$

The formula (7.4) for transforming the line element of a manifold on going over from one coordinate system to another enables us to conclude that $\sigma(E)$ does not depend on the choice of coordinate system in M.

We assume that E is an arbitrary Borel set in M. Then E can be represented as the union of a no more than countable set of pairwise disjoint Borel sets, each of which is contained in the domain of definition of an isothermal coordinate system of M. Let

$$E = \bigcup_k E_k$$

be such a representation of the set E. It is easy to show that the sum $\sum_k \sigma(E_k)$ does not depend on the choice of this representation of E. We put

$$\sum_k \sigma(E_k) = \sigma(E).$$

Thus, on the totality of all Borel sets of M there is defined a non-negative set function σ. It is easy to verify that σ is a measure in M. We shall call σ the area in M. We shall call $\sigma(E)$, where E is a Borel set in M, the *area of the set E.*

This definition of area is in some respects formal and relies on the analogy with the case of a Riemannian manifold. We now state some results that enable us to establish the geometrical meaning of the set function σ.

Theorem 8.1.9. (Aleksandrov and Zalgaller (1962)). *Let G be an open set in a two-dimensional manifold of bounded curvature, and (ρ_n), $n = 1, 2, \ldots$, a sequence of Riemannian metrics defined in G that converges to the metric ρ_G induced in G. We assume that the absolute curvatures of the Riemannian manifolds (G, ρ_n) are bounded in aggregate and let $\sigma_n(E)$ be the area of the Borel set E in the Riemannian metric ρ_n. Then as $n \to \infty$ the set functions σ_n converge weakly to the set function σ.*

8.2. A Theorem on the Contraction of a Cone. Angle between Curves. Comparison Theorems. Using the results of the previous part, we can first of all make the theorem on the contraction of a cone (Lemma 6.5.1) more precise. Namely, the following proposition is true.

Theorem 8.2.1. *Let M be a two-dimensional manifold of bounded curvature, and Γ a simple closed curve in M that bounds an open domain D homeomorphic to a disc. We assume that Γ is rectifiable and that $\Omega^+(D) < 2\pi$. Then there is a convex cone Q such that $\omega(Q) \leqslant \Omega^+(D)$ and Q has a map φ onto the set $D \cup \Gamma$ such that the following conditions are satisfied:*

1) *the boundary ∂Q of the cone is mapped one-to-one onto the curve Γ. The image of any simple arc $L \subset \partial Q$ is a simple arc of Γ whose length is equal to the length of L;*

2) *for any points $X, Y \in Q$ we have*

$$\rho_Q(X, Y) \geqslant \rho_D(\varphi(X), \varphi(Y)).$$

For the proof of the theorem it is sufficient to observe that the sequence of polyhedral metrics (ρ_n) considered in 6.5 can be chosen so that the set functions $\omega^+_{\rho_n}$ converge weakly to the function $\omega_0 = \Omega^+$. The existence of such an approximating sequence follows easily from the results of 8.1 and §7.

Let us mention some applications of Theorem 8.2.1.

Theorem 8.2.2 (the first comparison theorem, Aleksandrov and Zalgaller (1962)). *Let T be a triangle homeomorphic to a disc in a two-dimensional manifold of bounded curvature M, $\bar{\alpha}$ the upper angle at one vertex of T, and α_0 the angle at the corresponding vertex of the development of the triangle T onto the plane. Then*

$$\bar{\alpha} - \alpha_0 \leqslant \Omega^+(T^0), \tag{8.16}$$

where T^0 denotes the interior of the domain bounded by the sides of the triangle T.

A few words about the proof of Theorem 8.2.2. Let A, B, C be the vertices of the triangle T^0. The inequality (8.16) is obvious if $\Omega^+(T^0) \geqslant 2\pi$. We shall assume that $\Omega^+(T^0) < 2\pi$. Let us construct a convex cone Q and a map φ of Q onto T that corresponds in the sense of Theorem 8.2.1 to the domain T^0. We denote the points on the boundary of the cone corresponding to A, B, C by A', B', C' respectively. Let $K = [BC]$, $L = [CA]$, $M = [AB]$ be the sides of the triangle T, and $K' = [B'C']$, $L' = [C'A']$, $M' = [A'B']$ the arcs of the boundary of the cone Q that correspond to them. Each of the arcs K', L', M' is a shortest curve. For example, let R' be an arbitrary curve on the cone Q that joins the points B' and C', and let $R = \varphi(R')$. Then obviously $s(R) \leqslant s(R')$. The curve R joins the points B and C, and since K is a shortest curve, $s(K) \leqslant s(R) \leqslant s(R')$. We have $s(K) = s(K')$, and so $s(K') \leqslant s(R')$, that is, we see that any curve in Q that joins the points B' and C' has length at least $s(K')$, and so K' is a shortest curve.

Let α' be the angle of the boundary of the cone Q at the point A'. We show that

$$\bar{\alpha} \leqslant \alpha'. \tag{8.17}$$

Let $X \in L$ and $Y \in M$ be two arbitrary points on the sides of the triangle T starting from its vertex A, where $X \neq A$, $Y \neq A$. We put $x = \rho(A, X)$, $y = \rho(A, Y)$, $z = \rho(X, Y)$. We construct a planar triangle with sides equal to x, y, z, and let $\gamma(x, y)$ be the angle of this triangle lying opposite the side equal to z. We put $z' = \rho_Q(X', Y')$ and let R' be the shortest curve in Q that joins X' and Y', $R = \varphi(R')$. We have $s(R) \leqslant s(R') = z'$. On the other hand, we obviously have $z \leqslant s(R)$, so $z \leqslant z'$. Let us construct a planar triangle with sides x, y, z'; let $\gamma'(x, y)$ be its angle opposite the side equal to z'. Since $z' \geqslant z$ we have $\gamma(x, y) \leqslant \gamma'(x, y)$. Obviously $\gamma'(x, y) \to \alpha'$ as $x \to 0$, $y \to 0$. We have $\bar{\alpha} = \overline{\lim\limits_{x \to 0, y \to 0}} \gamma(x, y)$, and so the inequality (8.17) is proved.

Since $\Omega(Q) \leqslant \Omega^+(T^0)$, by virtue of (8.17) for the proof of Theorem 8.2.2 it is sufficient to establish that

$$\alpha' - \alpha_0 \leqslant \Omega(Q). \tag{8.18}$$

The simple arcs K', L', M' are shortest curves on the cone Q. The inequality (8.18) is therefore a special case of (8.16) when M is the convex cone Q. The proof of (8.18) is a problem of elementary geometry. Let us give a solution of it. If $\Omega(Q) = 0$, then $A'B'C'$ is an ordinary planar triangle: $\alpha' = \alpha_0$ and $\alpha' - \alpha_0 = 0 \leqslant \Omega(Q)$, so in this case (8.18) is true. We shall assume that $\Omega(Q) > 0$. Let O be the vertex of the cone Q. We join A' to O by a shortest curve and then draw from O the generator OZ' (where $Z' \in \partial Q$) of the cone Q such that the angles to the left and right of O between the shortest curves OA' and OZ' are equal. Let θ be their common value. We have $2\theta = 2\pi - \Omega(Q)$. The points A' and Z' are joined in Q by two shortest curves lying on opposite sides of Q (Fig. 39). On the cone Q we obtain two triangles $A'OZ'$. Since the vertex of the cone Q lies on the boundary of each of these triangles, the latter are isometric to planar triangles with the same lengths of sides. Let ξ and η be the angles of the triangles $A'OZ'$ at the points A' and Z' respectively. We have $\xi + \eta = \pi - \theta$, so $2\xi + 2\eta = 2\pi - 2\theta = \Omega(Q)$. In particular, $2\xi \leqslant \Omega(Q)$. The point Z' lies on the shortest curve $K' = [B'C']$. For if Z' were an interior point of the shortest curve $M' = [A'B']$, then we could shorten M', since the angle between the arc $Z'B'$ and the shortest curve $A'Z'$ on the other side of O with respect to the shortest curve M' is equal to $\pi - 2\eta < \pi$. Similarly we can establish that Z' cannot be an interior point of the

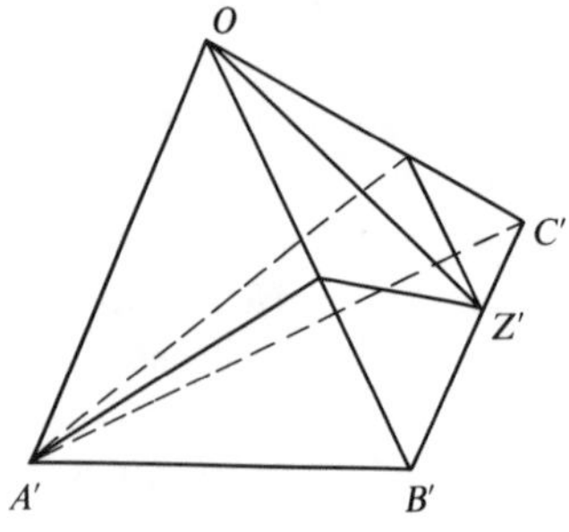

Fig. 39

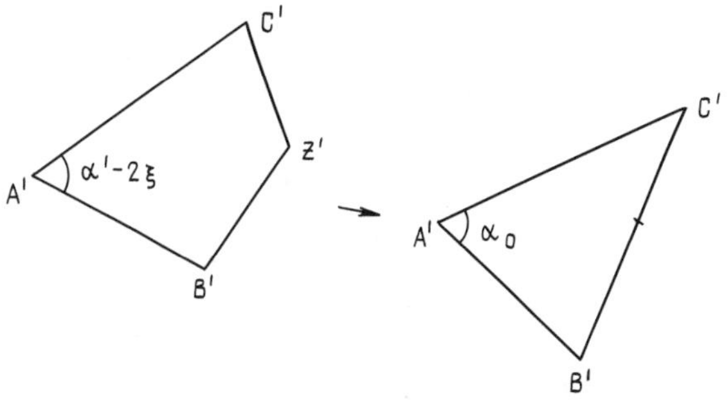

Fig. 40

shortest curve $[A'C']$. Cutting out from the cone Q the lune formed by the shortest curves $[A'Z']$ and pasting these shortest curves together, we obtain a planar quadrangle $A'B'Z'C'$ for which $|A'B'| = \rho(A, B)$, $|A'C'| = \rho(A, C)$, $|B'Z'| + |Z'C'| = \rho(B, C)$. The angle of this triangle at A' is equal to $\alpha' - 2\xi$. Straightening the planar polygonal line $B'Z'C'$ (Fig. 40), we obtain a planar triangle that is the development of the original triangle T, so the angle at its vertex corresponding to A' is equal to α_0. On straightening the polygonal line $B'Z'C'$ the angle at the vertex A' increases, and so we deduce that $\alpha' - 2\xi \leqslant \alpha_0$, so $\alpha' - \alpha_0 \leqslant 2\xi \leqslant \Omega(Q)$, and the inequality (8.18) and with it Theorem 8.2.2 are proved.

Theorem 8.2.3 (Aleksandrov and Zalgaller (1962)). *There is a definite angle between any two shortest curves in a two-dimensional manifold of bounded curvature starting from one point of the manifold.*

In other words, we assert the following. Let K and L be shortest curves in a manifold of bounded curvature with common starting point O, let $X \in K$ and $Y \in L$ be points of these shortest curves such that $\rho(O, X) = x > 0$, $\rho(O, Y) = y > 0$, and let $\gamma(x, y)$ be the angle of the development onto the plane of the triangle OXY corresponding to the vertex O. Then there exists the limit

$$\lim_{x \to 0, y \to 0} \gamma(x, y).$$

Here we shall not dwell on how Theorem 8.2.3 is proved. We only mention that in the proof of the first basic theorem on approximation we use a special case of Theorem 8.2.3. We did not mention this above, relating it to details of technical character. In the special case that is necessary for the proof of the first theorem on approximation, the required result is established by applying Lemma 6.3.1 (an estimate of the difference $\bar{\alpha} - \alpha_0$ in terms of $\bar{\nu}^+(OXY)$). In the general case the truth of Theorem 8.2.3 was established in Aleksandrov and Zalgaller (1962) by means of the first theorem on approximation.

A proof of Theorm 8.2.3 different from that given in Aleksandrov and Zalgaller (1962) can be obtained by using the estimate of Theorem 8.2.2.

Remark. In the conditions of Theorem 8.2.2 we denote by $\tilde{\alpha}$ the angle of the domain T^0 at the point A, and by α the angle of the triangle T at this point. Obviously $\alpha \leqslant \tilde{\alpha}$. The estimate of Theorem 8.2.2 can be strengthened, namely we have

$$\tilde{\alpha} - \alpha_0 \leqslant \omega^+(T^0). \tag{8.19}$$

The proof of (8.19) can also be established by applying Theorem 8.2.1.

Theorem 8.2.4 (the second comparison theorem, Aleksandrov and Zalgaller (1962)). *Let $T = ABC$ be a triangle homeomorphic to a disc in a two-dimensional manifold of bounded curvature, T^0 its interior, α the angle of the triangle T at the point A, and α_0 the corresponding angle of the development of the triangle. Then*

$$\alpha - \alpha_0 \geqslant -\omega^-(T^0) - \sum_{i=1}^{3} \kappa_i^-,$$

where κ_i^-, $i = 1, 2, 3$, are the negative parts of the turns of the sides of the triangle T from the side converted to the domain T^0.

Remark. In Aleksandrov and Zalgaller (1962) Theorems 8.2.2 and 8.2.4 were stated in more general form. This more general result can be obtained as a consequence of Theorems 8.2.2 and 8.2.4, given here.

The proof of Theorem 8.2.4 in its basic part relies on a proposition analogous to Lemma 6.3.1 which gives a lower estimate of the difference $\alpha - \alpha_0$ (in contrast to Lemma 6.3.1, which enables us to estimate the difference from above). Namely, the following lemma is true.

Lemma 8.2.1 (Aleksandrov and Zalgaller (1962)). *Suppose that for each pair of points X, Y on the sides $[AB]$ and $[AC]$ of the triangle $T = ABC$ in a two-dimensional manifold of bounded curvature we choose a definite shortest curve $[XY]$, where in all cases the left convergence (that is, with the side of the point A) $X_n \to X$ or $Y_n \to Y$ implies the convergence $[X_nY] \to [XY]$, $[XY_n] \to [XY]$. Then*

$$\alpha - \alpha_0 \geqslant -\nu_A^-, \tag{8.20}$$

where α is the angle of the triangle T at the point A, α_0 is the corresponding angle of the development, and ν_A^- is the greatest lower bound of the excesses of the triangles AXY.

The condition concerning the convergence of shortest curves in the statement of the lemma can be established if we understand by $[XY]$ the extreme (from the side of the point A) shortest curve joining X and Y.

We mention that Aleksandrov and Zalgaller (1962) stated a more general proposition relating to arbitrary metric spaces. In this case instead of α we need to take a lower limit. Excesses of triangles are understood in the corresponding way. The proof of Lemma 8.2.1 (also, like the more general proposition, given in Aleksandrov and Zalgaller (1962)) is carried out by means of arguments similar to those by means of which Lemma 6.3.1 is proved. The proof of Theorem 8.2.4 reduces to an estimate of ν_A^-.

Theorems 8.2.3 and 8.2.4 enable us to show, in particular, that a curve K belongs to the class $\tilde{\Delta}$ if and only if it has definite directions at its ends.

As a preliminary we indicate some auxiliary propositions that also have a definite independent interest.

Let O be a point in a two-dimensional manifold of bounded curvature M such that $\Omega(O) < 2\pi$. In a neighbourhood U of the point O we specify a definite orientation of the manifold M. Let $L_0 = [XO]$ be a simple arc with bounded variation of the turn lying in U, and $L = [OY]$ any other simple arc with bounded variation of the turn contained in U. We assume that L_0 and L have no points in common other than O. Then they form a curve K. Let us find the angle on the right of the curve K at the point O. Whatever the number $\varphi \in [0, \theta(X)]$, where $\theta(X) = 2\pi - \Omega(X)$, the arc $L = [OY]$ can be chosen so that the angle is equal to the given φ. It is easy to verify the truth of this assertion by using an isothermal coordinate system. We say that $\varphi \in [0, \theta(X)]$ is non-singular if there is a shortest curve $L = [OY]$ such that the angle on the right of the curve $K = L_0 \cup L$ at the point X is equal to φ.

Lemma 8.2.2. *In the notation introduced above, the set of non-singular values is everywhere dense in the interval* $[0, \theta(X)]$.

Lemma 8.2.3. *Let $L_0 = [AO]$, $L = [OB]$ be arbitrary shortest curves starting from a point O of a two-dimensional manifold of bounded curvature M. We assume that $\Omega(O) < 2\pi$ and that L_0 and L have no points in common other than O. Then for any $\varepsilon > 0$ there is a neighbourhood U of the point O such that for any triangle $T = OXY$ lying on one side of the simple arc $K = L_0 \cup L$ the angle α of the development of this triangle corresponding to the point O does not exceed $\varphi + \varepsilon$, where φ is the angle of the curve K at the point O from the side where the triangle T lies.*

The proof of Lemma 8.2.3 is carried out by a simple application of Theorem 8.2.4.

Let L be a curve of class $\tilde{\Delta}$, and O its starting point. In a neighbourhood of O we introduce an isothermal coordinate system $\varphi\colon U \to \mathbb{C}$ and let $K = \varphi(L)$. We assume that $\varphi(O) = 0$. The curve K has a half-tangent at O. Let K' be a segment of this half-tangent and L' the simple arc that goes over to K under the map φ. By Lemma 8.2.2 there are shortest curves L_1 and L_2 starting from O that lie on different sides of L and make an angle less than $\varepsilon/3$ with L'. The curves $K_1 = \varphi(L_1)$ and $K_2 = \varphi(L_2)$ are curves with bounded rotation and a small arc of the curve K is contained in the angle included between them. Lemma 8.2.3 enables us to conclude that for any triangle OXY, where $X \in L$, $Y \in L$, the angle $\gamma(X, Y) < \frac{2}{3}\varepsilon + \frac{\varepsilon}{3}$ if X and Y are sufficiently close to O. Hence it follows that $\lim\limits_{X \to 0, Y \to 0} \gamma(X, Y) = 0$, that is, L has a definite direction at O.

The verification that if the curve L has a definite direction at one of its end-points, then its image in an isothermal coordinate system has a half-tangent at the corresponding point, is easily carried out.

Theorem 8.2.5. *If the curvature of a two-dimensional manifold of bounded curvature M, as a set function, is non-positive, then the angles of any triangle in M homeomorphic to a disc do not exceed the corresponding angles of the development.*

Remark. The condition in the statement of the theorem that the triangle is homeomorphic to a disc cannot be omitted. In fact, let M be the surface of a right circular cylinder and Γ the section of M by a plane perpendicular to the generators. On Γ we mark points X, Y, Z that divide Γ into three equal arcs. The curve Γ is a geodesic, and the arcs into which it is split by the points X, Y, Z are shortest curves. We thus obtain a triangle XYZ. The angles at its vertices are all equal to π. At the same time the angles of the development of the triangle XYZ are all equal to $\pi/3 < \pi$.

The next result follows from Theorem 8.2.4.

Theorem 8.2.6. *Let M be a complete two-dimensional manifold of bounded curvature, where the curvature of M, as a set function, is non-negative. Then for any triangle in M the angles at its vertices are not less than the corresponding angles of the development.*

The given assertion follows from the fact that in a manifold of non-negative curvature the turn of a shortest curve is equal to zero, and any two shortest curves in such a manifold do not have points in common other than the end-points. Using the results in Aleksandrov (1948a) about the topological structure of complete two-dimensional manifolds of bounded curvature, it is not difficult to conclude that any triangle in such a manifold is homeomorphic to a disc (so long as its sides do not lie on one shortest curve). Theorem 8.2.6 is thus obtained by a direct application of Theorem 8.2.4.

Let M be a two-dimensional manifold of bounded curvature, K a real number, Ω the curvature and σ the area of the manifold. We say that M is a manifold of curvature not greater than K if for any Borel set E such that $\Omega(E)$ is defined we have $\Omega(E) \leqslant K\sigma(E)$. If $\Omega(E) \geqslant K\sigma(E)$ for any $E \subset M$ for which $\Omega(E)$ is defined, then M is called a two-dimensional manifold of curvature not less than K. Let Σ_K denote, in the case $K > 0$, a sphere of radius $1/\sqrt{K}$, in the case $K = 0$ an ordinary Euclidean plane, and finally in the case $K < 0$ a Lobachevskij plane with Gaussian curvature K. Let $T = XYZ$ be a triangle in an arbitrary metric space (M, ρ), and $x = \rho(Y, Z)$, $y = \rho(Z, X)$, $z = \rho(X, Y)$ the lengths of its sides. On the surface Σ_K we construct a triangle T' with side lengths x, y, z (in the case $K > 0$ this is possible only if $x + y + z \leqslant 2\pi/\sqrt{K}$). The triangle T' will be called the *K-development of the triangle T.* Theorems 8.2.5 and 8.2.6 have analogues for the cases when M is a two-dimensional manifold of curvature not greater than K (respectively, not less than K and instead of an ordinary development we consider a K-development (see Aleksandrov (1951), (1954), (1957b).

We note that for the case when M is a K-polyhedron (see § 6), where $K > 0$, an analogue of Theorem 8.2.1 was obtained in Belinskij (1975).

8.3. A Theorem on Pasting Together Two-Dimensional Manifolds of Bounded Curvature. In 3.2 we described a general procedure for pasting together two-dimensional manifolds with intrinsic metric. Here we mention a theorem that establishes conditions under which the pasting together of two-dimensional manifolds of bounded curvature gives as a result a manifold of bounded curvature.

Let M be a two-dimensional manifold with boundary, endowed with an intrinsic metric ρ_M. We shall say that M is a two-dimensional manifold of bounded curvature if M is isometric to a subset of some two-dimensional manifold M' without boundary that is endowed with an intrinsic metric and is a manifold of bounded curvature in the sense of the definitions in 6.1.

Similarly a two-dimensional manifold M with boundary will be called a Riemannian manifold if it is isometric to a domain in a two-dimensional Riemannian manifold in the sense of the definition given in § 4.

In order that the basic concepts of Riemannian geometry, the Gaussian curvature, the geodesic curvature of a curve, and so on, can be introduced for a Riemannian space it is sufficient that the metric tensor of the manifold should belong to the class C^∞. If in the conditions of 3.2 the pasted manifolds are Riemannian manifolds of class C^∞, then the manifold obtained as a result of the pasting will be a Riemannian manifold of the same smoothness class only if restrictions on the pasted arcs are satisfied; these are very severe even in the case when these arcs are "good" from the viewpoint of the geometry of the manifold. In contrast to this, the conditions under which as a result of pasting together manifolds of bounded curvature we obtain a manifold of the same type are substantially weaker.

Theorem 8.3.1 (Aleksandrov and Zalgaller (1962)). *Let (D_k), $k = 1, 2, \ldots$, be a finite or denumerable set of two-dimensional manifolds with boundary, each of which is endowed with an intrinsic metric. We assume that each of the manifolds D_k is a two-dimensional manifold of bounded curvature and that for each k any simple arc contained in ∂D_k is a curve with bounded variation of turn. We assume that for the manifolds D_k there is specified a law of pasting that satisfies conditions* 1)–5) *in* 3.2. *Then the two-dimensional manifold D obtained by pasting together the D_k is a manifold of bounded curvature.*

The next theorem gives a means of calculating the basic geometrical characteristics of a manifold D from the corresponding characteristics relating to the manifolds D_k.

Theorem 8.3.2. *Suppose that the manifolds (D_k), $k = 1, 2, \ldots$, satisfy all the conditions of the preceding theorem, and that D is the manifold obtained as a result of pasting them together. Let Δ_k be the set of points of D corresponding to points of D_k. Then the following assertions are true*:

A) *for any Borel set $E \subset D$ its area is equal to the sum of the areas of the sets $E_k \subset D_k$ from which E arises under the pasting*;

B) *if a point $x \in D$ belongs to Δ_k^0 for some k and $x' \in D_k$ is the point from which x is obtained under the pasting, then the total angle of D at x is equal to the total* *angle of D_k at x'. If the point $x \in D$ is obtained by pasting together the points $x_1 \in \partial D_{k_1}, x_2 \in \partial D_{k_2}, \ldots, x_m \in \partial D_{k_m}$, then the total angle of D at x is equal to the sum of the total angles of the manifolds $D_{k_1}, D_{k_2}, \ldots, D_{k_m}$ at the points $x_1, x_2, \ldots, x_m$ respectively;*

C) *if a simple arc $L \subset D$ is obtained by pasting together the simple arcs $L_1 \subset \partial D_{k_1}$, $L_2 \subset \partial D_{k_2}$, then L is a curve with bounded variation of turn. The turn of L on the side of a domain $\Delta_{k_i} \subset D$ is equal to the turn of L_i on the side of the manifold D_i. If $L \subset \Delta_k^0$ for some k, then the left and right turns of the arc L in the manifold D are equal to the left and right turns respectively of the arc $L' \subset D_k$ from which L is obtained by the pasting;*

D) *let $E \subset D$ be a Borel set. We assume that $E \subset \Delta_k^0$ for some k. Then the curvature of the set E in the manifold D is equal to the curvature in D_k of the set E' that goes over to E under the pasting.*

Theorems 8.3.1 and 8.3.2 indicate a method for constructing a large number of specific examples of two-dimensional manifolds of bounded curvature.

Example. A collection of flat domains whose pasting together gives the surface of the right circular cylinder shown in Fig. 6 satisfies the conditions of Theorem 8.3.1. Hence it follows that the surface of the cylinder is a manifold of bounded curvature. Assertions C and D of Theorem 8.3.2 enable us to conclude that the curvature, as a set function of this manifold, is positive and concentrated on the two circles that bound the base of the cylinder. If the set E is a simple arc lying on one of these circles, then $\omega(E)$ is equal to the rotation of this arc in the plane of the base (we assume that the plane of the base and the bounding circle are oriented so that the rotation of any arc of the circle is positive).

Another example. Let R be a convex open set on the plane $\mathbb{E}^2$ such that the set $M = \mathbb{E}^2 \setminus R$ is connected (this excludes the case when R is the strip included between two parallel lines). Pasting together the manifold M and a second copy of it, we obtain a two-dimensional manifold $\tilde{M}$ (the twice covered manifold M). By Theorem 8.3.1, $\tilde{M}$ is a manifold of bounded curvature. Theorem 8.3.2 enables us to conclude that its curvature, as a set function, is non-positive and concentrated on the curve that arises from the bounding curve of the domain R.

8.4. Theorems on Passage to the Limit for Two-Dimensional Manifolds of Bounded Curvature. Let (M, ρ) and (M_n, ρ_n), $n = 1, 2, \ldots$, be metric spaces with intrinsic metric. In accordance with the definition given above, we shall say that the spaces (M_n, ρ_n) converge to the space (M, ρ) if for each n there is specified a homeomorphism φ_n of M onto M_n such that $\rho_n(\varphi_n(X), \varphi_n(Y)) \to \rho(X, Y)$ uniformly as $n \to \infty$. Putting $\tilde{\rho}_n(X, Y) = \rho_n(\varphi_n(X), \varphi_n(Y))$, we obtain in this case a metric $\tilde{\rho}_n$ on the set M. Since φ_n is a homeomorphism, the metric $\tilde{\rho}_n$ is compatible with the topology of M. As $n \to \infty$ the metrics $\tilde{\rho}_n$ converge uniformly to the metric ρ. The general situation described above thus reduces to the following. There is a topological space M on which there are specified intrinsic metrics ρ and $\tilde{\rho}_n$, $n = 1, 2, \ldots$, where each of these metrics is compatible with the topology of M and $\tilde{\rho}_n \to \rho$ uniformly on the set $M \times M$ as $n \to \infty$.

Next we assume to be fixed a two-dimensional manifold M and intrinsic metrics ρ, ρ_n, $n = 1, 2, \ldots$, defined in M. The length of a curve $L \subset M$ with respect to the metric ρ_n is denoted by $s_n(L)$. The symbol $s(L)$ denotes the length of L with respect to the metric ρ. If each of the manifolds (M, ρ), (M, ρ_n) is a two-dimensional manifold of bounded curvature, then for any Borel set $E \subset M$ whose closure is compact there are defined quantities $\Omega(E)$ and $\Omega_n(E)$, its curvatures with respect to the metrics ρ and ρ_n respectively. If (M, ρ) is a manifold of bounded curvature, we shall say for brevity that ρ is a metric of bounded curvature in M.

Theorem 8.4.1. *If the metrics ρ_n on the manifold M converge to the metric ρ as $n \to \infty$, then for any sequence $\{x_n(t), a \leqslant t \leqslant b\}$ of parametrized curves that converges as $n \to \infty$ to a parametrized curve $x(t)$, $a \leqslant t \leqslant b$, we have*

$$s(x; a, b) \leqslant \varliminf_{n \to \infty} s_n(x_n; a, b). \tag{8.21}$$

A set $E \subset M$ will be said to be bounded if its closure is compact. The totality of all Borel sets $E \subset M$ is denoted by $\mathfrak{B}(M)$. Henceforth $\mathfrak{B}_0(M)$ denotes the totality of all bounded Borel subsets of M. A measure in M will be any totally additive real set function defined either on $\mathfrak{B}(M)$ or on $\mathfrak{B}_0(M)$.

A function $f: M \to \mathbb{R}$ is said to be compactly supported if there is a compact set $A \subset M$ such that $f(x) = 0$ for all $x \notin A$. The totality of all continuous compactly supported functions $f: M \to \mathbb{R}$ is denoted by $C_0(M)$. We shall denote by $C(M)$ the set of all bounded continuous functions $f: M \to \mathbb{R}$.

If the measure α has the totality of sets $\mathfrak{B}(M)$ as domain of definition, then for any function $f \in C(M)$ the integral

$$\int_M f(x)\alpha(dx) \tag{8.22}$$

is defined and finite. In the case when the measure α has the totality of sets $\mathfrak{B}_0(M)$ as domain of definition, that is, $\alpha(E)$ makes sense only for a bounded Borel set E, the integral (8.22) is defined for any compactly supported continuous function.

Let $(\alpha_n: \mathfrak{B}(M) \to \mathbb{R})$, $n = 1, 2, \ldots$, be a sequence of measures in M, each of which is defined on all Borel subsets of M. Then we say that the sequence (α_n), $n = 1, 2, \ldots$, converges weakly to the measure $\alpha: \mathfrak{B}(M) \to \mathbb{R}$ as $n \to \infty$ if for any function $f \in C(M)$

$$\int_M f(x)\alpha(dx) = \lim_{n \to \infty} \int_M f(x)\alpha_n(dx). \tag{8.23}$$

We shall say that the sequence of measures $\alpha_n: \mathfrak{B}_0(M) \to \mathbb{R}$ converges locally weakly to the measure $\alpha: \mathfrak{B}_0(M) \to \mathbb{R}$ as $n \to \infty$ if (8.23) holds for any function $f \in C_0(M)$.

Above we defined the concept of weak convergence for a sequence of non-negative measures. In this case the definition given here is equivalent to the previous one.

Theorem 8.4.2. *We assume that the metrics ρ_n converge to the metric ρ as $n \to \infty$ and that (M, ρ_n) are two-dimensional manifolds of bounded curvature, where the quantities $|\Omega_n|(M)$ are bounded in aggregate,*

$$|\Omega_n|(M) \leqslant C = \text{const} < \infty \tag{8.24}$$

for all n. Then as $n \to \infty$ the set functions Ω_n converge locally weakly to the function Ω.

For the proof see Aleksandrov and Zalgaller (1962).

Let (ρ_n), $n = 1, 2, \ldots$, be a sequence of metrics of bounded curvature in a two-dimensional manifold M. We shall say that the sequence (ρ_n) converges tamely to the metric ρ if $\Omega_n^+ \to \Omega^+$, $\Omega_n^- \to \Omega^-$ as $n \to \infty$ in the sense of weak convergence.

Theorem 8.4.3. *We assume that the space (M, ρ) is isometric to a domain in a two-dimensional manifold of bounded curvature that has compact closure and is endowed with the induced metric. Then in M there is a sequence of Riemannian metrics ρ_n that converges tamely to ρ as $n \to \infty$. Under the same assumptions it is also possible to construct a sequence of polyhedral metrics (ρ_n) in the manifold M that converges tamely to ρ as $n \to \infty$.*

A proof of the theorem can be obtained, for example, by means of the theorems on analytic representation of a two-dimensional manifold of bounded curvature stated in §7.

We note that the construction by means of which Theorem 6.2.1 on the approximability of the metric of a manifold of bounded curvature by polyhedral metrics was proved leads to a tamely convergent sequence of polyhedral metrics (see Aleksandrov and Zalgaller (1962)).

Theorem 8.4.4. *We assume that ρ_n are metrics of bounded curvature that converge as $n \to \infty$ to a metric ρ, and that the quantities $|\Omega_n|(M)$ are bounded in aggregate, $|\Omega_n|(M) \leqslant C = \text{const} < \infty$ for all n. Let (L_n), $n = 1, 2, \ldots$, be a sequence of simple arcs in the manifold M that converges to a simple arc L. We assume that L_n for each n is a curve with bounded variation of turn in the manifold (M, ρ_n). Let $|\kappa_l|(L_n)$ denote the total variation of left turn of the arc L_n in the metric ρ_n. Then if the sequence $(|\kappa_l|(L_n))$, $n = 1, 2, \ldots$, is bounded and the curve L does not contain peak points in the sense of the metric ρ, it follows that L is a curve with bounded variation of turn. We have*

$$s(L) = \lim_{n\to\infty} s_n(L_n). \tag{8.25}$$

Remark. Under the conditions of the theorem the sequence of variations of right turns of the curves L_n in the metrics ρ_n is also bounded.

For a proof of Theorem 8.4.4 see Aleksandrov and Zalgaller (1962).

In Burago and Zalgaller (1965) some results are cited that touch on the relations between the variation of turn of a limiting curve and the variations of turn of the curves that converge to it. In the same paper an analogue of Theorem 8.4.4 was established for the case of curves that do not have simple arcs.

In Aleksandrov and Zalgaller (1962), Ch. VIII, § 2, there is discussed the question of the convergence of domains, each of which is a polygon with a fixed number of sides in the manifold (M, ρ_n). (We recall that by a polygon we understand a domain whose boundary consists of finitely many shortest curves, and each connected component of the boundary is the homeomorphic image of either a circle or an open interval.)

The theorems stated here can be extended to the case when convergence of metrics is understood in a weaker sense. We assume that in the manifold M there is specified a sequence of domains (M_n), $n = 1, 2, \ldots$, such that for each n the closure of M_n is compact and contained in M_{n+1} and

$$\bigcup_{n=1}^{\infty} M_n = M.$$

Let us assume that for each n there is specified a metric ρ_n. We shall say that the sequence of metrics (ρ_n), $n = 1, 2, \ldots$, converges locally to the metric ρ if for any n the sequence of metrics (ρ_m), $m = n, n + 1, \ldots$, converges to ρ on the set M_n, that is, $\rho_m(X, Y) \to \rho(X, Y)$ as $m \to \infty$ uniformly when $X, Y \in M_n$, whatever the value of n.

The results of this section admit an extension to the case of a sequence of metrics (ρ_n) that converges locally to the metric ρ as $n \to \infty$. The best possible results are obtained as a simple consequence of the theorems given above. All the necessary details can be found in Aleksandrov and Zalgaller (1962).

In the paper Burago (1965b) the author considered the question of the structure of metric spaces that are not two-dimensional manifolds and at the same time admit an approximation by two-dimensional Riemannian manifolds with absolute integral curvatures that are bounded in aggregate. It is easy to construct simple examples of such spaces. Let P be a plane in the three-dimensional space $\mathbb{E}^3$, let $O \in P$, and let $[OA]$ be an interval perpendicular to P. We put $M = P \cup [OA]$ and introduce in M the intrinsic metric induced from $\mathbb{E}^3$. Obviously the set M is not a manifold. Let Γ_n be the circle in the plane P with centre O and radius $1/n$. We denote by M_n the surface in $\mathbb{E}^3$ formed by the part of P outside Γ_n and the lateral surface of the right circular cone with vertex A for which Γ_n is the base circle (see Fig. 41). It is natural to regard M as the limit of a sequence of spaces M_n. Each of them is a manifold of bounded curvature and the sequence $|\Omega|(M_n)$ is bounded. Another example of this kind occurs when M

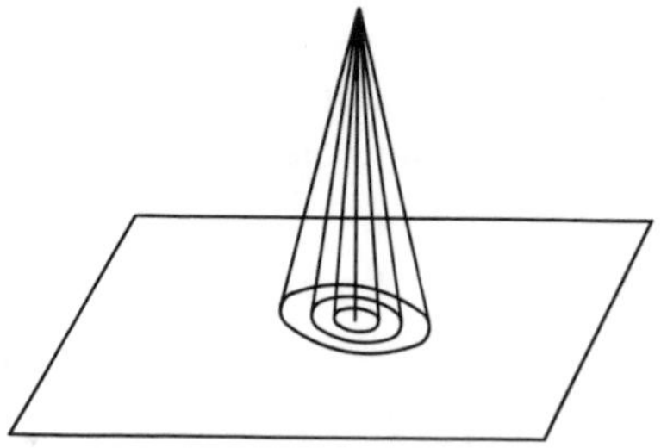

Fig. 41

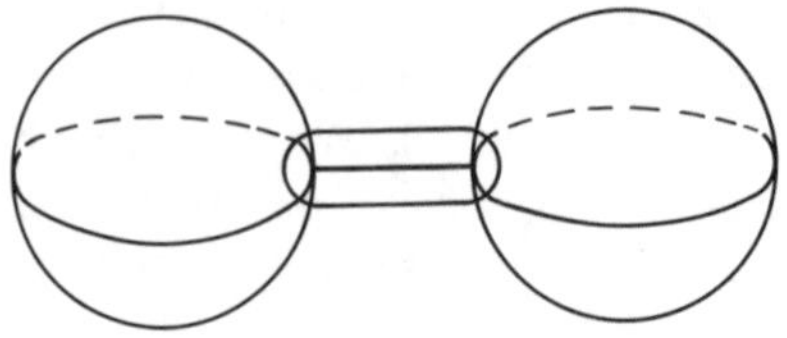

Fig. 42

is formed by two spheres S_1 and S_2 in $\mathbb{E}^3$ and an interval $[AB]$ joining them (see Fig. 42). Replacing the interval by the cylindrical hub of width $1/n$ on whose axis $[AB]$ lies, we obtain a sequence of two-dimensional manifolds of bounded curvature M_n, which is naturally regarded as approximating M, for which the sequence $|\Omega|(M_n)$ is also bounded.

Let us refine the statement of the problem considered in Burago (1965b). The function $r(x, y)$ of pairs of points of the two-dimensional manifold M is said to be semimetric if $r(x, x) = 0$, $r(x, y) = r(y, x)$ and $r(x, y) \leqslant r(x, z) + r(z, y)$ for any $x, y, z \in M$. Obviously $r(x, y)$ is non-negative. In contrast to the metric, $r(x, y) = 0$ does not necessarily imply that $x = y$.

Such concepts as the length of a curve, a shortest curve, an intrinsic metric, a triangle, and a triangle with convex boundary are defined for a semimetric word for word the same as for an ordinary metric. The definition of the concept of the upper angle between curves is also similar to the case of curves in a metric space.

A degenerate manifold of bounded curvature is a two-dimensional manifold with intrinsic semimetric for which an axiom of boundedness of curvature is satisfied that is similar to the corresponding axiom for the case of manifolds with intrinsic metric with the difference, however, that it is necessary to require that the sums of the absolute values of the excesses of the triangles are bounded (this is due to the importance of the matter). The main result of Burago (see Burago (1965b)) relating to degenerate manifolds of bounded curvature is contained in the following theorem.

Theorem 8.4.5. *Let M be a two-dimensional manifold with boundary, homeomorphic to a closed disc, and r a semimetric in M. Then in order that (M, r) should be a degenerate two-dimensional manifold of bounded curvature it is necessary and sufficient that there should be a sequence of metrics (ρ_n), $n = 1, 2, \ldots$, defined in M that converges uniformly to r as $n \to \infty$ and is such that (M, ρ_n) is a two-dimensional manifold of bounded curvature for each n, and the quantities $|\kappa_n|(\partial M) + |\Omega_n|(M)$ are bounded in aggregate, where $|\kappa_n|(\partial M)$ is the turn of the boundary of M and $|\Omega_n|(M)$ is the absolute curvature of M in the metric ρ_n.*

Let M be a two-dimensional manifold endowed with a semimetric $r(x, y)$. In M we introduce an equivalence relation by putting $x \sim y$ if $r(x, y) = 0$. Let $\tilde{M}$ be the set of equivalence classes with respect to this relation. For $X \in \tilde{M}$, $Y \in \tilde{M}$ we put $\rho(X, Y) = r(x, y)$, where $x \in X$, $y \in Y$. Generally speaking, the metric space $(\tilde{M}, \rho)$ is not a manifold. In Burago (1965b) the author gave an exhaustive description of the topological structure of metric spaces, to which we add by the method described here the concept of a degenerate two-dimensional manifold of bounded curvature. In that paper Burago established some compactness theorems for semimetrics of bounded curvature in a two-dimensional manifold, and proved the closure of the class of degenerate two-dimensional manifolds of bounded curvature with respect to some naturally defined passages to the limit.

8.5. Some Inequalities and Estimates. Extremal Problems for Two-Dimensional Manifolds of Bounded Curvature.

Application of the theorem on contraction of a cone to the solution of extremal problems. Theorems 5.6.1 and 5.6.2 on the contraction of a convex cone enable us to distinguish a class of extremal problems with respect to polyhedra, for each of which the solution is a convex cone.

Consider the set of all two-dimensional polyhedra homeomorphic to a closed disc. A function $F(M)$ defined on this set is said to be *increasing* if for any polyhedra M_1 and M_2 such that M_2 admits a contracting map onto M_1 we have $F(M_1) \leqslant F(M_2)$. If equality holds if and only if M_1 and M_2 are isometric, then F is said to be *strictly increasing.*

Theorem 8.5.1. *Let $\mathfrak{M} = \mathfrak{M}(l_1, l_2, \ldots, l_m, \omega_0)$ be the totality of all polyhedra homeomorphic to a closed disc for which the boundary links have lengths equal to $l_1, l_2, \ldots, l_m$ respectively, and the positive part of the curvature does not exceed a number ω_0, $0 \leqslant \omega_0 < 2\pi$, and $F(M)$ is an increasing function of the polyhedra. If F has a maximum in $\mathfrak{M}$, then among the manifolds $M \in \mathfrak{M}$ for which it takes its greatest value there is at least one convex cone. If F is a strictly increasing function, then the desired polyhedron that realizes the maximum can only be a convex cone.*

By Theorem 8.5.1 the solution of the problem about finding the maximum of F comes down to consideration of the case when the polyhedron M is a convex cone. Theorem 8.5.1 is an obvious consequence of Theorem 5.6.1.

From Theorem 8.2.1 there follows in an obvious way a result analogous to Theorem 8.5.1, but relating to arbitrary two-dimensional manifolds of bounded curvature.

An increasing (and even strictly increasing) function of a polyhedron is its area $\sigma(M)$. Another example is the diameter $d(M)$ of the polyhedron M, that is, the least upper bound of he distances between points of the polyhedron.

The solution of the extremal problem in the class $\mathfrak{M}(l_1, l_2, \ldots, l_m, \omega_0)$ in the case when $F(M)$ is the area, $F(M) = \sigma(M)$, is given by Theorem 6.5.3. Theorem 8.2.1 enables us to extend it to the case of arbitrary two-dimensional manifolds of bounded curvature. Namely, the following proposition is true.

Theorem 8.5.2. *Let G be a domain in a two-dimensional manifold of bounded curvature M, homeomorphic to a closed disc. We assume that the boundary of G is rectifiable and let p be its length. If $\omega_0 = \Omega^+(G^0) < 2\pi$, then*

$$\sigma(G) \leqslant \frac{p^2}{2(2\pi - \omega_0)}. \tag{8.26}$$

Here the equality sign holds if and only if the domain G with the induced metric is isometric to the lateral surface of a right circular cone.

Theorem 8.5.2 is due to Aleksandrov (Aleksandrov (1945)). We note that for the special case of this theorem when M is Riemannian the inequality (8.26) was obtained earlier by Fiala (Fiala (1941)).

The assertion of Theorem 8.5.2 for the case of equality in (8.26) follows from the fact that the area is a strictly increasing functional.

The extremal problem corresponding to the case when $F(M)$ is the diameter of the polyhedron, $F(M) \equiv d(M)$, was solved by Aleksandrov and Strel'tsov (Aleksandrov and Strel'tsov (1965)). The best possible result is given by the following theorem.

Theorem 8.5.3. *Let G be a domain in a two-dimensional manifold of bounded curvature, homeomorphic to a closed disc, $p < \infty$ the length of the boundary of G, $\Omega^+ = \Omega^+(G^0)$, and $d(G)$ the intrinsic diameter of G, that is, the exact limit of the distances between points of G in the metric induced in G. Then when $0 \leqslant \Omega^+ \leqslant \pi$*

$$d(G) \leqslant \frac{p}{2}, \tag{8.27}$$

and when $\pi < \Omega^+ < 2\pi$

$$d(G) \leqslant \frac{p}{2 \sin \frac{\Omega^+}{2}}. \tag{8.28}$$

In both cases the inequality is exact. In (8.27) the equality sign holds in many cases, for example, for lunes between two meridians on a sphere. In (8.28) equality is attained if and only if G with respect to its intrinsic geometry is an isosceles triangle, pasted together along the lateral sides, with base p and angle at the vertex $2\pi - \Omega^+$.

Theorem 8.2.1 and its special case – Theorem 5.6.1 – can also be applied as an auxiliary means of investigation in the study of extremal problems that do not immediately reduce to it. From the results proved by means of it we give some theorems on the estimate of the length of a curve in a two-dimensional manifold of bounded curvature.

Let L be a simple arc with bounded variation of turn in a two-dimensional manifold of bounded curvature. In certain cases instead of the variation of left and right turns of the arc L it is advisable to consider the quantity

$$\kappa^*(L) = \frac{1}{2}(|\kappa_r|(L) + |\kappa_l|(L) - |\Omega|(L^0)). \tag{8.29}$$

Here $|\kappa_l|$ and $|\kappa_r|$ are the variations of left and right turns of the curve L, and L^0 is the totality of all interior points of L. The quantity $\kappa^*(L)$ is called the characteristic turn or winding of the simple arc L. It does not depend on the choice of orientation of L. We assume that L is split by an interior point X of it into two arcs L_1 and L_2, and let θ_1 and θ_2 be the left and right angles of L at the point X. We put $\kappa^*(X) = \frac{1}{2}(|\pi - \theta_1| + |\pi - \theta_2| - |2\pi - \theta_1 - \theta_2|)$. Then

$$\kappa^*(L) = \kappa^*(L_1) + \kappa^*(L_2) + \kappa^*(X).$$

Using the given property of winding, we can define it for curves that are not simple arcs. Namely, we shall say that L is a curve of bounded variation of turn if it can be formed from finitely many simple arcs $L_1, L_2, \ldots, L_m$, each of which is a curve of bounded variation of turn, and the end of L_i is the beginning of L_{i+1}, $i = 1, 2, \ldots, m - 1$. Let X_i be this common point of L_i and L_{i+1}. Then there is defined the quantity $\kappa_L^*(X_i) = \frac{1}{2}(|\pi - \theta_i'| + |\pi - \theta_i''| - |2\pi - \theta_i' - \theta_i''|)$, where θ_i' and θ_i'' are the angles of the sectors into which L_i and L_{i+1} split a small neighbourhood of X_i. We put

$$\kappa^*(L) = \sum_{i=1}^{m} \kappa^*(L_i) + \sum_{i=1}^{m-1} \kappa^*(X_i). \tag{8.30}$$

It is easy to show that the sum on the right does not depend on the choice of splitting into simple arcs of the curve L.

Theorem 8.5.4 (Aleksandrov and Strel'tsov (1965)). *Let G be a domain homeomorphic to a closed disc in a two-dimensional manifold, and L a curve lying in G. If $\Omega^+(G) + \kappa^*(L) < 2\pi$, then L either does not have multiple points or it consists of a loop (a curve without multiple points and with coincident ends) and one or two simple arcs extending the ends of the loop that do not have other points in common with the loop and lie outside the region bounded by it.*

Theorem 8.5.5 (Aleksandrov and Strel'tsov (1965)). *Let G be a domain homeomorphic to a closed disc in a two-dimensional manifold of bounded curvature. We assume that the length of the boundary of G is finite. Then for any curve $L \subset G$ the length $s(L)$ is bounded above by the quantity $S_0 = s(p, \Omega^+(G^0), \kappa^*(L)) < \infty$, where p is the length of the boundary of G.*

The next theorem gives an exact estimate of the quantity $s(L)$, which is true, however, only under certain restrictions.

Theorem 8.5.6. *Suppose that the conditions of Theorem* 8.5.5 *are satisfied. We put $\Omega^+(G^0) + \kappa^*(L) = 2\omega_0$. If $2\omega_0 \leqslant \pi$, then*

$$s(L) \leqslant \frac{p}{1 + \cos \omega_0},$$

and if $\pi \leqslant 2\omega_0 < 2\pi$, then

$$s(L) \leqslant \frac{p}{\sin \omega_0}.$$

In both cases the estimate for $s(L)$ is exact.

The proofs of Theorems 8.5.5 and 8.5.6 are contained in Aleksandrov and Strel'tsov (1965).

In Aleksandrov and Strel'tsov (1965) Theorem 8.5.6 was established for the case when G is a domain of more general type than in the statement of the theorem given here. The description of this class turns out, however, to be rather cumbersome; this is why we do not give Theorem 8.5.6 in the most general form here. We note that in Aleksandrov and Strel'tsov (1965) cases are also described in which each of the inequalities of Theorem 8.5.6 becomes an equality.

Inequalities of isoperimetric type. The inequality of Theorem 8.5.2 is a special case of a whole class of inequalities known as isoperimetric.

Henceforth M denotes a two-dimensional manifold of bounded curvature. Let G be a domain, that is, a connected open set in M. We denote by $\tilde{G}$ the metric space that arises from G if we cut M along ∂G. The set G is embedded in a natural way in $\tilde{G}$. We put $\partial\tilde{G} = \tilde{G}\setminus G$. We shall say that G is simple if $\tilde{G}$ is compact and the set $\partial\tilde{G}$ is non-empty and consists of finitely many connected components, each of which is a simple closed curve of finite length. We denote the sum of the lengths of the connected components of $\partial\tilde{G}$ by $p(G)$ and call it the *perimeter* of G. The boundary of a simple domain is the union of finitely many rectifiable curves and the perimeter of G can be understood as the length of the boundary, taking account of multiplicity. We denote by $\chi = \chi(\tilde{G})$ the Euler characteristic of $\tilde{G}$. Since $\tilde{G}$ is connected and the boundary of $\tilde{G}$ is non-empty, $\chi \leqslant 1$.

Let k be a real number. We put $\Omega_k(E) = \Omega(E) - k\sigma(E)$. Defined as a measure, Ω_k is called the curvature with respect to a k-plane. We shall also consider the measures Ω_k^+ and Ω_k^-, the upper and lower variations of Ω.

If M is Riemannian and $K(x)$ denotes the Gaussian curvature at a point $x \in M$, we have

$$\Omega_k(E) = \int_E (K(x) - k)\sigma(dx),$$

$$\Omega_k^+(E) = \int_E (K(x) - k)^+\sigma(dx), \quad \Omega_k^-(E) = \int_E (K(x) - k)^-\sigma(dx).$$

Theorem 8.5.7. *Let G be a simple domain in a two-dimensional manifold of bounded curvature M, $p = p(G)$, $\Omega_k^+ = \Omega_k^+(G)$, and $F = \sigma(G)$ the area of G. Then for any k we have*

$$p^2 + 2(\Omega_k^+ - 2\pi\chi)F + kF^2 \geqslant 0. \tag{8.31}$$

For the proof of Theorem 8.5.7 see Burago (1989) and Ionin (1969a), (1969b), (1972a), (1972b). The given theorem contains as a special case many other well-known results. For example, Theorem 8.5.2 follows from Theorem 8.5.7 if we put $k = 0$.

In the article Aleksandrov and Strel'tsov (1965) and the monograph Burago and Zalgaller (1980) other extremal problems of the theory of two-dimensional manifolds of bounded curvature are also considered. Some unsolved problems relating to the same range of equations are stated in the same article and monograph. We should also mention the articles Aleksandrov (1945c), Aleksandrov, Borisov and Rusieshvili (1975), Aleksandrov and Strel'tsov (1953), and Rusieshvili (1957), which are devoted to extremal problems.

§9. Further Results. Some Additional Remarks

In conclusion we give a survey of further research on the theory of two-dimensional manifolds of bounded curvature.

A circle, an equidistant, and other special sets in a two-dimensional manifold of bounded curvature. Let us specify arbitrarily a two-dimensional manifold of bounded curvature M. Let $A \subset M$ be an arbitrary non-empty subset of M. For $r > 0$ we put $S(A, r) = \{x \in M \mid \rho(X, A) = r)$. Sets $S(A, r)$ corresponding to different r are called *equidistants* of the set A. In the special case when A consists of a unique point $OA = \{O\}$, $S(A, r)$ is the circle with centre O and radius r. The question of the structure of a circle in an arbitrary two-dimensional manifold of bounded curvature was considered in the articles Burgo and Stratilatova (1965) and Zalgaller (1950b). The case of manifolds of positive curvatures was studied in the second of these articles, and the general case was considered in the first. In the case when A is a simple arc with bounded variation of turn, not containing peak points of M, the question of the structure of the equidistant $S(A, r)$ for small r was considered in Borisov (1955) and Borisov (1965).

We give here some results about the structure of a circle and an equidistant of a set, obtained in Burago and Stratilatova (1965). Apart from the significance of a result, we are also guided in our choice by the simplicity of the statements.

In the general case a circle on a two-dimensional manifold of bounded curvature can be arranged in a rather complicated way. In particular, it may not be a simple closed curve. Moreover, a circle may even be a disconnected set.

Let G be a domain homeomorphic to an open disc in a two-dimensional manifold of bounded curvature, and O an arbitrary point of G. We shall consider circles with centre O contained in G.

It is natural to pose the following question: under which conditions is the circle $S(O, r)$ a simple closed curve? The answer is contained in the following theorem.

Theorem 9.1 (Burago and Stratilatova (1965)). *If the circle $S(O, r)$ lies in a domain G_0 homeomorphic to a disc, and $\Omega^+(G_0 \setminus \{O\}) < \pi$, then $S(O, r)$ is a simple closed curve.*

Theorem 9.2 (Burago and Stratilatova (1965)). *If a connected component Γ of a circle $S(O, r) \subset G$ is not a one-point set, then it is a continuous image of the circle $S_1 = S(0, 1)$ in $\mathbb{R}^2$. There is a continuous map x: $S_1 \to M$ such that $x(S_1) = \Gamma$ and the following conditions are satisfied:*

1) *the path x: $S_1 \to M$ does not have essential intersections, that is, by an arbitrarily small deformation of the map x we can obtain a topological map of S_1 into M;*

2) *let $R_l(t)$ ($R_r(t)$) be the leftmost (rightmost) shortest curve joining the points O and $x(t)$, where $t \in S_1$. Then, if t runs through S_1 in one direction, the radius $R_l(t)$ ($R_r(t)$) rotates monotonically around the point O.*

Remark 1. The rotation of the radius $R_l(t)$ of which we spoke in condition 2 of the theorem may take place in a discontinuous way.

Remark 2. We shall call the map $x: S_1 \to \Gamma$ mentioned in Theorem 9.2 a canonical parametrization of the component Γ of the circle $S(O, r)$.

Other properties of a circle are described by the following theorem.

Theorem 9.3 (Burago and Stratilatova (1965)). *If the circle $S(O, r)$ lies in a domain G of the manifold M that is homeomorphic to a disc and is a connected set, then it is a curve with bounded variation of turn. We have*

$$|\kappa|(S(O, r)) \leqslant \theta(O) + \Omega^-[B(O, r)\backslash\{O\}],$$

$$\kappa^-(S(O, r)) \leqslant \Omega^+[B(O, r)\backslash\{O\}].$$

(Here κ denotes the turn from the side of the domain $B(O, r)$ – the disc bounded by the given circle.)

Remark 1. Since in the conditions of Theorem 9.3 the circle $S(O, r)$ may not be a simple closed curve, explanations of how its turn should be understood are necessary. Let $x(t)$, $t \in S_1$, be an arbitrary parametrization of the circle $S(O, r)$. We take an arc $[a, b]$ of S_1 arbitrarily. We shall say that any such arc defines an arc L of the circle $S(O, r)$. The turn of this is defined as follows. On the arc $[a, b]$ we can specify a sequence of points $t_0 = a < t_1 < \cdots < t_{m-1} < t_m = b$ such that $L_i = x([t_{i-1} t_i])$, $i = 1, 2, \ldots, m$, is a simple arc with bounded variation of turn that lies on the boundary of the disc $B(O, r)$. The turn of the arc L is defined as the sum of the turns of the arcs L_i and the turns from the side of the arc at the points $X_i = x(t_i)$, $i = 1, 2, \ldots, m - 1$.

Theorem 9.3 admits an extension to the case when the circle is disconnected. The first of the estimates of Theorem 9.3 extends to this case.

In Burago and Stratilatova (1965) the authors gave a formula for the length of the circle $S(O, r)$. They established that the length of $S(O, r)$ is a function of bounded variation with respect to the variable r on any interval $[0, l]$, where $l < \infty$. In that article they also studied the question of circles in metrics of bounded curvature that converge to some limit metric. They established some conditions under which the lengths of these circles converge to the length of a circle in the limit metric.

Concerning equidistants in a two-dimensional manifold of bounded curvature we give just the following result.

Theorem 9.4. *Let L be a simple arc with bounded variation of turn in a two-dimensional manifold of bounded curvature M that does not contain peak points. Then there is a $\delta > 0$ such that when $0 < r < \delta$, $S(L, r)$ is a simple closed curve with bounded variation of turn that bounds a domain $B(L, r)$ homeomorphic to a disc.*

We have

$$|\kappa|[S(L,r)] \leqslant |\kappa_l|(L) + |\kappa_r|(L) + |\Omega|[B(L,r)\setminus L] + \theta(A) + \theta(B),$$

where A and B are the end-points of the curve L.

Theorem 9.4 is a consequence of the estimates contained in Theorem 9.3 and a theorem established in Borisov (1965). We do not give a statement of the latter theorem because it is rather cumbersome.

Along with a circle and an equidistant in a two-dimensional manifold of bounded curvature, we can also consider various other loci. For example, we can study the locus of points such that the sum of the distances from them to two given points is a constant. The following questions arise: 1) under what conditions is a given locus a simple arc or a simple closed curve? 2) will the curve have bounded variation of turn? 3) how do we estimate the variation of turn or the length of the curve?

In Stratilatova (1965) propositions were proved that establish sufficient conditions under which the answer to the questions posed above is positive.

Quasigeodesic. As we know, a geodesic in a Riemannian manifold can be characterized as a curve for which the geodesic curvature at each point is zero. There naturally arises the idea of considering curves characterized by similar properties in an arbitrary two-dimensional manifold of bounded curvature. For a curve the concepts of left and right turns are defined. In contrast to the case of a Riemannian manifold, their sum may be non-zero. In this connection we first introduce a characterization of a curve in which the left and right turns of the curve occur in an equivalent way.

Let L be a simple arc with bounded variation of turn. We put

$$\kappa^*(L) = \frac{1}{2}\{|\kappa_r|(L) + |\kappa_l|(L) - |\Omega|(L^0)\}$$

(L^0 is the arc L with the end-points excluded). The quantity $\kappa^*(L)$ is called the characteristic turn or *winding of the arc L.*

The concept of winding is also defined for curves with multiple points (see Aleksandrov and Burago (1965)).

A curve L is called a *quasigeodesic* if its winding is equal to zero. In the case when M is a Riemannian manifold the concepts of quasigeodesic and geodesic coincide. In the general case this is not so. For example, let M be a cone with vertex O, where the curvature at this vertex is positive. We assume that the curve L is composed of segments of generators of the cone that start from O and split the total angle at M in half. Let $\theta = \theta(O)$, $\theta < 2\pi$. Then the left and right turns of the curve L at the point O are equal to $\pi - \theta/2$, $\Omega(L) = 2\pi - \theta$, and it is easy to see that $\kappa^*(L) = 0$. The curve L is thus a quasigeodesic. No arc L inside which O lies is a shortest curve, so L is not a geodesic.

The concept of a quasigeodesic was introduced by Aleksandrov in Aleksandrov (1949b). Some known results about geodesics admit an extension to the case of two-dimensional manifolds of bounded curvature if instead of geodesics we consider quasigeodesics. In particular, generalizations of this kind have been obtained for a famous theorem of Lyusternik and Shnirel'man on three closed geodesics (Aleksandrov (1950a)), a theorem of Cohn-Vossen (Cohn-Vossen (1959)) on the structure of a geodesic in a domain whose curvature is small (Aleksandrov and Burago (1965)), and some others.

We shall say that M is a *two-dimensional manifold of specific curvature bounded above* if there is a number K such that for any Borel set $E \subset M$ we have $\Omega(E) \leqslant K\sigma(E)$.

Theorem 9.5 (Aleksandrov and Burago (1965)). *In a manifold of specific curvature bounded above, any quasigeodesic is a geodesic.*

Theorem 9.6 (Aleksandrov and Burago (1965)). *If the metrics* (ρ_n), $n = 1, 2, \ldots$, *of bounded curvature in a two-dimensional manifold converge tamely to the metric* ρ *and the quasigeodesics in the metric* ρ *converge to some curve* L_0, *then* L_0 *is a quasigeodesic in the metric* ρ.

Therem 9.7 (Aleksandrov and Burago (1965)). *From each point* X *of a two-dimensional manifold of bounded curvature in each direction we can draw a quasigeodesic. The quasigeodesic can be extended indefinitely beyond any of its ends.*

Special coordinate systems in a two-dimensional manifold of bounded curvature. Apart from an isothermal coordinate system in a two-dimensional manifold of bounded curvature, we can also consider some others.

Let $\mathbb{R}^2_+$ be the quadrant $\{(x, y) \in \mathbb{R}^2 | x > 0, y > 0\}$ of the plane $\mathbb{R}^2$. We assume that from $\mathbb{R}^2_+$ there is cut out a polygon P_0 homeomorphic to a disc and that in its place there is pasted a polygon P, also homeomorphic to a disc, cut out from a two-dimensional manifold of bounded curvature. The manifold $M = (\mathbb{R}^2_+ \setminus P_0) \cup P$ formed under this pasting is in turn a two-dimensional manifold of bounded curvature.

Henceforth we shall assume that $|\Omega|(M) < \pi/2$. This condition will be ensured, in particular, if for a polygon P taken on a manifold of bounded curvature we have $|\Omega|(P) + |\kappa|(\partial P) < \pi/2$.

Theorem 9.8 (Bakel'man (1961), Bakel'man (1965)). *Let* M *be the manifold defined as above. Then there is a topological map* φ *of* M *onto* $\mathbb{R}^2_+$ *such that if we introduce in* $\mathbb{R}^2_+$ *the line element*

$$ds^2 = du^2 + 2 \cos \tau(u, v)\, du\, dv + dv^2, \tag{9.1}$$

where $\tau(u, v) = \pi/2 - \Omega[\varphi^{-1}(D_{uv})]$, *and* D_{uv} *is the rectangle* $\{(x, y) \in \mathbb{R}^2 | 0 < x \leqslant u, 0 < y \leqslant v\}$, *then* φ *is an isometric map of* M *onto the square* $\mathbb{R}^2_+$ *endowed with the metric generated by the line element* (9.1).

From Theorem 9.8 it follows that for any point X of a two-dimensional manifold of bounded curvature such that $|\Omega|(X) < \pi/2$ we can find a circle in which the metric of the manifold is defined by the line element (9.1). The coordinate system in the given manifold obtained in this way is called a *Chebyshev coordinate system.*

The proof of Theorem 9.8 is based on an approximation of the metric of the manifold M by polyhedral metrics. It is much simpler than the proof of the theorem on isometric coordinates given in § 7.

The condition $|\Omega|(M) < \pi/2$ in the statement of the theorem cannot be discarded.

Theorem 9.8 has a converse. Namely, we assume that in $\mathbb{R}^2_+$ we are given a metric ω such that $|\omega|(\mathbb{R}^2_+) < \pi/2$. We put

$$\tau(u, v) = \pi/2 - \omega(D_{uv})$$

and let ρ be the metric determined in $\mathbb{R}^2_+$ by the line element (9.1). Then, as Bakel'man proved in Bakel'man (1965), the manifold $\mathbb{R}^2_+$ with metric ρ is a manifold of bounded curvature and the function ω is its curvature.

A so-called *equidistant coordinate system* was constructed in Borisov (1965). Let L be a simple arc with finite variation of turn that does not contain peak points of the manifold. The domain of definition of the equidistant coordinate system is the set G bounded by the arc L, two shortest curves starting from its ends, and an arc of the equidistant $S(L, h)$, where $h > 0$. We assume that when $0 < r \leqslant h$ the set $S(L, r)$ is a simple closed curve. With each point $X \in G$ there are associated two numbers r and t. Here $r = \rho(X, L)$. The number t is determined from X in a more complicated way. We shall not describe the construction given in Borisov (1965), because it is rather cumbersome. We just mention that in the case when the manifold M is Riemannian, and the curve L is sufficiently smooth, t is determined as follows. Let $\xi(t)$, $0 \leqslant t \leqslant 1$, be a parametrization of L such that for any $t_1 t_2 \in [0, 1]$ the length of the arc $[\xi(t_1)\xi(t_2)]$ is equal to $s(L)|t_2 - t_1|$. Let $\pi(X)$ be the point of L that is closest to X. Then t is defined by the condition $\pi(X) = \xi(t)$.

As an application of the equidistant coordinate system a formula was given in Borisov (1965) for the variation of the length of an arc with finite variation of turn in a two-dimensional manifold of bounded curvature.

Suppose we are given numbers K_1 and K_2, where $K_1 \leqslant K_2$. We shall say that the specific curvature of a manifold of bounded curvature M lies between K_1 and K_2 if for any Borel set $E \in \mathfrak{B}_0(M)$ we have $K_1\sigma(E) \leqslant \Omega(E) \leqslant K_2\sigma(E)$.

Theorem 9.9. *If a two-dimensional manifold of bounded curvature is such that its specific curvature at each point lies in some interval* $[K_1, K_2]$, *then for any point* $O \in M$ *there is a number* $r_0 > 0$ *such that any point* X *lying in the disc* $B(O, r_0)$ *is joined to* O *by a unique shortest curve. We fix the radius* $[OX_0]$ *of the disc* $B(O, r_0)$ *arbitrarily, and for an arbitrary point* $X \in B(O, r_0)$ *let* $r = \rho(X, O)$ *and let* θ *be the angle between the shortest curves* $[OX_0]$ *and* $[OX]$, *taken in a definite direction. Then the metric of* M *in the disc* $B(O, r_0)$ *is defined by the line element of the form* $ds^2 = dr^2 + b^2(r, \theta)\, d\theta^2$, *where the function* $b(r, \theta)$ *is such that* $b(0, \theta) = 0$, $\dfrac{\partial b}{\partial r}(0, \theta) = 1$ *for all* $\theta \in [0, 2\pi)$, *for almost all* (r, θ) *such that* $0 < r < r_0$, $0 < \theta < 2\pi$ *there is a finite second derivative* $\dfrac{\partial^2 b}{\partial r^2}(r, \theta)$, *where*

$$K_1 b(r, \theta) \leqslant -\frac{\partial^2 b}{\partial r^2}(r, \theta) \leqslant K_2 b(r, \theta),$$

and for an arbitrary Borel set E

$$\Omega(E) = -\iint_E \frac{\partial^2 b}{\partial r^2}(r, \theta)\, dr\, d\theta.$$

A proof of Theorem 9.9 for the case when $K_1 \geqslant 0$ was given in Aleksandrov (1948a). However, for the general case the arguments do not differ from those in Aleksandrov (1948a).

Proportional convergence. In certain questions a concept of convergnce of metrics different from the one used above turns out to be useful. We assume that on the set M there are specified metrics ρ, ρ_n, $n = 1, 2, \ldots$. We shall say that as $n \to \infty$ the metrics ρ_n converge proportionally to the metric ρ if for any $\varepsilon > 0$ we can find a number n_0 such that for any $n \geqslant n_0$

$$\frac{1}{1+\varepsilon}\rho(X, Y) \leqslant \rho_n(X, Y) \leqslant (1+\varepsilon)\rho(X, Y)$$

for any $X, Y \in M$.

Let M be a two-dimensional manifold with intrinsic metric, and A an arbitrary point of M. We shall say that M has a *tangent cone* at A if there is a cone $Q(\theta, r)$ that admits a topological map φ onto a neighbourhood of A such that A corresponds to the vertex O of the given cone and for any $X, Y \in Q(\theta, r)$ such that $X \neq Y$

$$\frac{\rho_Q(X, Y)}{\rho_M[\varphi(X), \varphi(Y)]} \to 1$$

as $X \to O$, $Y \to O$.

Theorem 9.10 (Burago (1965c), Zalgaller (1950b)). *A two-dimensional manifold of bounded curvature has a tangent cone at any point A of it that is not a peak point.*

In Zalgaller (1950b) this theorem was proved for the case of manifolds whose curvature is non-negative. In Burago (1965c) a proof was given for the general case.

Theorem 9.11 (Reshetnyak (1959)). *Let M be a compact two-dimensional manifold with intrinsic metric ρ. We assume that M has a tangent cone at each point of it. Then there is a sequence of polyhedral metrics (ρ_n), $n = 1, 2, \ldots$, defined in M, that converges proportionally to the metric of the manifold.*

Classes of surfaces with metric of bounded curvature. Let F be a two-dimensional surface in three-dimensional Euclidean space. We assume that any two points $X, Y \in F$ can be joined by a rectifiable curve lying on F and that the intrinsic metric defined in a natural way on M is compatible with the natural topology of F as a subset of $\mathbb{E}^3$. This condition will be satisfied under rather weak assumptions about F. For example, it is satisfied if F is a smooth surface of class C^1. There arises the following question: under what additional conditions is a surface F endowed with an intrinsic metric a two-dimensional manifold of bounded curvature? Let us quote some results in this direction.

First of all we mention that any convex surface in $\mathbb{E}^3$ is a two-dimensional manifold of bounded curvature. The curvature of the manifold as a set function is non-negative. As Aleksandrov showed (see Aleksandrov (1948a)) any two-dimensional manifold with non-negative curvature homeomorphic to a sphere is isometric to a closed convex surface. (For further generalizations of the given theorem see Pogorelov (1957), Pogorelov (1969).) This theorem enables us to conclude, in particular, that any two-dimensional manifold of non-negative curvature can be realized at least locally in the form of a convex surface.

By analogy with the case of convex surfaces it is natural to pose the following problem: to find a class $\mathfrak{R}$ of surfaces in $\mathbb{E}^3$ determined by certain outwardly geometrical conditions and such that if a surface F in $\mathbb{E}^3$ belongs to the class $\mathfrak{R}$, then it is a two-dimensional manifold of bounded curvature, and conversely, given any manifold of bounded curvature, for any point of it we can find a neighbourhood isometric to some surface $F \in \mathfrak{R}$. This problem seems rather difficult, and probably does not have an effective solution. Some particular classes of surfaces are known that satisfy the first of the conditions in the statement of the problem, namely such that the surfaces that occur in this class are manifolds of bounded curvature.

In this connection we mention, first of all, the *surface representable by the difference of convex surfaces*, introduced by A. D. Aleksandrov (Aleksandrov (1949a)). We shall say that F is a surface representable by the difference of convex surfaces if in some orthogonal Cartesian coordinate system F can be specified by the equation

$$z = f_1(x, y) - f_2(x, y),$$

where $(x, y) \in G$, a convex domain of $\mathbb{R}^2$, and $f_1(x, y)$ and $f_2(x, y)$ are convex functions. (A function is said to be convex if the segment joining two arbitrary points of its graph lies above the graph of the function.) It was proved in Aleksandrov (1949a) that if F is a surface representable by the difference of convex surfaces, then F is a two-dimensional manifold of bounded curvature.

Another class consists of the *surfaces of bounded extrinsic curvature*, introduced by Pogorelov in Pogorelov (1956). Let F be a smooth surface in $\mathbb{E}^3$, and $\tilde{F}$ its convex hull, that is, the smallest convex set containing F. Let E be the totality of all internal points of F that belong to the boundary of $\tilde{F}$. If $X \in E$, then X is a smooth point of the boundary of the convex body $\tilde{F}$, and in it there is defined a vector $\nu(X)$, the outward normal vector of the convex body $\tilde{F}$ at the point X. Laying off the vector $\nu(X)$ from a fixed point O in $\mathbb{E}^3$, we obtain a map ν of the set E into the unit sphere $S(O, 1)$ of the space. We denote by $\tilde{\Theta}(E)$ the area of the set $\nu(E)$ on the sphere $S(O, 1)$, $\tilde{\Theta}(E) = \sigma[\nu(E))$. We shall say that F is a surface of bounded extrinsic curvature if there is a constant $\Theta_0 < \infty$ such that for any finite system of pairwise non-overlapping domains $F_1, F_2, \ldots, F_m$ of F we have

$$\sum_{k=1}^{n} \tilde{\Theta}(F_k) \leqslant \Theta_0.$$

Any surface of bounded extrinsic curvature is a two-dimensional manifold of bounded curvature, as Pogorelov showed in Pogorelov (1956).

Smooth surfaces of class W_2^2 were considered in Bakel'man (1956). Let F be a smooth surface in $\mathbb{E}^3$. We shall say that F is smooth of class W_2^2 if any point of it has a neighbourhood that admits a parametrization $f(u, v) = [x_1(u, v), x_2(u, v), x_3(u, v)]$, $(u, v) \in B(0, 1)$, such that the functions x_i have generalized (in the sense of Sobolev) derivatives of the first and second orders in the disc $B(0, 1)$, where the derivatives $\partial f/\partial u$, $\partial f/\partial v$ are continuous and linearly independent at each point $(u, v) \in B(0, 1)$, and the second derivatives are square integrable in $B(0, 1)$. It was shown in Bakel'man (1956) that any smooth surface of class W_2^2 is a two-dimensional manifold of bounded curvature. This class of surfaces is interesting in that most of the results in a classical course of differential geometry admit an extension to surfaces that belong to this class.

The class of saddle surfaces in $\mathbb{E}^3$ was introduced in Shefel' (1964). A surface F is called a *saddle surface* if no plane can cut off a crust from it, that is, for any plane P, any connected component of the set $F \setminus P$ contains boundary points of F. In the regular case the class of saddle surfaces coincides with the class of surfaces of non-positive Gaussian curvature. The main result of Shefel' (1964) consists in proving that a saddle surface is a manifold of bounded curvature, and its curvature as a set function is non-positive. A more detailed discussion of the paper Shefel' (1964), and of some others that are an extension of it, is given in one of the books in the present series (Burago (1989)).

Generalizations of two-dimensional manifolds of bounded curvature. Apart from Riemannian geometry on a two-dimensional manifold, we can specify various others, for example, pseudo-Riemannian geometry, the geometry of affine connection, and Finsler geometry. In each of these geometries the concept of curvature is defined and there arises the question of the construction for them of an analogue of the concept of a two-dimensional manifold of bounded curvature. For the case of pseudo-Riemannian geometry the solution of this problem is known. Namely, in Gurevich (1979a), (1979b), (1981a), (1981b) the concept of plane kinematics of bounded curvature is introduced; this is a generalization of the concept of a manifold of bounded curvature to the pseudo-Riemannian case. (The term "kinematics" is connected with a known physical interpretation of pseudo-Riemannian geometry.)

Another question is how to regard the multidimensional analogue of the theory presented in this article. In this setting there are individual results. Namely, Aleksandrov introduced the concepts of metric spaces of curvature no greater than some number K and a space of curvature bounded below by K.

Some questions of the theory of spaces of unilaterally bounded curvature are considered in the article by Berestovskii and Nikolkaev (Part II of this volume).

In the general case a theory of multidimensional manifolds that can be regarded as the complete analogue of two-dimensional manifolds of bounded curvature has not been constructed.

On isothermal coordinates. The possibility of using isothermal coordinates to solve certain problems of the theory of surfaces was observed quite a long time ago. In this connection we mention the papers Lozinskij (1944), Beckenbach and Rado (1933), Beurling (1952).

References*

Aleksandrov, A.D. [1944]: Intrinsic metric of a convex surface in a space of constant curvature. Dokl. Akad. Nauk SSSR *45*, 3–6 [Russian], Zbl.61,376

Aleksandrov, A.D. [1945a]: Isoperimetric inequalities on curved surfaces. Dokl. Akad. Nauk SSSR *47*, 235–238 [Russian], Zbl.61,376

Aleksandrov, A.D. [1945b]: Convex surfaces as surfaces of positive curvature. Dokl. Akad. Nauk SSSR *50*, 27–30 [Russian]

Aleksandrov, A.D. [1945c]: One isoperimetric problem. Dokl. Akad. Nauk SSSR *50*, 31–34 [Russian]

Aleksandrov, A.D. [1947]: The method of gluing in the theory of surfaces. Dokl. Akad. Nauk *57*, 863–865 [Russian], Zbl.29,228

Aleksandrov, A.D. [1948a]: Intrinsic Geometry of Convex Surfaces. Gostekhizdat, Moscow-Leningrad [Russian], Zbl.38,352. German transl.: Academie-Verlag. Berlin 1955, Zbl.65,151

Aleksandrov, A.D. [1948b]: Foundations of the intrinsic geometry of surfaces. Dokl. Akad. Nauk SSSR *60*, 1483–1486 [Russian], Zbl.38,351

Aleksandrov, A.D. [1948c]: Curves on manifolds of bounded curvature. Dokl. Akad. Nauk SSSR *63*, 349–352 [Russian], Zbl.41,508

Aleksandrov, A.D. [1949a]: Surfaces representable by the difference of convex functions. Izv. Akad. Nauk KazSSR *60*, Ser. Mat. Mekh. 3, 3–20 [Russian]

Aleksandrov, A.D. [1949b]: Quasigeodesics. Dokl. Akad. Nauk SSSR *69*, 717–720 [Russian], Zbl.39,390

Aleksandrov, A.D. [1950a]: Quasigeodesics on manifolds homeomorphic to a sphere. Dokl. Akad. Nauk SSSR *70*, 557–560 [Russian], Zbl.39,391

Aleksandrov, A.D. [1950b]: Surfaces representable by differences of convex functions. Dokl. Akad. Nauk SSSR *72*, 613–616 [Russian], Zbl.39,180

Aleksandrov, A.D. [1950c]: Convex Polyhedra. Gostekhizdat, Moscow-Leningrad [Russian], Zbl.41,509

Aleksandrov, A.D. [1951]: A theorem on triangles in a metric space and some applications of it. Tr. Mat. Inst. Steklova *38*, 5–23 [Russian], Zbl.49,395.

---

*For the convenience of the reader, references to reviews in Zentralblatt für Mathematik (Zbl.), compiled using the MATH database, and Jahrbuch über die Fortschritte der Mathematik (Jbuch) have, as far as possible, been included in this bibliography.

Aleksandrov, A.D. [1954]: Synthetic methods in the theory of surfaces. Convegno Int. Geometria Differenziale, Italia 1953, 162–175, Zbl.55,409.

Aleksandrov, A.D. [1957a]: Ruled surfaces in metric spaces. Vestn. Leningr. Univ. *12*, No. 1, 5–26 [Russian], Zbl.96,166

Aleksandrov, A.D. [1957b]: Über eine Verallgemeinerung der Riemannschen Geometrie. Schr. Forschungsinst. Math. *1*, 33–84, Zbl.77,357

Aleksandrov, A.D., Borisov, Yu.F. and Rusieshvili, G.I. [1975]: An extremal property of cones in Lobachevskij space. Tr. Ped. Inst. GruzSSR *2*, 3–27 [Russian]

Aleksandrov, A.D. and Burago, Yu.D. [1965]: Quasigeodesics. Tr. Mat. Inst. Steklova *76*, 49–63. English transl.: Proc. Steklov Inst. Math. *76*, 58–76 (1965), Zbl.142,204

Aleksandrov, A.D. and Reshetnyak Yu.G. [1989]: General Theory of Irregular Curves, Kluwer Academic Publishers, Dordrecht, Boston-London, Zbl.691.53002

Aleksandrov, A.D. and Strel'tsov, V.V. [1953]: Estimate of the length of a curve on a surface. Dokl. Akad. Nauk SSSR *93*, 221–224 [Russian], Zbl.52,182

Aleksandrov, A.D. and Strel'tsov, V.V. [1965]: The isoperimetric problem and estimates of the length of a curve on a surface. Tr. Mat. Inst. Steklova *76*, 67–80. English transl.: Proc. Steklov Inst. Math. *76*, 81–99 (1965), Zbl.147,224

Aleksandrov, A.D. and Zalgaller, V.A. [1962]: Two-dimensional manifolds of bounded curvature. Tr. Mat. Inst. Steklov *63*, English transl.: Intrinsic geometry of surfaces. Transl. Math. Monographs *15*, Am. Math. Soc. (1967), Zbl.122,170

Alekseevskij, D.V., Vinogradov, A.M. and Lychagin, V.V. [1988]: Basic ideas and notions of differential geometry. Itogi Nauki Tekh., Ser. Sovrem. Probl. Mat., Fundam. Napravleniya *28*, 1–299. English transl. in: Encycl. Math. Sci. *28*, Springer, Berlin-Heidelberg-New York 1991, Zbl.675.53001

Bakel'man, I.Ya. [1956]: The differential geometry of smooth non-regular surfaces. Usp. Mat. Nauk *11*, No. 2, 67–124 [Russian], Zbl.70,392

Bakel'man, I.Ya. [1961]: Generalized Chebyshev nets and manifolds of bounded curvature. Dokl. Akad. Nauk SSSR *138*, 506–507. English transl.: Sov. Math., Dokl. *2*, 631–632, Zbl.104,168

Bakel'man, I.Ya. [1965]: Chebyshev nets in manifolds of bounded curvature. Tr. Mat. Inst. Steklova *76*, 124–129. English transl.: Proc. Steklov Inst. Math. *76*, 154–160 (1967), Zbl.171,432

Beckenbach, E.F. and Rado, T. [1933]: Subharmonic functions and surfaces of negative curvature. Trans. Amer. Math. Soc. *35*, 662–674, Zbl.7,130

Belinskij, S.P. [1975]: A mapping of a cone onto a polyhedron for the Lobachevskij plane. Sib. Mat. Zh. *16*, 203–211. English transl.: Sib. Math. J. *16*, 159–164, Zbl.324.53044

Beurling, A. [1952]: Sur la géométrie métrique des surfaces à courbure totale 0. Comm. Sém. Math. Univ. Lund, Suppl., 7–11, Zbl.48,388

Borisov, Yu.F. [1949]: Curves on complete two-dimensional manifolds with boundary. Dokl. Akad. Nauk SSSR *64*, 9–12 [Russian], Zbl.36,234

Borisov, Yu.F. [1950]: Manifolds of bounded curvature with boundary. Dokl. Akad. Nauk SSSR *74*, 877–880 [Russian], Zbl.39,180

Borisov, Yu.F. [1955]: The geometry of the semi-neighbourhood of a curve in a two-dimensional manifold of bounded curvature. Dokl. Akad. Nauk SSSR *103*, 537–539 [Russian], Zbl.67,402

Borisov, Yu.F. [1965]: Semi-neighbourhoods and variation of the length of a curve on a surface. Tr. Mat. Inst. Steklova *76*, 26–48. English transl.: Proc. Steklov Inst. Math. *76*, 31–57 (1967), Zbl.173,504

Burago, Yu.D. [1965a]: Curves in convergent spaces. Tr. Mat. Inst. Steklova *76*, 5–25. English transl.: Proc. Steklov Inst. Math. *76*, 3–29 (1967), Zbl.142,203

Burago, Yu.D. [1965b]: Closure of the class of manifolds of bounded curvature. Tr. Mat. Inst. Steklova *76*, 141–147. English transl.: Proc. Steklov Inst. Math. *76*, 175–183 (1967), Zbl.168,200

Burago, Yu.D. [1965c]: On the proportional approximation of a metric. Tr. Mat. Inst. Steklova *76*, 120–123. English transl.: Proc. Steklov Inst. Math. *76*, 149–153 (1967), Zbl.142,204

Burago, Yu.D. [1978]: On the radius of injectivity on a surface with curvature bounded from above. Ukr. Geom. Sb. *21*, 10–14 [Russian], Zbl.438.53052

Burago, Yu.D. [1989]: Geometry of surfaces in Euclidean spaces. Itogi Nauki Tekh., Ser. Sovrem. Probl. Math., Fundam. Napravleniya *48*, 5–97. English. transl. in: Encycl. Math. Sc. *48*, Springer, Berlin-Heidelberg-New York 1992, Zbl.711.53003

Burago, Yu.D. and Stratilatova, M.B. [1965]: A circle on a surface. Tr. Mat. Inst. Steklova *76*, 88–114. English transl.: Proc. Steklov Inst. Math. *76*, 109–141 (1967), Zbl.142,203

Burago, Yu.D. and Zalgaller, V.A. [1965]: The isometric problem for domains with restricted width on a surface. Tr. Mat. Inst. Steklova *76*, 81–87. English transl.: Proc. Steklov Inst. Math. *76*, 100–108 (1967)

Burago, Yu.D. and Zalgeller, V.A. [1980]: Geometric Inequalities. Nauka, Leningrad. English transl.: Springer, Berlin-Heidelberg-New York 1988, Zbl.436.52009

Cohn-Vossen, S.E. [1959]: Some Problems of Differential Geometry. Gosudarstv. Izdat. Fiz.-Mat. Lit., Moscow [Russian], Zbl.91,341

Fiala, F. [1941]: Le problème des isoperimètres sur les surfaces ouvertes à courbure positive. Comment. Math. Helv. *13*, 293–346, Zbl.25,230.

Gurevich, V.L. [1970]: Rotation of a longest curve in kinematics of bounded curvature. Deposited at VINITI, No. 361–79 [Russian], Zbl.423.53007

Gurevich, V.L. (1979a): Kinematics of bounded curvature. Sib. Mat. Zh. *20*, No. 1, 37–48. English transl.: Sib. Math. J. *20*, 26–33 (1979), Zbl.412.53029

Gurevich, V.L. [1979b]: Two-dimensional pseudo-Riemannian manifolds of bounded curvature. Dokl. Akad. Nauk SSSR *247*, 22–25. English transl.: Sov. Math., Dokl. *20*, 638–641 (1979), Zbl.428.53030

Gurevich, V.L. [1981a]: Two-dimensional kinematics of bounded curvature. I. Sib. Mat. Zh. *22*, No. 5, 3–30. English transl.: Sib. Math. J. *22*, 653–672 (1982), Zbl.492.53013

Gurevich, V.L. [1981b]: Two-dimensional kinematics of bounded curvature. II. Sib. Mat. Zh. *22*, No. 6, 41–53; English transl.: Sib. Math. J. *22*, 836–845 (1982), Zbl.492.53014

Hayman W.K., Kennedy P.B. [1976]: Subharmonic Functions, vol. I. Academic Press, London-New York-San Francisco, Zbl.419.31001

Huber, A. [1954]: On the isoperimetric inequality on surfaces of variable Gaussian curvature. Ann. Math., II. Ser. *60*, 237–247, Zbl.56,158

Huber, A. [1957]: Zur isoperimetrischen Ungleichung auf gekrümmten Flächen. Acta Math. *97*, 95–101, Zbl.77,358

Huber, A. [1960]: Zum potentialtheoretischen Aspekt der Alexandrowschen Flächentheorie. Comment. Math. Helv. *34*, 99–126, Zbl.105,161

Ionin, V.K. [1969a]: Riemannian spaces with Euclidean isoperimetric inequality. Dokl. Akad. Nauk SSSR *188*, 990–992. English transl.: Sov. Math. Dokl. *10*, 1225–1227 (1970), Zbl.208.248

Ionin, V.K. [1969b]: Isoperimetric and certain other inequalities for a manifold of bounded curvature. Sib. Mat. Zh. *10*, 329–342. English transl.: Sib. Math. J. *10*, 233–243, Zbl.188,275

Ionin, V.K. [1972a]: Isoperimetric inequalities in simply-connected Riemannian spaces of nonpositive curvature. Dokl. Akad. Nauk SSSR *203*, 282–284. English transl.: Sov. Math. Dokl. *13*, 378–381 (1972), Zbl.258.52011

Ionin, V.K. [1972b]: Isoperimetric inequalities for surfaces of negative curvature. Sib. Mat. Zh. *13*, 933–938. English transl.: Sib. Math. J. *13*, 650–653 (1973), Zbl.241.52002

Ionin, V.K. [1982]: Upper estimate of the radius of the circumscribed disc of a domain on a manifold of bounded curvature. Symposium on geometry in the large and foundations of the theory of relativity. Novosibirsk, 49–50 [Russian], Zbl.489.53001

Kobayashi, S. and Nomizu, K. [1963]: Foundations of Differential Geometry. Vol. I. Interscience, New York-London-Sydney, Zbl.119,375.

Kuratowski, K. [1966]: Topology. Academic Press, New-York-London, Zbl.158,408

Lewy, H. [1935]: A priori limitations for solutions of Monge-Ampère equatons. Trans. Am. Math. Soc. *37*, 417–434, Zbl.11,350

Lozinskij, S.M. [1944]: Subharmonic functions and their applications to the theory of surfaces. Izv. Akad. Nauk SSSR, Ser. Mat. *8*, 175–194 [Russian], Zbl.61,241

Men'shov, D.E. [1948]: Sur une généralisation d'un théorème de M.H. Bohr. Mat. sb. (2) *44*, 339–356

Petrovskij, I.G. [1961]: Lectures on Partial Differential Equations. 3rd ed. Gosudarstv. Izdat. Fiz.-Mat. Lit., Moscow, Zbl.115,81. English transl.: Iliffe Books Ltd., London 1967, Zbl.163,117

Pogorelov, A.V. [1949]: Quasigeodesic lines on a convex surface. Mat. Sb., Nov. Ser. *25*, 275–306. English transl.: Am. Math. Soc., Transl., I. Ser. *6*, 430–473, Zbl.41,89

Pogorelov, A.V. [1956]: Surfaces of Bounded Extrinsic Curvature. Izdat. Khar'kov. Gos. Univ., Khar'kov [Russian], Zbl.74,176

Pogorelov, A.V. [1957]: Some Questions of Geometry in the Large in Riemannian Space. Izdat. Khar'kov. Gos. Univ., Khar'kov, Zbl.80,149. German transl.: VEB Deūtscher Verlag der Wissenschaften, Berlin 1960, Zbl.90,125.

Pogorelov, A.V. [1961]: Some Results on Geometry in the Large. Izdat. Khar'kov. Gos. Univ., Khar'kov. English transl.: Adv. Math. *1*, 191–264 (1964), Zbl.129,357

Pogorelov, A.V. [1969]: Extrinsic Geometry of Convex Surfaces. Nauka, Moscow. English transl.: Transl. Math. Monographs *35*, Am. Math. Soc., Providence 1973, Zbl.311.53067

Privalov, I.I. [1937]: Subharmonic Functions. Moscow-Leningrad [Russian]

Rado, T. [1925]: Über den Begriff der Riemannschen Flächen. Acta Szeged. *2*, 101–121, Jbuch 51,273

Reshetnyak, Yu.G. [1954]: Isothermal coordinates in manifolds of bounded curvature. Dokl. Akad. Nauk SSSR *94*, 631–633 [Russian], Zbl.58,384

Reshetnyak, Yu.G. [1959]: Investigation of manifolds of bounded curvature by means of isothermal coordinates. Izv. Sib. Otd. Akad. Nauk SSSR *1959*, No. 10, 15–28 [Russian], Zbl.115,164

Reshetnyak, Yu.G. [1960]: Isothermal coordinates in manifolds of bounded curvature. I, II. Sib. Mat. Zh. *1*, 88–116, 248–276 [Russian], Zbl.108,338

Reshetnyak, Yu.G. [1961a]: A special map of a cone onto a manifold. Mat. Sb., Nov. Ser. *53*, 39–52 [Russian], Zbl.124,152

Reshetnyak, Yu.G. [1961b]: On the isoperimetric property of two-dimensional manifolds of curvature not greater than K. Vestn. Leningr. Univ. *16*, No. 19, 58–76, Zbl.106,365

Reshetnyak, Yu.G. [1962]: On a special map of a cone into a two-dimensional manifold of bounded curvature. Sib. Mat. Zh. *3*, 256–272 [Russian], Zbl.124,153

Reshetnyak, Yu.G. [1963a]: The length of a curve in a manifold of bounded curvature with an isothermal line element. Sib. Mat. Zh. *4*, 212–226 [Russian], Zbl.128,159

Reshetnyak, Yu.G. [1963b]: Rotation of a curve in a manifold of bounded curvature with an isothermal line element. Sib. Mat. Zh. *4*, 870–911 [Russian], Zbl.145,193

Rusieshvili, G.I. [1957]: Some extremal problems of the intrinsic geometry of manifolds in Lobachevskij space. Vestn. Leningr. Univ. *12*, No. 1, 76–79 [Russian], Zbl.94,352

Shefel', S.Z. [1964]: On the intrinsic geometry of saddle surfaces. Sib. Mat. Zh. *5*, 1382–1396 [Russian], Zbl.142,189

Stratilatova, M.B. [1965]: Certain geometrical loci of points on a surface. Tr. Mat. Inst. Steklova *76*, 115–119. English transl.: Proc. Steklov Inst. Math. *76*, 142–148 (1965), Zbl.142,204

Strel'tsov, V.V. [1952a]: Estimate of the length of an open polygon on a polyhedron. Izv. Akad. Nauk KazSSR *1*, 3–36 [Russian].

Strel'tsov, V.V. [1952b]: Some extremal problems of the intrinsic geometry of polyhedra. Izv. Akad. Nauk KazSSR *2*, 37–63 [Russian].

Strel'tsov, V.V. [1953]: General points of geodesics. Izv. Akad. Nauk KazSSR *3*, 89–103 [Russian], Zbl.55,410

Strel'tsov, V.V. [1956]: A map of a surface of negative curvature. Izv. Akad. Nauk KazSSR *5*, 29–44 [Russian], Zbl.72,174

Toponogov, V.A. [1969]: An isoperimetric inequality for surfaces whose Gaussian curvature is bounded above. Sib. Mat. Zh. *10*, 144–157. English transl.: Sib. Math. J. *10*, 104–113 (1969), Zbl.186,558

Verner, A.L. [1965]: A semigeodesic coordinate net on tubes of non-positive curvature. Tr. Mat. Inst. Steklova *76*, 130–140. English transl.: Proc. Steklov Inst. Math. *76*, 161–173 (1967), Zbl.156,430

Zalgaller, V.A. [1950a]: Curves with bounded variation of rotation on convex surfaces. Mat. Sb., Nov. Ser. *26*, 205–214 [Russian], Zbl.39,181

Zalgaller, V.A. [1950b]: The circle on a convex surface. The local almost-isometry of a convex surface to a cone. Mat. Sb., Nov. Ser. *26*, 401–424 [Russian], Zbl.37,250

Zalgaller, V.A. [1952]: On the class of curves with bounded variation of rotation on a convex surface. Mat. Sb. Nov. Ser. *30*, 59–72 [Russian], Zbl.49,395

Zalgaller, V.A. [1956]: Foundations of the theory of two-dimensional manifolds of bounded curvature. Dokl. Akad. Nauk SSSR *108*, 575–576 [Russian], Zbl.70,392

Zalgaller, V.A. [1965]: Curves on a surface near to cusp points. Tr. Mat. Inst. Steklova *76*, 64–66. English transl.: Proc. Steklov Inst. Math. *76*, 77–80 (1967), Zbl.156,430

II. Multidimensional Generalized Riemannian Spaces

V.N. Berestovskij and I.G. Nikolaev

Translated from the Russian
by E. Primrose

Contents

Introduction

0.1. Riemannian Spaces. A Riemannian space is usually understood as a space such that in small domains of it Euclidean geometry holds approximately, to within infinitesimals of higher order in comparison with the dimensions of the domain.

Distances in a Riemannian space are specified by means of a metric tensor: from the metric tensor we calculate the lengths of piecewise-differentiable curves, and the distance between a pair of points of the Riemannian space is taken to be equal to the length of the "shortest curve" of this type that joins these points.

Riemannian geometry traditionally considers spaces whose geometry differs from Euclidean geometry by an infinitesimal quantity of the second order. The difference in the second order of distances of Riemannian space from the corresponding Euclidean distances is controlled by the sectional curvature, which plays a key role in Riemannian geometry.

Classical Riemannian geometry assumes that the metric tensor has sufficiently good differential properties. The minimal requirements on the smoothness of the metric – the fact that it is twice differentiable – are due to the fact that the sectional curvature of a Riemannian space is calculated in terms of the second derivatives of the metric tensor.

Henceforth a Riemannian space, or as we shall more often call it, a Riemannian manifold, will be understood as a Riemannian space whose metric tensor is at least twice continuously differentiable.

In this article we shall be dealing with Riemannian metrics of inferior smoothness that are "generalized" with respect to the traditional point of view. We shall also call such spaces Riemannian manifolds, each time making precise the smoothness class of their metric tensor.

0.2. Generalized Riemannian Spaces. The main concepts of Riemannian geometry are "distance" and "curvature", which estimates the difference between the distances of a Riemannian space and Euclidean distances. However, "curvature" can be defined by starting from distances. This gives the possibility of defining the "boundedness of curvature" of a metric space. The "generalized Riemannian spaces" thus obtained are the main object of study in this article.

Thus, the starting point is a metric space. To avoid pathological cases we shall consider only those spaces whose metric is intrinsic. This means that the distance between points of the space coincides with the length (measured in the metric of the space) of the "shortest curve" joining these points.

At the basis of the definition of boundedness of curvature lies the following observation.

It is known that the sectional curvature of a Riemannian manifold at a point in a two-dimensional direction can be calculated in terms of the excesses of geodesic triangles in the following way:

$$K_\sigma(P) = \lim_{T \xrightarrow{\sigma} P} \delta(T)/\sigma(T), \tag{0.1}$$

where $\delta(T)$ is the excess of the triangle T, that is, the number equal to the sum of its angles minus π, $\sigma(T)$ is the "area" of the triangle T, that is, the number equal to the area of a Euclidean triangle with the same lengths of sides as T, and the triangles T are contracted to a point P so that in the limit they are tangent to the plane element s at the point P (Cartan (1928)).

For general metric spaces there is the concept of the angle between shortest curves (the "upper angle", which always exists; for the definition see 1.4.1). To calculate the angle between curves in the metric space, we need to do the following, roughly speaking. We consider points lying on curves other than the origin. From the distances between these points and the origin we calculate the angle as if the curves were situated in Euclidean space, and then proceed to the limit, making the points tend to the origin.

The role of shortest geodesic in a metric space is played by a shortest curve, that is, a curve whose length is equal to the distance between its beginning and end.

Thus, in a metric space we can consider triangles formed from shortest curves.

By (0.1), it is natural to define the boundedness above of the curvature of a metric space $\mathfrak{M}$ by requiring that for an arbitrary sequence of triangles contracting to any point we have

$$\overline{\lim_{T \to P}}\ \bar{\delta}(T)/\sigma(T) \leqslant K, \tag{0.2}$$

where $\bar{\delta}(T)$ and $\sigma(T)$ have in this case the same meaning as in (0.1) (it is necessary to specify the case of degenerate triangles in (0.2) separately). If in addition we require that

$$\underline{\lim}_{T \to P}\ \bar{\delta}(T)/\sigma(T) \geqslant K',$$

then we specify the bilateral boundedness of the curvature of the metric space.

In the first case we arrive at the so-called spaces of curvature $\leqslant K$, and in the second case at the spaces of curvature both $\leqslant K$ and $\geqslant K'$.

These spaces were introduced by A.D. Aleksandrov in Aleksandrov (1951).

The theory of spaces of curvature $\leqslant K$ (and $\geqslant K'$) is based on those ideas that lie at the basis of the intrinsic geometry of convex surfaces (Aleksandrov (1948)). Here the important features are the *axiomatic* method, in which the spaces under consideration are defined by a minimal set of properties of their metric, and the *synthetic* method, which goes back to Euclid and is based on geometrical constructions in the spaces, and consideration of curves, triangles and other figures in them.

0.3. Riemannian Geometry and Generalized Riemannian Spaces. In classical Riemannian geometry, when the existence of any necessary analytic properties of a Riemannian metric is implied, a series of naturally occurring metrics falls out of consideration. Among those not included in classical Riemannian geometry are the metrics considered in this article of curvature $\leqslant K$ or of curvature both $\leqslant K$ and $\geqslant K'$. Apart from the fact that they arise in a natural geometric way, we should mention that consideration of them is useful in the framework of Riemannian geometry itself.

0.3.1. Problems of Riemannian Geometry in which Generalized Riemannian Spaces Arise. First of all there are problems in which we need to consider "limits" of Riemannian metrics (extremal problems, problems connected with the concept of stability, and so on).

In this article we mention one of these problems – Berger's theorem (Berger (1983)) – connected with "loosening" the conditions of the rigidity theorem (15.2 and 6.1). In Berger's theorem the question is whether we can assert that if the curvature of a compact simply-connected Riemannian manifold of even dimension has the estimate $\frac{1}{4} - \varepsilon \leqslant K_\sigma \leqslant 1$ (in the rigidity theorem $\frac{1}{4} \leqslant K_\sigma \leqslant 1$), then if the manifold is not homeomorphic to a sphere, it is diffeomorphic to one of the symmetric spaces of rank one (in the rigidity theorem isometry was asserted).

The method by which Berger proved this theorem is noteworthy. The proof of the theorem reduces to the investigation of the "limit" space for a sequence of Riemannian manifolds with estimates for the curvature $\frac{1}{4} - \varepsilon_n \leqslant K_\sigma \leqslant 1$, where $\varepsilon_n \to 0$ as $n \to \infty$. Using the specific character of the situation, one can prove that the limit space (and this is precisely a space of curvature $\leqslant 1$ and $\geqslant \frac{1}{4}$ in the sense of Aleksandrov) is a C^∞-smooth Riemannian manifold, and so the rigidity theorem can be applied to it, from which the required result is obtained.

We mention that in this theorem the crucial step is to use a theorem proved by Gromov on the compactness of a class of Riemannian manifolds (Gromov (1981)).

We observe that such a use of generalized Riemannian spaces recalls the situation in differential equations when we first turn to a "generalized solution", and then establish that it is a classical solution.

0.3.2. Generalized Riemannian Spaces and the "Closure" of Classes of Riemannian Manifolds with Bounded Curvatures. The set of Riemannian metrics on a given manifold forms a metric space (we can also consider the space of metrics specified on different manifolds; see § 15). This space is not complete – its completion may be rather poor. However, under natural restrictions, namely if bounds for the sectional curvatures $K' \leqslant K_\sigma \leqslant K$ (K', K = const) are satisfied, and for the radius of injectivity of the manifold we have $i(\mathfrak{M}) \geqslant \text{const} > 0$, the completion is obtained by adjoining to the Riemannian manifolds under consideration Aleksandrov spaces of curvature $\leqslant K$ and $\geqslant K'$.

In exactly the same way we can consider classes of Riemannian manifolds with upper bounds on the sectional curvatures, $K_\sigma \leqslant K$, and a lower bound on the radius of injectivity. The completion of such spaces leads to spaces of curvature $\leqslant K$.

We emphasize that in the first case it has been proved that the completion consists exactly of spaces of curvature both $\leqslant K$ and $\geqslant K'$ (more details about this are given in 0.4).

0.3.3. Synthetic Description of Riemannian Geometry. Spaces of curvature both $\leqslant K$ and $\geqslant K'$ give a synthetic description of the completion of the class of Riemannian manifolds with bounds on the sectional curvatures $K' \leqslant K_\sigma \leqslant K$. The problem of a synthetic coordinate-free description of Riemannian geometry is also very important.

In order to obtain such a description, a new concept is necessary for metric spaces that is an analogue of the sectional curvature of a Riemannian manifold. The corresponding concept ("non-isotropic Riemannian curvature" in § 17 of this article) is copied from Riemannian geometry by means of formula (0.1). The role of plane element is played by a pair of "directions" of the metric space, which can be defined in the most general case (see § 3).

Briefly, the most important axioms imposed on a metric space so that it is a C^2-smooth Riemannian manifold are that at each point of the metric space for arbitrary pairs of directions at this point (provided that the angle between them is greater than zero and less than π) there should exist a non-isotropic Riemannian curvature and it should have a "good" modulus of continuity (for example, it satisfies a Hölder condition; see § 18). In the definition of the "continuity" of the non-isotropic Riemannian curvature we need to introduce a function that specifies the distances between the directions at different points of the metric space (see § 16).

It is curious that in this way we can give the synthetic axiomatics of C^m-smooth ($m = 2, 3, 4, \ldots$) and thereby C^∞-smooth Riemannian manifolds (see § 18).

We also consider the isotropic case when the "curvature" of a metric space does not depend on the directions. Such a curvature, defined in the metric space in accordance with (0.1), we call an isotropic Riemannian curvature (§ 17).

We note that the isotropic Riemannian curvature that we have introduced coincides with the Wald curvature, well known in distance geometry, which is defined in terms of the embeddability into a two-dimensional space of constant curvature of quadruples of points of a metric space that contain "linear" triples of points.

Metric spaces at each point of which there is an isotropic Riemannian curvature we shall call isotropic metric spaces. Of course, it makes sense to consider the very concept of isotropy only in the multidimensional case.

The main result for isotropic metric spaces is the generalization of a well-known theorem of Schur: under additional assumptions, which are satisfied if the metric space under consideration is a manifold of dimension greater than two, an isotropic metric space is isometric to a space of constant curvature.

0.4. A Brief Characterization of the Article by Chapters. In the first chapter we describe the main concepts of intrinsic geometry: intrinsic metric, shortest curve, angle, triangle, and excess of a triangle. Here we give the construction of the space tangent to an arbitrary metric space.

The main demands on the concepts that we introduce are that they should be associated with an arbitrary metric space and that in the Riemannian case they should lead to standard concepts.

Chapter 2 is devoted to the study of spaces of curvature $\leqslant K$. At the end of the chapter we consider spaces of curvature both $\leqslant K$ and $\geqslant K'$, and we also discuss the case when the curvature is only bounded below.

The method by which the spaces under consideration are investigated in this chapter is purely synthetic.

In Chapter 2 we discuss "boundedness of curvature" in some detail. We mention different (but equivalent) points of view on the definition of boundedness of curvature of a metric space and obtain consequences that imply the boundedness of curvature of a metric space. In the main these consequences are described by properties that are in common with Riemannian spaces: local uniqueness of the shortest curve joining a given pair of points, convexity of a small ball, the existence of an angle between shortest curves, and so on.

As an illustration of the methods developed we mention the solution of Plateau's problem on the existence of a surface of minimal area and establish an isoperimetric inequality that generalizes Carleman's well-known inequality to the case of spaces with curvature bounded above.

Without dwelling on these results in detail, we give the two most important assertions for spaces of curvature $\leqslant K$, on which (and on the analogue of them for spaces of curvature both $\leqslant K$ and $\geqslant K'$) is based the whole synthetic theory of the generalized Riemannian spaces considered in this article and which, properly speaking, make possible the study of these spaces.

The first is Aleksandrov's theorem of comparison of the angles of a triangle:

For an arbitrary sufficiently small triangle in a space of curvature $\leqslant K$ each of its upper angles has the bound

$$\bar{\alpha} \leqslant \alpha_K,$$

where α_K is the corresponding angle in a triangle with the same lengths of sides as the original but on a surface of constant curvature K.

The second is the property of K-concavity:

Let L and M be two shortest curves in a small domain of a space of curvature $\leqslant K$, starting from a point O. Consider points X and Y lying on the shortest curves L and M respectively. We denote by $\gamma_{LM}^K(x, y)$ $(x = OX, y = OY)$ the angle of a triangle on a surface of constant curvature K with sides x, y, z $(z = XY)$, lying opposite the side of length z.

K-concavity means that for any two such shortest curves L and M (lying in a small domain of the space) the angle $\gamma_{LM}^K(x, y)$ is a non-decreasing function of x and y.

We observe that formally the property of K-concavity follows easily from the theorem on comparison of angles. Its importance, however, is based on the following three features: firstly, K-concavity enables us to characterize in a new way the upper boundedness of curvature, including the case of a Riemannian manifold; secondly, most of the properties of spaces of curvature $\leqslant K$ have K-concavity as their basis; thirdly, there are situations (for example, the case of a space of curvature $\geqslant K'$ in § 11) when the analogues of K-concavity (K'-convexity) do not follow only from the theorem on comparison of angles (but, for example, for spaces of curvature $\geqslant K'$ significant results are based on K'-convexity).

Chapter 3 is devoted to spaces of curvature both $\leqslant K$ and $\geqslant K'$.

More significant assertions are obtained under the assumption that the space of curvature $\leqslant K$ and $\geqslant K'$ under consideration is a topological manifold. It is interesting that there is no necessity to assume in advance that the metric space is a manifold. The condition of local compactness, local extendability of a shortest curve (12.1) and bilateral boundedness of curvature already guarantee that the space under consideration is a manifold (12.2.1). Such spaces we shall call *spaces with bounded curvature* for short.

The central result proved for spaces of bounded curvature is the following:

1) A space with bounded curvature is a differentiable manifold; the intrinsic metric of a space with bounded curvature is specified by means of a certain metric tensor, specified on this manifold.

2) In a neighbourhood of each point of a space with bounded curvature we can introduce a so-called harmonic coordinate system (for the definition see 14.1.3). Harmonic coordinate systems specify in a space with bounded curvature an atlas of class $C^{3,\alpha}$, where for α we can take an arbitrary number lying in the interval (0, 1).

3) The components of a metric tensor in an arbitrary harmonic coordinate system are continuous functions that have generalized second derivatives in the sense of Sobolev (Sobolev (1950)) that are summable with an arbitrarily large degree. In particular, the first derivatives of the components of the metric tensor satisfy a Hölder condition with arbitrary $\alpha \in (0, 1)$.

Example 12.1 given in the text shows that we cannot expect two-fold continuous differentiability of the components of a metric tensor in a space with bounded curvature.

The importance of the theorem we have proved (in the text it is Theorem 14.1) is that it gives information on the second derivatives of the metric tensor of a space with bounded curvature. The enables us to state a theorem (it follows from Theorem 14.1) which gives the possibility of regarding metrics of bounded curvature as limits of Riemannian metrics (of class C^∞) with sectional curvatures bounded in aggregate (see Theorem 15.1).

All this enables us to look on spaces with bounded curvature as Riemannian spaces, but with "worse" differential properties of the metric tensor compared with the generally accepted ones, and at the same time as spaces with metrics that are the limits of smooth Riemannian metrics, where for them we preserve the assertions of "smooth" Riemannian geometry that "depend on the maximum modulus of sectional curvatures"; compare with 0.3.2.

The investigation of spaces with bounded curvature in this chapter has mixed synthetic-analytic character. The synthetic part, which is essentially used in the proof of the main results, is related to parallel displacement, which is introduced in Chapter 3 in a geometric way. Consideration of it has independent interest. The analytic part is based on rather subtle results of the theory of functions with generalized derivatives and the theory of elliptic equations.

In Chapter 4 it is a question of the synthetic description of Riemannian geometry and of isotropic metric spaces. We have already discussed this in 0.3.3. We only note that these results have a main theorem which asserts the existence in a definite sense of "almost everywhere" non-isotropic Riemannian curvature. More precisely, this theorem (in the text it is Theorem 17.1) asserts that there is a set $\mathcal{O} \in \mathfrak{M}$ of zero n-dimensional Hausdorff measure ($n = \dim M$) such that for each point $P \in \mathfrak{M} \backslash \mathcal{O}$ and arbitrary "admissible" pair of directions ξ, $\zeta \in \Omega_P \mathfrak{M}$ the sectional curvature $K_\sigma(P)$, which is formally calculated from the metric tensor at the point P in the direction of the two-dimensional element σ of $\mathfrak{M}_P$, specified by the bivector $\xi \wedge \zeta$, can also be calculated as the limit of the ratios $\bar{\delta}(T_m)/\sigma(T_m)$ for some sequence of triangles that contract to the point P "in the direction of the pair (ξ, ζ)".

0.5. In what Sense Do the Stated Results Have Multidimensional Character? The answer to this question is given by the theory of so-called two-dimensional manifolds of bounded integral curvature, in which restrictions are imposed not on the Gaussian curvature but the integral curvature (Aleksandrov and Zalgaller (1962), Reshetnyak (1960a)) (the case of two-dimensional spaces for which "curvature exists" at each point was considered by Aleksandrov in Aleksandrov (1948)), since spaces of curvature $\leqslant K$ and spaces of curvature $\leqslant K$ and $\geqslant K'$ in the two-dimensional case have bounded integral curvature in the sense of Aleksandrov.

For the multidimensional case we have not succeeded in introducing the concept of a manifold of bounded integral curvature, which in the two-dimensional case has much greater generality than a space of curvature $\leqslant K$.

0.6. Final Remarks on the Text. As a basis for writing the present article we have taken the survey Aleksandrov, Berestovskij and Nikolaev (1986). However, we have added sections with remarks and examples that comment on the material set forth. In addition, in § 15 and Chapter 4 we place results not stated in Aleksandrov, Berestovskij and Nikolaev (1986).

The contents of Chapters 1 and 2 reflect the fundamental work of Aleksandrov. Chapter 3 follows the results of Berestovskij on the introduction of a Riemannian structure in spaces with bounded curvature (Berestovskij (1975)) and of Nikolaev on the geometrical definition of parallel displacement (Nikolaev (1980), (1983a)), and on the smoothness of the metric of a space with bounded curvature (Nikolaev (1980), (1983b)).

Chapter 4 was written as a result of the work of Nikolaev, announced earlier in Nikolaev (1987), (1989).

In conclusion, the authors express their gratitude to Yu.D. Burago and V.A. Zalgaller for the attention they have shown to the work and their valuable advice.

Chapter 1
Basic Concepts Connected with the Intrinsic Metric

It is well known that such "Euclidean" concepts as straight line, angle, triangle, and so on, make sense and are very useful in Riemannian geometry.

In this chapter we show how basic "Euclidean" concepts of this kind carry over to the case of general metric spaces, and we describe their properties.

The main demands on the concepts introduced here are that they must admit a definition in an arbitrary metric space and in the case of a Riemannian manifold they must coincide with the corresponding concepts that exist there.

Henceforth the material of Chapter 1 will be a kind of "linguistic" basis for the construction of the synthetic axiomatics of Riemannian geometry.

§ 1. Intrinsic Metric, Shortest Curve, Triangle, Angle, Excess of a Triangle

1.1. Intrinsic Metric. We recall that in a Riemannian manifold $\mathfrak{M}$ the lengths of piecewise-differentiable curves are calculated by means of the metric tensor. If $\mathfrak{M}$ is connected, we can introduce a standard metric in the Riemannian manifold $\mathfrak{M}$. The distance between points of $\mathfrak{M}$ with respect to this metric is taken to be equal to the infimum of the lengths of all piecewise-differentiable curves $L \subset \mathfrak{M}$ joining the points under consideration. The metric on $\mathfrak{M}$ introduced in this way will be *intrinsic* in accordance with the definition given below.

We now consider an arbitrary metric space with metric ρ. It will often be denoted by $(\mathfrak{M}, \rho)$. For an arbitrary curve L in $\mathfrak{M}$ its *length* in the metric ρ is defined:

$$l_\rho(L) = \sup \sum_{i=1}^{m-1} \rho(X_i, X_{i+1}),$$

where $X_1, X_2, \ldots, X_m$ is an arbitrary sequence of points of L, numbered in the order of their position on the curve, and the upper bound is taken over all such sequences of points.

The metric ρ is called *intrinsic* if for any X, $Y \in \mathfrak{M}$, $\rho(X, Y)$ is equal to the greatest lower bound of the lengths (measured in the metric ρ) of curves joining X and Y.

From the definition it follows, in particular, that $\mathfrak{M}$ is a *metrically connected space*. This means that any two points of it are joined by a curve of finite length.

1.2. Shortest Curve. It is well known that a geodesic curve in a Riemannian manifold is locally the shortest curve among all curves joining the same points. In this connection the following definition is natural.

A curve L in a metric space $(\mathfrak{M}, \rho)$ joining points $A, B \in \mathfrak{M}$ is called a *shortest curve* if its length is equal to $\rho(A, B)$. Both $\rho(A, B)$ and the shortest curve with ends A and B are often denoted by AB.

In what follows the concept of convexity turns out to be useful for us. A subset V in $(\mathfrak{M}, \rho)$ is said to be *convex* if any two points of it are joined by a shortest curve in $\mathfrak{M}$ and any such shortest curve lies in V.

We observe that often, in contrast to the definition of convexity that we have given, the second condition is replaced by the condition that at least one shortest curve joining points of the set lies in the set.

1.3. Triangle. As in Riemannian manifolds, in metric spaces we can consider triangles.

A *triangle* $T = ABC$ in a metric space $(\mathfrak{M}, \rho)$ (where $A, B, C \in \mathfrak{M}$) is a set consisting of points of the shortest curves AB, BC, AC, called the *sides* of T. The points A, B, C are called the *vertices* of T. The *perimeter* of the triangle is the sum $AB + BC + AC$. A short notation for the triangle is ΔABC.

Later on we define the measure of the difference between a triangle in a metric space and a planar triangle with the same lengths of sides – the excess of the triangle – which enables us to estimate the "distortion" of the metric space, as is done in the case of Riemannian manifolds. However, it is sometimes convenient to regard as the "model" space not only a Euclidean plane, but other spaces of constant curvature. In this connection we introduce the following notation. We denote by S_K the Euclidean plane when $K = 0$, the Lobachevskij plane of curvature K when $K < 0$, and a two-dimensional open hemisphere of radius $1/\sqrt{K}$ when $K > 0$; we denote the n-dimensional analogue of S_K by S_K^n.

Let $T = ABC$ be a triangle in a metric space; we denote by T^K the triangle $A^K B^K C^K$ on S_K that has sides of the same length as T does (degeneracy of the triangle T^K into an interval is allowed). When $K \leqslant 0$ the triangle T^K always exists on account of the triangle inequality for the metric. When $K > 0$ it is necessary to require that the perimeter of the triangle T is less than $2\pi/\sqrt{K}$. When $K > 0$ it will be assumed everywhere in this chapter that this requirement is satisfied.

1.4. Angle. In order to calculate the angle between geodesics of a Riemannian manifold starting from a point O it is sufficient, as we know, to consider their tangent vectors at O and by means of the metric tensor to calculate from the well-known cosine formula the angles between these vectors. However, if we take into account the "infinitesimal Euclidean property" of a Riemannian manifold at a point, we can calculate the same quantity by starting only from the metric of the manifold. For this we can consider geodesic triangles, one vertex of which is at O and the other two lie on geodesics starting from O. Next we calculate from the sides of these triangles the angle at the vertex O by means of the formulae of Euclidean geometry and proceed to the limit, contracting the triangles to the point O. Now we give the definition of angle according to Aleksandrov (Aleksandrov (1957b)) in the metric space.

Let L and M be two curves in the metric space $(\mathfrak{M}, \rho)$ that have common starting point O. On L and M respectively we choose arbitrary points X and Y, and let $x = OX$, $y = OY$, $z = XY$. We denote by $\gamma_{LM}^K(x, y)$ the angle opposite to the side of length z in the triangle $T^K = O^K X^K Y^K$ on S_K corresponding to the triangle $T = OXY$.

1.4.1. Upper Angle, Angle. The *upper angle* between the curves L and M is by definition the quantity

$$\bar{\alpha}(L, M) = \varlimsup_{x,y \to 0} \gamma_{LM}^K(x, y).$$

We observe that the upper angle between L and M always exists and does not depend on K. We obtain the definition of the *lower angle* between curves if in the definition given above the upper limit is replaced by the lower.

We shall say that there is an angle α between L and M if $\bar{\alpha} = \underline{\alpha} = \alpha$.

1.4.2. Strong Upper and Lower Angles, Angle in the Strong Sense. A central role in our presentation will be played by the upper angle and the angle. However, there are situations when such an approach to the definition of an angle turns out to be unsatisfactory. Namely, in order to obtain a meaningful definition of lower boundedness of the curvature it is necessary to invoke the concepts given in this subsection of strong lower angle and angle in the strong sense (see 4.2 and 11.3).

The *strong upper angle* between L and M is defined as the upper limit of $\gamma_{LM}^K(x, y)$ on condition that x or y tends to zero. In particular, we allow the situation when x tends to zero but y is non-zero, or vice versa. Similarly we define the *strong lower angle* (in the definition of the strong upper angle the upper limit is replaced by the lower). The upper, lower and strong upper and lower angles will be denoted by $\bar{\alpha}$, $\underline{\alpha}$, $\bar{\alpha}_s$, $\underline{\alpha}_s$ respectively. Obviously,

$$\underline{\alpha}_s \leqslant \underline{\alpha} \leqslant \bar{\alpha} \leqslant \bar{\alpha}_s.$$

The existence of an angle in the strong sense α_s is determined in the same way as the existence of an angle: $\underline{\alpha}_s = \bar{\alpha}_s$.

To conclude 1.4 we note again that in a Riemannian manifold there is an angle between any two geodesics. The existence of an angle in the strong sense can be guaranteed in the general case only for small geodesics (in contrast to $\bar{\alpha}$ and $\underline{\alpha}$, the strong lower angle $\underline{\alpha}_s$ is not a local concept); see § 4, where the angles we have introduced will be discussed in more detail. In an arbitrary metric space, for shortest curves the angles α and α_s may not exist.

1.5. Excess. We now introduce a quantity that characterizes the difference between a triangle in a metric space and the corresponding planar triangle. We need the following notation. If $T = ABC$ is a triangle in a metric space, then the angle $\bar{\alpha}$ at its vertex A is the upper angle between its sides AB and AC. We denote by $\bar{\beta}$ and $\bar{\gamma}$ the angles of ΔABC at the vertices B and C, and by α_K. β_K, γ_K the angles of the corresponding triangle $T^K = A^K B^K C^K$ at the vertices A^K, B^K, C^K.

The *K-area* $\sigma_K(T)$ or $\sigma_K(\Delta ABC)$ of the triangle T is understood as the area of the triangle T^K; $\sigma_0(T)$ will be called the area of T and denoted by $\sigma(T)$.

The *absolute excess of the triangle* $T = ABC$ (or simply the excess) is the quantity

$$\bar{\delta}(T) = \bar{\alpha} + \bar{\beta} + \bar{\gamma} - \pi.$$

We recall that by the Gauss-Bonnet theorem the excess and the K-area of the triangle T on S_K are connected by the equality

$$\delta(T) = K \cdot \sigma_K(T).$$

It is also convenient to use the concept of the excess of a triangle with respect to S_K, or briefly the K-excess. The *K-excess* of a triangle T is the quantity

$$\bar{\delta}_K(T) = (\bar{\alpha} + \bar{\beta} + \bar{\gamma}) - (\alpha_K + \beta_K + \gamma_K).$$

We mention the following equalities, which will be useful later:

$$\bar{\delta}(T) = \bar{\delta}_0(T), \quad \bar{\delta}_K(T) = \bar{\delta}(T) - \delta(T^K). \tag{1.1}$$

§2. General Propositions about Upper Angles

The general propositions about upper angles stated in this section relate to general metric spaces.

Proposition 2.1 (Aleksandrov (1948), Aleksandrov and Zalgaller (1962)). *Let $\bar{\alpha}_{12}$, $\bar{\alpha}_{13}$, $\bar{\alpha}_{23}$ be the upper angles between curves L_1, L_2, L_3 starting from one point. Then*

$$\bar{\alpha}_{13} \leqslant \bar{\alpha}_{12} + \bar{\alpha}_{23}.$$

If L_1 and L_3 are branches of the same shortest curve, then $\bar{\alpha}_{13} = \pi$. Therefore the next result follows from Proposition 2.1.

Proposition 2.2. *The sum of the upper adjacent angles is not less than π.*

It turns out that for shortest curves the concepts of upper angle and strong upper angle coincide.

Proposition 2.3 (Aleksandrov (1957b)). *The upper angle $\bar{\alpha}$ between two shortest curves L and M starting from a common point is equal to the strong upper angle between them, that is,*

$$\bar{\alpha} = \overline{\lim_{x \to 0}}\, \gamma^K_{LM}(x, y).$$

The proof is based on the following lemma.

Lemma (Aleksandrov (1951), (1957b)). *For any number K*

$$\cos \gamma_{LM}^K(x, y) = (y - z)/x + \varepsilon,$$

where $\varepsilon \to 0$ as $x/y \to 0$.

We note that, generally speaking, the strong lower angle does not coincide with the lower angle.

Theorem 2.1, stated below, is a kind of foundation for the further development of the theory. It enables us, under very general assumptions, to estimate the difference between the upper angle in a triangle $T = ABC$ and the corresponding angle α_K in the triangle T^K on S_K. The only condition on T is that any two points on its sides can be joined by a shortest curve.

Theorem 2.1 (Aleksandrov and Zalgaller (1962)). *Let $\bar{\alpha}$ be the upper angle at the vertex A of the triangle ABC, and α_K the corresponding angle in the triangle on the K-plane with the same lengths of sides (when $K > 0$ the perimeter of ΔABC must be less than $2\pi/\sqrt{K}$). We assume that any two points on the sides of ΔABC can be joined by a shortest curve. Then*

$$\bar{\alpha} - \alpha_K \leqslant \bar{\nu}_{KA}^+,$$

where $\bar{\nu}_{KA}^+ = \sup\{\inf\{\bar{\delta}_K(\Delta AXY)\}\}$; here the supremum is taken over all possible points $X \in AB$, $Y \in AC$, and the infimum is taken over all possible shortest curves XY joining X and Y.

The next theorem gives an estimate of the angles of a triangle that is in a certain sense opposite to the estimate of Theorem 2.1.

Theorem 2.2 (Aleksandrov (1957b)). *Let ABC be a triangle in some metric space, any two points of which can be joined by a unique shortest curve, where between these shortest curves and the corresponding segments of the sides there is an angle in the strong sense. Suppose also that K' is an arbitrary number and $\mu_{K'}$ is the lower bound of K'-excesses of triangles AXY with vertices X and Y on the sides AB and AC of the triangle ABC. Under these assumptions, for the angle α of ΔABC at the vertex A and the corresponding angle $\alpha_{K'}$ of the triangle $A'B'C'$ with the same lengths of sides on the K'-plane we have*

$$\alpha - \alpha_{K'} \geqslant \mu_{K'}.$$

§ 3. The Space of Directions at a Point. K-Cone. Tangent Space

In a Riemannian manifold $\mathfrak{M}$ the *space of directions* at a given point $P \in \mathfrak{M}$ is naturally assumed to be the set of unit vectors touching the manifold at P (that is, the unit sphere in $\mathfrak{M}_P$). At the same time the direction at a point P of a

Riemannian manifold $\mathfrak{M}$ can be represented as a set of paths in $\mathfrak{M}$ beginning at P and having initially a given unit velocity vector. This leads us to the definition of the space of directions at a given point of a metric space in the sense of Aleksandrov. Having the space of directions at a given point of a metric space, we can easily construct the tangent cone to the metric space at the given point. The concept of K-cone introduced here is auxiliary.

3.1. Direction (Aleksandrov and Zalgaller (1962), Aleksandrov (1957b)). The following definition of the space of directions is related to any metric space $\mathfrak{M}$.

Let us fix an arbitrary point $P \in \mathfrak{M}$. A curve $L \subset \mathfrak{M}$ starting from P *has a definite direction* at this point if the upper angle that it makes with itself at P is equal to zero. We shall also say that two curves L and M beginning at P have the same direction at this point if the upper angle between them at P is zero (we observe that in this case it follows from the inequality $\bar{\alpha}_{11} \leqslant \bar{\alpha}_{12} + \bar{\alpha}_{21}$ that L and M have a definite direction at P).

On the set of curves that have a definite direction at P we introduce an equivalence relation by calling two curves equivalent if they have the same direction at this point (by Proposition 2.1, the given relation is an equivalence relation). Now we define the set of directions of $\mathfrak{M}$ at P as the set of classes of equivalent curves that have a direction at this point. The next result follows easily from the triangle inequality for upper angles (Proposition 2.1).

Proposition 3.1. *The directions at a given point form a metric space in which the distance between directions is defined as the upper angle between them.* Notation: $(\Omega_P M, \bar{\alpha})$.

3.2. *K*-Cone (Aleksandrov, Berestovskij and Nikolaev (1986)). At the basis of the definition there lies (depending on K) the cosine theorem of hyperbolic, Euclidean or spherical geometry.

3.2.1. $K \leqslant 0$. The K-cone $C_K\mathfrak{M}$ over the metric space $\mathfrak{M}$ is homeomorphic to the factor space of the space $\mathfrak{M} \times \mathbb{R}_+$ ($\mathbb{R}_+ = [0, +\infty)$) obtained by contracting the set $\mathfrak{M} \times 0$ to one point O, called the vertex of the cone $C_K\mathfrak{M}$. If $(X, t) \in \mathfrak{M} \times \mathbb{R}_+$, then $[X, t]$ denotes the corresponding point in $C_K\mathfrak{M}$. The metric in $C_K\mathfrak{M}$ is introduced as follows:

If $K = 0$, then for $X, Y \in \mathfrak{M}$, $t, s \in \mathbb{R}_+$

$$\rho_{C_K}([X, t], [Y, s]) = \begin{cases} [t^2 + s^2 - 2ts\cos\rho(X, Y)]^{1/2} & \text{if } \rho(X, Y) \leqslant \pi, \\ (t + s) & \text{if } \rho(X, Y) > \pi. \end{cases}$$

In the case $K < 0$, for $k = \sqrt{-K}$, $\rho_{C_K}([X, t], [Y, s])$ is defined by the relations

$$k^{-1} \cdot \operatorname{arcosh}(\cosh kt \cdot \cosh ks - \sinh kt \cdot \sinh ks \cdot \cos\rho(X, Y)) \qquad \text{if } \rho(X, Y) \leqslant \pi,$$

$$(t + s) \qquad \text{if } \rho(X, Y) > \pi.$$

3.2.2. $K > 0$. Topologically $C_K\mathfrak{M}$ is the factor space of the space $\mathfrak{M} \times [0, \pi/\sqrt{K}]$ obtained by contracting the set $\mathfrak{M} \times 0$ to one point, and the set $\mathfrak{M} \times \pi/\sqrt{K}$ to another (that is, suspension over $\mathfrak{M}$).

For $k = \sqrt{K}$, $\rho_{C_K}([X, t], [Y, s])$ is defined by the relations

$$k^{-1} \arccos(\cos kt \cdot \cos ks + \sin kt \cdot \sin ks \cdot \cos \rho(X, Y)) \qquad \text{if } \rho(X, Y) \leqslant \pi.$$

$$\min\{(t + s), 2\pi/\sqrt{K} - (t + s)\} \qquad \text{if } \rho(X, Y) > \pi.$$

3.3. Tangent Space (Aleksandrov, Berestovskij and Nikolaev (1986)). The tangent space $\mathfrak{M}_P$ to the metric space $\mathfrak{M}$ at a point $P \in \mathfrak{M}$ is by definition the O-cone over the space of directions $\Omega_P\mathfrak{M}$ to $\mathfrak{M}$ at the point P, that is,

$$\mathfrak{M}_P = C_O\Omega_P\mathfrak{M}.$$

The vertex of the O-cone is called the origin in $\mathfrak{M}_P$. Sometimes $\mathfrak{M}_P$ is also called a tangent cone.

§4. Remarks, Examples

4.1. Intrinsic Metric, Shortest Curve, Angles.

4.1.1. Extrinsic and Intrinsic Metric of a Surface. The concept of intrinsic metric arose originally in the study of the metric properties of surfaces, where it was particularly intuitive. The fact is that on a surface there is already a metric – the extrinsic metric of the ambient space. However, it is natural to measure the distance between given points of the surface along the surface, that is, to understand it as the greatest lower bound of the lengths of all curves on the surface that join given points, and this is the intrinsic metric. We should observe that metrically disconnected spaces fall out of consideration, since on them it is impossible to specify the intrinsic metric from an existing one. An example of such a space is a cylinder with directrix in the form of a curve that is not rectifiable on any part. We also observe that in the case of a metrically connected space the topologies determined by the extrinsic metric and the intrinsic metric constructed from it may differ substantially. As an example we can consider a cone with directrix in the form of a curve that is not rectifiable on any part. From the viewpoint of its intrinsic metric this cone is isometric to a continuum of isolated segments with one common end-point (Aleksandrov and Zalgaller (1962)).

4.1.2. Shortest Curves in a Space with Intrinsic Metric. The example of an open non-convex domain in E_n shows that, generally speaking, any two points of a metric space with intrinsic metric are not necessarily joined by a shortest curve. However, any two points of a compact metric space with intrinsic metric are obviously joined by at least one shortest curve, in particular, if the metric space with intrinsic metric is a manifold, then at least locally any two of its points are joined by a shortest curve. In the general case the shortest curve joining the points is not unique.

Example 4.1 (Busemann (1955)). Suppose that on the set $\mathbb{R}^2 = \{(x, y) | x, y \in \mathbb{R}\}$ we introduce the norm $\|(x, y)\|_1 = |x| + |y|$. Then any curve $(x(t), y(t))$ in $(\mathbb{R}^2, \| \ \|_1)$, each "coordinate" of which $x(t)$, $y(t)$ varies monotonically, is a shortest curve in $(\mathbb{R}^2, \| \ \|_1)$. In particular, the curves $y = x^P$, $p > 0$, in $\mathbb{R}^2$ are shortest curves joining the points (0, 0) and (1, 1).

4.1.3. The Triangle Inequality for Lower Angles. The Existence of α and α_s. In $(\mathbb{R}^2, \| \ \|_1)$ we consider the rays $e_1 = (t, t)$, $e_2 = (t, 0)$, $e_3 = (0, t)$, $t \geqslant 0$. We denote by $\bar{\alpha}_{ij}$ the upper angle between the rays e_i and e_j, $i, j = 1, 2, 3$. We denote by $\underline{\alpha}_{ij}$, $\bar{\alpha}^s_{ij}$, $\underline{\alpha}^s_{ij}$ the lower angle and the strong upper and lower angles between the corresponding rays. A simple calculation shows that

$$\bar{\alpha}_{12} = \bar{\alpha}_{13} = \bar{\alpha}^s_{12} = \bar{\alpha}^s_{13} = \pi/2; \quad \underline{\alpha}_{12} = \underline{\alpha}_{13} = \underline{\alpha}^s_{12} = \underline{\alpha}^s_{13} = 0;$$

$$\underline{\alpha}^s_{23} = \underline{\alpha}_{23} = \bar{\alpha}_{23} = \bar{\alpha}^s_{23} = \pi.$$

Hence it is obvious that for lower angles (and strong lower angles) between the shortest curves e_1, e_2, e_3 the triangle inequality if not satisfied (compare with Proposition 2.1):

$$\pi = \underline{\alpha}^s_{23} = \underline{\alpha}_{23} > \underline{\alpha}_{13} + \underline{\alpha}_{12} = \underline{\alpha}^s_{13} + \underline{\alpha}^s_{12} = 0.$$

We also see that there is a strong angle between the shortest curves e_2 and e_3, but no angle or strong angle between e_1 and e_2 or between e_1 and e_3. Later in Example 4.2 it will be shown that a strong angle may not exist even between shortest curves on a sphere.

4.1.4. Strong lower angle. The upper and lower angles are local concepts. If it is a question of the angle between shortest curves, then by Proposition 2.3 the strong upper angle is equal to the upper angle, and so this is also a local concept. However, this does not hold for the strong lower angle. In Example 4.3 given below the lower angle at the vertex B is equal to π, and the strong lower angle at this vertex is equal to ψ, where $0 < \psi < \pi/3$. Thus, the definition of strong lower angles has the defect that it is connected not with the movement of the shortest curves L and M in a small neighbourhood of their origin O, but with the properties of whole parts of L or M. Moreover, in the general case the shortest curve is not unique and the strong lower angle now describes something stronger than the angle between shortest curves (see Fig. 1). We also observe that in two-dimensional manifolds of bounded curvature and even in smooth manifolds, as Example 4.2 shows, the strong angle between certain shortest curves may not exist. In this connection we note that there is another definition of the strong lower angle $\underline{\alpha}'_s$ (and correspondingly the strong angle α'_s: $\underline{\alpha}'_s = \bar{\alpha}'_s$). This differs from the fact mentioned above that the limit is considered not over all sequences of points X_n, Y_n such that X_n or Y_n tends to O, but only over those such that as $X_n \to O$ there are shortest curves X_n, Y_n that converge to

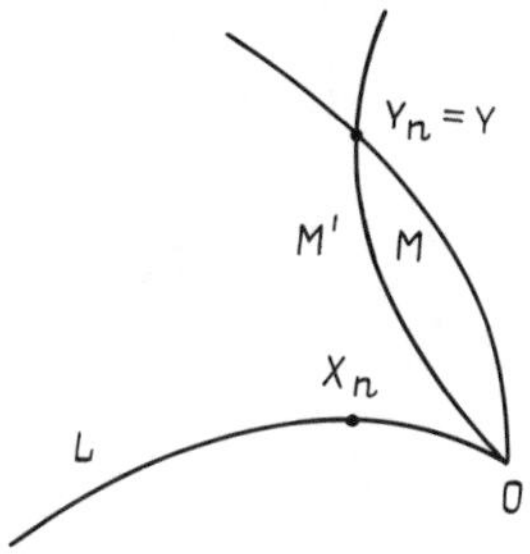

Fig. 1

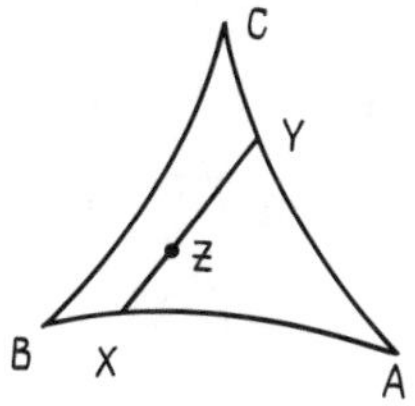

Fig. 2

part of the shortest curve M, and as $Y_n \to O$ they converge to part of the shortest curve L (it is assumed that at least one such sequence exists).

Example 4.2 (Aleksandrov and Zalgaller (1962)). On a sphere we consider two shortest curves starting from a point O and forming an angle φ, $0 < \varphi < \pi/2$, and the opposite ends of the shortest curves coincide at the point diametrically opposite to O. Then obviously $\underline{\alpha}_s = 0$, but $\underline{\alpha}'_s = \underline{\alpha} = \overline{\alpha} = \overline{\alpha}'_s = \varphi$. Here between the shortest curves under consideration there is an angle α'_s but not α_s.

We note that in two-dimensional manifolds of bounded curvature there is an angle α'_s between shortest curves, but not α_s (Aleksandrov and Zalgaller (1962)).

4.2. An Assertion Completely Dual to Theorem 2.1, that is, the Corresponding Lower Bound for the Lower Angle Does not Hold. Let us give the corresponding example.

Example 4.3 (Aleksandrov and Zalgaller (1962)). On a plane we consider a concave curvilinear triangle with angles at the vertices A, B, C equal to ψ, where $0 < \psi < \pi/3$ (see Fig. 2). We now consider the metric space consisting of the three threads AB, BC, CA of equal length l, and infinitely many threads joining points $X \in AB$ and $Y \in AC$ having the same length as the corresponding shortest curve XY in Fig. 2. The metric is determined from the lengths of curves. Then in the resulting metric space, for example, the distance from A to a point $Z \in XY$ close to Y is equal to $YZ + YA$, and ABC is a triangle. At its vertex A we have $\underline{\alpha} = \psi$, $\alpha_0 = \pi/3$, $\underline{\alpha} - \alpha_0 < 0$. However, for any triangle AXY we have $\delta(\Delta AXY) = (\varphi + \pi + \pi) - \pi > 0$. Thus, the inequality $\underline{\alpha} - \alpha_0 \geqslant \inf \delta_0(\Delta AXY)$, $X \in AB$, $Y \in AC$, is impossible.

We observe that the lower bound can be obtained if we calculate the excesses of the triangles AXY from the strong lower angles $\underline{\alpha}_s$ (the same is true for strong lower angles $\underline{\alpha}'_s$) (Aleksandrov and Zalgaller (1962)).

Other definitions of angles occur in the literature. For this we refer the reader to the appendix to Chapter 2 of Aleksandrov and Zalgaller (1962).

4.3. Tangent Space, Space of Directions. The concept of the tangent space (cone) to an arbitrary metric space at a point $P \in \mathfrak{M}$ assumes that at P the metrics $\mathfrak{M}$ and $\mathfrak{M}_P$ coincide infinitesimally. However, in the general case this is not so. Consider two shortest curves L, $M \subset \mathfrak{M}$ with common origin P, and points $X \in L$, $Y \in M$. Then

$$\rho = XY = (PX^2 + PY^2 - 2PX \cdot PY \cdot \cos \gamma_{LM}^0(x, y))^{1/2}.$$

In $\mathfrak{M}_P$ the corresponding distance ρ' is

$$\rho' = (PX^2 \dotplus PY^2 - 2PX \cdot PY \cdot \cos \bar{\alpha})^{1/2},$$

where $\bar{\alpha}$ is the upper angle between L and M. We see that $\rho = \rho' + o(\rho_{\mathfrak{M}}(X, Y))$ if and only if there is an angle between L and M.

4.3.1. The space of directions of an arbitrary metric space is not necessarily arranged more simply than the original space. Consider an arbitrary metric space $\mathfrak{M}$ of diameter less than π. Then obviously the space of directions of the K-cone over $\mathfrak{M}$ at the vertex of this cone is isometric to $\mathfrak{M}$: $\Omega_O C_K \mathfrak{M} = \mathfrak{M}$.

4.3.2. In a normed vector space the existence of a direction of a curve at a point does not always imply the existence of the corresponding half-tangent at this point. A curve L in Euclidean space E_n with origin at a point $P \in E_n$ has a definite direction at P if and only if there is a right derivative of it at P, that is, in this case L has a right half-tangent at P. In an arbitrary normed space this is not so.

Example 4.4. On a Euclidean plane we consider an orthogonal coordinate system Oxy and the disc B: $x^2 + y^2 \leqslant 1$. We denote by $C(B)$ the space of continuous functions on B with the usual sup-norm. It is well known that the map that associates with an arbitrary point $(x_0, y_0) \in B$ the function

$$f_{(x_0,y_0)}(x, y) = [(x - x_0)^2 + (y - y_0)^2]^{1/2}$$

is isometric. Consider in $C(B)$ the image of the segment l: $x = t$, $y = 0$, $0 \leqslant t \leqslant 1$:

$$f_t(x, y) = [(x - t)^2 + y^2]^{1/2}.$$

Obviously the resulting curve in $C(B)$ is a shortest curve, and so it has a definite direction at the point $f_0(x, y) \in C(B)$. However, the curve $t \mapsto f_t \in C(B)$ does not have a right derivative at $t = 0$, and so it does not have a right half-tangent. For if it existed, then it could be calculated as a pointwise limit

$$d_{\text{rt.}} f_t(x, y)|_{t=0} = \lim_{t \to 0^+} \{[(x - t)^2 + y^2]^{1/2} - (x^2 + y^2)^{1/2}\}/t.$$

It is easy to calculate that then $d_{\text{rt.}} f_t(x, y)|_{t=0} = -x/(x^2 + y^2)^{1/2}$ when $x^2 + y^2 \neq 0$ and is equal to 1 when $x^2 + y^2 = 0$. However, the resulting function is discontinuous at zero (that is, it does not belong to $C(B)$) and so there cannot be a right derivative in $C(B)$.

We observe that a similar example can also be given in a finite-dimensional normed space.

Chapter 2
Spaces of Curvature $\leqslant K$ (and $\geqslant K'$)

In this chapter we shall consider metric spaces whose curvature is in a certain sense either bounded above or bilaterally bounded. These spaces, namely spaces of curvature $\leqslant K$ and spaces of curvature both $\leqslant K$ and $\geqslant K'$, were introduced by Aleksandrov (Aleksandrov (1951), (1957b)). As an example of the situation in which spaces of curvature $\leqslant K$ arise, we can consider metrics that are limits of smooth Riemannian metrics with bound on the sectional curvatures $K_\sigma^m \leqslant K$ (see 11.1.3).

A way of specifying restrictions on the curvature is based on a known expression for the sectional curvature of a Riemannian manifold in terms of the excesses of geodesic triangles (see (0.1), (0.2)).

The upper boundedness of the curvature, that is, just the right inequality being satisfied, enables us to prove the existence of many geometrical properties that Riemannian manifolds have. We mention that the results presented in this chapter have a purely synthetic character, and they have as their basis the angle comparison theorem (Theorem 5.1) and the property of K-concavity (Theorem 5.2), which are satisfied in the general case in a small domain R_K of a space of curvature $\leqslant K$. At the end of the chapter we consider spaces of curvature both $\leqslant K$ and $\geqslant K'$.

§ 5. Spaces of Curvature $\leqslant K$. The Domain R_K and its Basic Properties

5.1. Definition of a Space of Curvature $\leqslant K$. A space of curvature $\leqslant K$ can also be characterized by the fact that for each sufficiently small triangle T its K-excess $\bar{\delta}_K(T)$ is non-positive. A space of curvature $\leqslant K$ was originally defined in this form in Aleksandrov (1951). Later, in § 7, we shall prove that the definition given at the beginning of the chapter and the one given here are equivalent. Particular significance in what follows is played by a domain of space of curvature $\leqslant K$ in which the bounds for K-excesses of triangles mentioned above are satisfied. We shall denote it by R_K. Most of the results presented in this chapter refer to the domain R_K. A precise definition of a space of curvature $\leqslant K$ is as follows:

The *domain* R_K is a metric space with the following properties:

a) any two points in R_K can be joined by a shortest curve;

b) each triangle in R_K has non-positive K-excess;

c) if $K > 0$, then the perimeter of each triangle in R_K is less than $2\pi/\sqrt{K}$.

By a space of curvature $\leqslant K$ we understand a metric space, each point of which is contained in some neighbourhood of the original space, which is the domain R_K.

5.2. Basic Properties of the Domain R_K.

5.2.1. The Angle Comparison Theorem for Triangles in R_K. The next theorem follows immediately from Theorem 2.1.

Theorem 5.1 (Aleksandrov (1957b)). *The upper angles $\bar{\alpha}$, $\bar{\beta}$, $\bar{\gamma}$ of an arbitrary triangle T in R_K are not greater than the corresponding angles α_K, β_K, γ_K of the triangle T^K on S_K.*

In what follows, an important role is played by the following assertion of elementary geometry.

Lemma 5.1 (Aleksandrov (1957b)). *Let $Q = ABCD$ be a quadrangle on S_K, bounded by the triangle ABC (when $K > 0$ the perimeter of Q is less than $2\pi/\sqrt{K}$). Then there is a triangle T on S_K obtained from Q by rectifying the polygonal line CDB, that is, a triangle with side lengths AB, AC, $BD + DC$. If Q is not equal to T, then:*

1. *the angles of Q at the vertices A, B, C are less than the corresponding angles of T;*
2. *the area of Q is less than the area of T.*

5.2.2. *K*-concavity (Aleksandrov (1957b)). From Theorem 5.1 by means of Lemma 5.1 we can obtain the following theorem.

Theorem 5.2 (Aleksandrov (1957b)). *For any two shortest curves L and M in R_K, starting from one point O, the angle $\gamma^K_{LM}(x, y)$ (see § 1) is a non-decreasing function of x and y.*

The proof of Theorem 5.2 consists in the following. Suppose that the shortest curves L and M start from the point O. We take points Y and Y_1 on M and a point X on L so that $OY = y$, $OY_1 = y_1$, $OX = x$, $0 < y_1 < y$, $x > 0$ (see Fig. 3). Let $T = \Delta OXY$, $T_1 = \Delta OXY_1$, $T_2 = \Delta XY_1Y$; $T_1^K = \Delta O'X'Y_1'$, $T_2^K = \Delta X'Y_1'Y'$ be triangles on S_K corresponding to T_1 and T_2 such that T_1^K and T_2^K lie on opposite sides of $X'Y_1'$ (when $K > 0$ the existence of such triangles follows from condition c); see 5.1).

In consequence of Proposition 2.2 and Theorem 5.1, the angle of the quadrangle Q composed of the triangles T_1^K and T_2^K at the vertex Y_1' is either equal to π or reentrant, that is, Q is bounded by the triangle $O'X'Y'$. By rectifying the polygonal line $O'Y_1'Y'$ we obtain a triangle T^K. By Lemma 5.1 the angle of Q at the vertex O' (that is, $\gamma^K_{LM}(x, y_1)$) is not greater than the corresponding angle of the triangle T^K (that is, $\gamma^K_{LM}(x, y)$), as required.

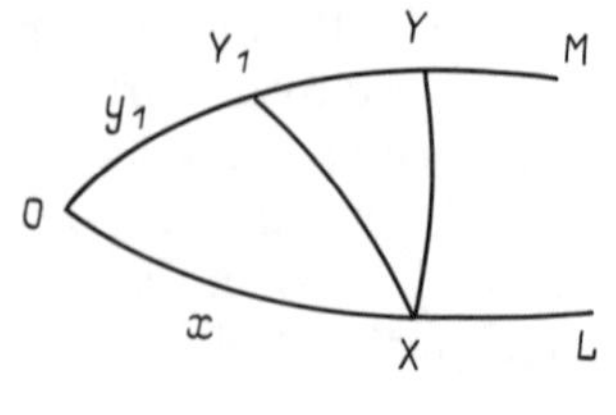

Fig. 3

5.2.3. Consequences of K-Concavity.

Proposition 5.1 (Aleksandrov (1957b)). *Let X and Y be points on the sides AB and AC of the triangle $T = ABC$ in R_K, and X' and Y' the corresponding points on the sides of the triangle $T^K = A'B'C'$ (that is, $A'X' = AX$, $A'Y' = AY$). Then $XY \leqslant X'Y'$.*

The next proposition, which follows immediately from the monotonicity of the angle $\gamma_{LM}^K(x, y)$, is particularly important for us. In Riemannian manifolds locally between shortest curves there is an angle in the strong sense. However, in Finsler spaces (even with infinitely differentiable metric tensor) the proposition stated below does not hold in any domain of the space (see 4.1.3).

Proposition 5.2 (Aleksandrov (1957b)). *Between any two shortest curves in R_K starting from one point there is an angle in the strong sense.*

We now state propositions that describe the local behaviour of shortest curves in R_K. It turns out that shortest curves in R_K have a series of properties common to the local properties of geodesics in Riemannian manifolds.

Proposition 5.3 (Aleksandrov (1957b)). *In R_K shortest curves depend continuously on their ends.*

Proof. Consider the shortest curves AB and A_nB_n in R_K, $A_n \to A$, $B_n \to B$. On AB we take an arbitrary point C and on A_nB_n points C_n so that

$$AC/AB = A_nC_n/A_nB_n.$$

We now consider the shortest curve AB_n (see condition a)) and take on it the point D_n that divides AB_n in the same ratio (see Fig. 4). Applying Proposition 5.1 to the triangle ABB_n and taking into account that $AC/AB = AD_n/AB_n = t$, we deduce that $CD_n \to 0$ as $BB_n \to 0$. Similarly $D_nC_n \to 0$ as $A_nA \to 0$. Since $CC_n \leqslant CD_n + D_nC_n$, we obtain the required result.

Proposition 5.4 (Aleksandrov (1957b)). *Any two points of R_K are joined by a unique shortest curve.*

Proposition 5.5 (Aleksandrov, Berestovskii and Nikolaev (1986)). *An open ball in R_K is convex (if its radius is less than $\pi/2\sqrt{K}$ when $K > 0$).*

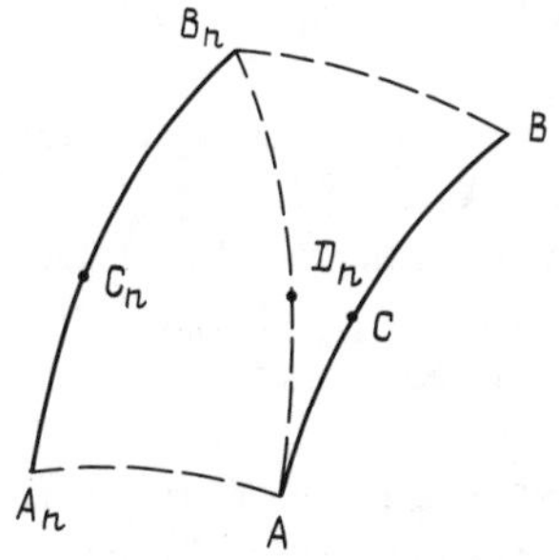

Fig. 4

5.3. The Domain P_K (Reshetnyak (1960b), (1968), Nikolaev (1979)). We give a modification of the concept of the domain R_K that occurs in the literature and is useful in what follows. The domain P_K is a metric space with the conditions a) of 5.1 and

b′) for any triangle contained in P_K whose perimeter is less than $2\pi/\sqrt{K}$ when $K > 0$ the K-excess is non-positive.

Obviously the concepts of the domains P_K and R_K differ only when $K > 0$ (the closure of S_K when $K > 0$ is P_K but not R_K).

§ 6. The Operation of Gluing

In this section we shall describe the operation of gluing for spaces of curvature $\leqslant K$ along compact convex sets and give some examples.

6.1. Gluing of Metric Spaces with Intrinsic Metric. We consider only the situation when the metric spaces that are glued contain isometric convex sets and the gluing is carried out along these sets. The general case is quite complicated and we do not discuss it.

Let $(\mathfrak{M}_1, \rho_1)$ and $(\mathfrak{M}_2, \rho_2)$ be metric spaces with intrinsic metric, $V_1 \subseteq \mathfrak{M}_1$ and $V_2 \subseteq \mathfrak{M}_2$ isometric convex sets in these spaces, and $\varphi: \mathfrak{M}_1 \to \mathfrak{M}_2$ the corresponding isometry. Consider the set $\mathfrak{M}_1 \cup \mathfrak{M}_2$ and specify an equivalence relation on it. Namely,

$$\begin{array}{ll} X \text{ is equivalent to } X & \text{if } X \in \mathfrak{M}_1 \setminus V_1, \\ Y \text{ is equivalent to } Y & \text{if } Y \in \mathfrak{M}_2 \setminus V_2, \\ Z \text{ is equivalent to } \varphi(Z) & \text{if } Z \in V_1. \end{array}$$

We denote by $\mathfrak{M}_1 \cup_\varphi \mathfrak{M}_2$ the set of classes of equivalent elements of $\mathfrak{M}_1 \cup \mathfrak{M}_2$. We now introduce an intrinsic metric on $\mathfrak{M}_1 \cup_\varphi \mathfrak{M}_2$:

$$\rho_\varphi(\bar{X}, \bar{Y}) = \begin{cases} \rho_1(X, Y) & \text{if } X, Y \in \mathfrak{M}_1, \\ \rho_2(X, Y) & \text{if } X, Y \in \mathfrak{M}_2, \\ \inf\limits_{Z \in V_1} \{\rho_1(X, Z) + \rho_2(Y, \varphi(Z))\} & \text{if } X \in \mathfrak{M}_1,\ Y \in \mathfrak{M}_2, \end{cases}$$

where X, Y denote the points of $\mathfrak{M}_1 \cup \mathfrak{M}_2$ contained in the corresponding classes of equivalent elements $\bar{X}, \bar{Y} \in \mathfrak{M}_1 \cup_\varphi \mathfrak{M}_2$.

The metric space with intrinsic metric $(\mathfrak{M}_1 \cup_\varphi \mathfrak{M}_2, \rho_\varphi)$ is the result of gluing the metric spaces $(\mathfrak{M}_1, \rho_1)$ and $(\mathfrak{M}_2, \rho_2)$ with intrinsic metric along the convex sets V_1 and V_2 by means of the isometry φ.

6.1.1. Gluing of Spaces of Curvature $\leqslant K$. It turns out that the result of gluing spaces of curvature $\leqslant K$ along isometric compact convex sets is a space of curvature $\leqslant K$. More precisely, the following theorem holds.

Theorem 6.1 (Reshetnyak (1960b)). *If the spaces $\mathfrak{M}_1$ and $\mathfrak{M}_2$, which are domains of P_K, are glued together along compact convex sets $V_1 \subseteq \mathfrak{M}_1$, $V_2 \subseteq \mathfrak{M}_2$ by means of some isometry φ, then the space $\mathfrak{M}_1 \cup_\varphi \mathfrak{M}_2$ that arises as a result of the gluing is also a domain of P_K.*

6.1.2. Example. Let $\mathfrak{M}_1$ be a closed convex set in $S^\infty_{K_1}$ (for $S^m_{K_1}$ see § 1), $\mathfrak{M}_2$ a closed convex set in $S^m_{K_2}$, where $K_1, K_2 \leqslant K$, and $P \in \mathfrak{M}_1$ and $Q \in \mathfrak{M}_2$ certain points. Theorem 6.1 asserts that the space obtained by gluing $\mathfrak{M}_1$ and $\mathfrak{M}_2$ at the points P and Q is a space of curvature $\leqslant K$. Fig. 5 shows a space of curvature $\leqslant 0$ obtained by gluing two Euclidean triangles at the vertices O' and O''.

6.1.3. *K*-Fan (Reshetnyak (1968)). A *K-fan* is a space of curvature $\leqslant K$ obtained as a result of successive gluing of finitely many triangles of S_K. In this subsection, by a triangle we understand the corresponding convex domain of S_K bounded by a triangle in the original understanding of this term, and the triangle may be degenerate.

Let $T_i \subset S_K$, $i = 1, 2, \ldots, m$, be a set of triangles. For them we introduce the notation $T_i = A_iO_iB_i$. We assume that in T_i and T_{i+1} the sides O_iB_i and $O_{i+1}A_{i+1}$ have different lengths. We denote by φ_i the isometry of O_iB_i onto $O_{i+1}A_{i+1}$ under which the point O_i goes over to O_{i+1}. Then the K-fan formed from the triangles $T_1, \ldots, T_m$ is a metric space with intrinsic metric, glued together from the T_i along O_iB_i and $O_{i+1}A_{i+1}$, $i=1, \ldots, m-1$, that is, $T_1 \cup_{\varphi_1} T_2 \cup_{\varphi_2} \cdots \cup_{\varphi_{m-1}} T_m$. The resulting space is a space of curvature $\leqslant K$. Obviously all the vertices O_i are glued together at one point.

If we reject the condition $\varphi_i(O_i) = O_{i+1}$ in the definition given above, we arrive at the concept of a *generalized K-fan*. Figs. 6 and 7 show a K-fan and a generalized K-fan respectively, glued together from four triangles.

§ 7. Equivalent Definitions of Upper Boundedness of the Curvature

A definition of a space of curvature $\leqslant K$ was given in 5.1. In this section we give different versions of the definition of upper boundednes of the curvature: we

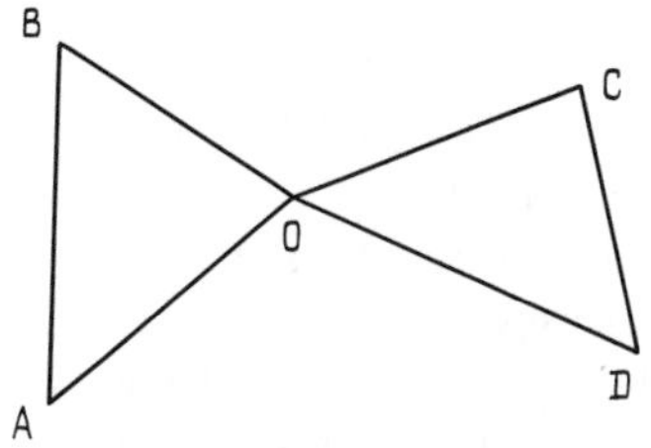

Fig. 5

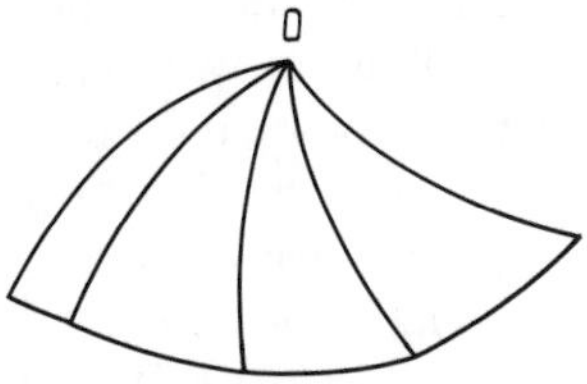

Fig. 6

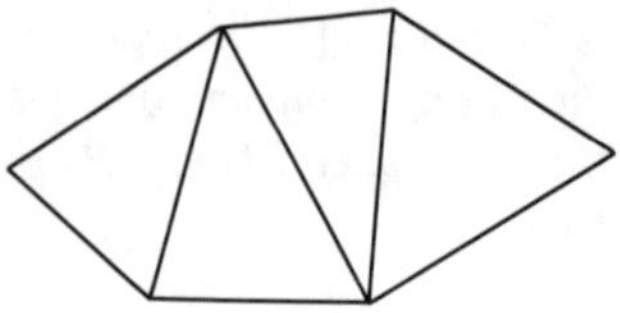

Fig. 7

give the connection with the usual Riemannian definition of curvature, we state a theorem about non-expanding maps in a space of curvature $\leqslant K$ that is characteristic for these spaces, and we give a definition of boundedness of curvature from the positions of distance geometry. We note that these definitions in the case of Riemannian manifolds describe the upper boundedness of the sectional curvatures of a Riemannian manifold by the number K.

7.1. Conditions under which a Space of Curvature $\leqslant K$ is the Domain R_K. In the definition of the domain R_K we require that the K-excesses of arbitrary triangles are non-positive. It would be desirable that this condition should be imposed only on small triangles. Naturally, this condition is necessary, but generally speaking insufficient. For example, a circular cylinder in this sense has non-positive curvature, and at the same time it is not the domain R_0, since some pairs of points on it are joined by two shortest curves. If we require in addition that a shortest curve depends continuously on its ends, we obtain conditions that specify the domain R_K.

Proposition 7.1 (Aleksandrov (1957b)). *Suppose that a domain G of a metric space satisfies the following conditions*:

1) *G is convex, that is, any two points of G are joined by a shortest curve, and any such shortest curve lies in G*;

2) *shortest curves in G depend continuously on their ends*;

3) *each point in G is contained in a neighbourhood in which the K-excess of any triangle is non-positive*;

4) *when* $K > 0$ *the perimeter of any triangle in G is less than* $2\pi/\sqrt{K}$.

Then G is the domain R_K.

Let us give a plan of the proof. From the conditions of Proposition 7.1 it follows that in G the conditions that determine the domain R_K are locally satisfied, and so in small domains G the results of § 5 are valid. In particular, between any two shortest curves lying in G the angle exists, and for small triangles the angle comparison theorem for triangles is satisfied.

Now let $T = ABC$ be an arbitrary triangle in the domian G. We prove that each angle of T is not greater than the corresponding angle in the triangle T^K with the same lengths of sides. To this end we split T by shortest curves AD_i into "narrow" triangles $T_i = AD_iD_{i+1}$, as shown in Fig. 8. This is possible by virtue of the conditions 1) and 2). In turn we split each "narrow" triangle T_i into small

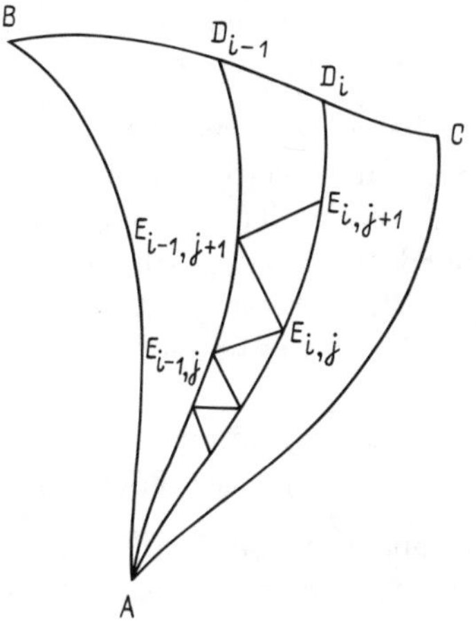

Fig. 8

triangles $T_{ij} = E_{i-1,j}E_{i,j}E_{i-1,j+1}$ (see Fig. 8) so that all the small triangles have non-positive excess $\bar{\delta}_K$. On S_K we consider the triangles T_{ij}^K corresponding to T_{ij}. By gluing together the T_{ij}^K for fixed i in the order of adjoining the triangles T_{ij} corresponding to them, we obtain a "narrow" generalized K-fan (6.1.3). We observe that the sum of the angles adjoining one vertex (lying inside AD_i or AD_{i-1}) is not less than π (Theorem 5.1, Proposition 2.2). Straightening the corresponding reentrant angles of the K-fan, we deduce from Lemma 5.1, applied successively under each straightening, that the angles of the "narrow" triangle T_i are not greater than the angles of the corresponding triangle T_i^K on S_K. We now glue together the K-fans $A'B'D'_1, \ldots, D'_nC'A'$ from the triangles T_i^K. Again straightening the reentrant angles at $D'_1, D'_2, \ldots, D'_n$ and using Lemma 5.1, we obtain the required assertion.

7.2. Connection with the Riemannian Definition of Curvature. Let us establish the connection between the definition of a space of curvature $\leqslant K$ and the definition of sectional curvature, known from Riemannian geometry, as the limit of the ratio of the excesses of geodesic triangles to their "areas".

Theorem 7.1 (Aleksandrov (1957b)). *In order that a metric space should be a space of curvature $\leqslant K$ it is necessary and sufficient that the following conditions are satisfied*:

1) *each point has a neighbourhood, any two points of which are joined by a shortest curve*;

2) *for any sequence of triangles T_m contracted to a point*

$$\overline{\lim_{m\to\infty}}\ \bar{\delta}(T_m)/\sigma(T_m) \leqslant K, \tag{7.1}$$

where $\bar{\delta}(T_m)$ is the excess of the triangle T_m, and $\sigma(T_m)$ is its area (see 1.5).

Remark. When $\sigma(T_m) = 0$ it is assumed that $\bar{\delta}(T_m) \leqslant 0$, so (7.1) makes sense in this case.

The proof of necessity is based on (1.1) and a known expression for the excess of a triangle on S_K in terms of its area (the Gauss-Bonnet theorem). The plan of the proof of sufficiency is as follows. If (7.1) is satisfied, it is not difficult to show that each point is contained in a neighbourhood $R_{K+\varepsilon}$, where ε is any preassigned positive number. In $R_{K+\varepsilon}$ the shortest curves depend continuously on their ends (Proposition 5.3). It remains to apply Proposition 7.1 and the limiting process as $\varepsilon \to 0$.

Starting from the geometrical meaning of the curvature of a Riemannian manifold and Theorem 7.1, we obtain the following result.

Corollary 7.1. *A Riemannian manifold with sectional curvature* $\leqslant K$ *(at all points in all two-dimensional directions) is a space of curvature* $\leqslant K$.

7.3. Definition of Upper Boundedness of Curvature. In a space in which locally any two points are joined by a shortest curve, the condition $\bar{\delta}_K(T) \leqslant 0$, which is satisfied for sufficiently small triangles T, is equivalent to the angle comparison theorem for triangles (that is, Theorem 5.1) for sufficiently small triangles or the property of K-concavity in the small (5.2.2). Thus, each of these properties can be adopted as the definition of upper boundedness of the curvature of the space.

7.4. Non-Expanding Maps in Spaces of Curvature $\leqslant K$. Let $(\mathfrak{M}_1, \rho_1)$ and $(\mathfrak{M}_2, \rho_2)$ be metric spaces. We say that a map $\varphi: \mathfrak{M}_1 \to \mathfrak{M}_2$ is *non-expanding* (Reshetnyak (1968)) if

$$\rho_2(\varphi(X), \varphi(Y)) \leqslant \rho_1(X, Y), \qquad X, Y \in \mathfrak{M}_1.$$

φ maps a rectifiable curve $\Gamma_1 \subset \mathfrak{M}_1$ equilengthwise into a rectifiable curve $\Gamma_2 \subset \mathfrak{M}_2$ if it takes any arc of Γ_1 into an arc of Γ_2 of the same length.

A convex domain V on S_K with bounding curve L majorizes a rectifiable curve Γ in the metric space $\mathfrak{M}$ (it is assumed that the length of Γ is equal to the length of L) if there is a non-expanding map of V into $\mathfrak{M}$ that maps L equilengthwise onto Γ.

The following theorem holds for non-expanding maps in a space of curvature $\leqslant K$.

Theorem 7.2 (Reshetnyak (1968)). *For any closed rectifiable curve* Γ *in* P_K *whose length is less than* $2\pi/\sqrt{K}$ *when* $K > 0$ *there is a convex domain on* S_K *that majorizes the curve* Γ.

Corollary 7.2. *In order that a metric space should be a space of curvature* $\leqslant K$ *it is necessary and sufficient that condition* 1) *of Theorem* 7.1 *should be satisfied and that*

2′) *for each sufficiently small triangle* T *there is a convex domain on* S_K *that majorizes* T *(obviously the boundary of this domain is a triangle with the same lengths of sides as* T *has).*

Theorem 7.2 is proved as follows. The condition on the length of the curve Γ enables us to span on Γ a ruled surface $\mathscr{S}$ consisting of shortest curves OX joining a fixed point $O \in \Gamma$ and points $X \in \Gamma$. In the curve Γ there is inscribed a sequence of polygonal lines $\Gamma_n = OX_1X_2 \ldots X_{m_n}O$, $n = 1, 2, \ldots$, that converges to Γ. We then construct an "almost" non-expanding map ψ_n of the K-fan $\mathscr{U}_n = OX_1X_2 \ldots X_{m_n}$ into $\mathfrak{M}$, under which the contour of the fan $\mathscr{U}_n$ is mapped equilengthwise onto the polygonal line Γ_n. By means of Lemma 5.1 there is established the existence of a majorizing map φ_n of the convex domain V_n in S_K onto the fan $\mathscr{U}_n$. The composition $\psi_n \circ \varphi_n$ "almost" majorizes the polygonal line Γ_n. There is established the existence of a subsequence $\psi_n \circ \varphi_n$ that converges to some map, majorizing Γ, of the convex domain $V \subseteq S_K$ onto the ruled surface $\mathscr{S}$.

7.5. Boundedness of the Curvature from the Viewpoint of Distance Geometry. There is a connection between the theory of generalized Riemannian spaces and the so-called distance geometry (Blumenthal (1970)). Let us give the necessary definitions.

We shall say that in a metric space $(\mathfrak{M}, \rho)$ a point B lies between points A and C if

$$\rho(A, C) = \rho(A, B) + \rho(B, C), \quad \rho(A, B), \rho(B, C) > 0.$$

A *quadruple* of distinct points in the metric space is said to be *exceptional* if one of these points lies between distinct pairs of points of the quadruple, but the quadruple does not have rectilinear arrangement, that is, it is not embedded isometrically into the number line. Otherwise the quadruple is said to be *ordinary*. Let us explain what we have said by an example. We turn to the example of 6.1.2 (see Fig. 5). In this metric space, obtained by gluing together two planar triangles at the vertices, the quadruple of points A, O, C, D is obviously exceptional.

The next theorem characterizes the boundedness of curvature in terms of the isometric embeddability of quadruples in the model space S_K^3.

Theorem 7.3 (Berestovksij (1986)). *A metric space is a space of curvature* $\leqslant K$ *if and only if it satisfies condition* 1) *of Theorem* 7.1 *and*

2″) *each point has a neighbourhood* $\mathscr{U}$ *such that each ordinary quadruple of points of* $\mathscr{U}$ *is embedded isometrically in* $S_{K'}^3$ (*for the definition of* $S_{K'}^3$ *see* 1.3) *with* $K' \leqslant K$, *depending on the chosen quadruple.*

We note that from the isometric embeddability of ordinary quadruples in S_K^3 there easily follows the property of K-concavity in $\mathscr{U}$.

§ 8. Space of Directions, Tangent Space at a Point of a Space of Curvature $\leqslant K$

The definitions of a space of directions and a tangent space were given in 3.1 and 3.3. In this section we discuss the properties of a space of directions and a tangent space at an arbitrary point of a space of curvature $\leqslant K$. In order to obtain significant results it is necessary to make additional assumptions. These assumptions are satisfied, for example, in the case when the space of curvature $\leqslant K$ is a topological manifold. The main results of this section are that the space of directions at a point of a space of curvature $\leqslant K$ is a space of curvature $\leqslant 1$, and the corresponding tangent space at the point has curvature $\leqslant 0$ in the sense of Aleksandrov.

8.1. Conditions under which a Shortest Curve Goes out in each Direction. The next proposition follows easily from the property of K-concavity in R_K (see 5.2.2).

Proposition 8.1 (Aleksandrov (1957b). *If in R_K the shortest curves L_n, M_n, $n = 1, 2, \ldots$, going out from a point P converge to shortest curves L, M, then the angles $\alpha(L_n, M_n)$ converge to the angle $\alpha(L, M)$. In particular, if $L_n \to L$, then $\alpha(L_n, L) \to 0$.*

Remark. In the case when the shortest curves L_n, M_n have starting point $P_n \neq P$ and converge to shortest curves L, M respectively, we have

$$\alpha(L, M) \geqslant \varlimsup_{n\to\infty} \alpha(L_n, M_n).$$

From Proposition 8.1 it follows, in particular, that the directions specified by shortest curves in R_K with common starting point depend continuously on the shortest curves.

We can now state a condition under which a shortest curve goes out in each direction.

Proposition 8.2 (Aleksandrov (1957b). *If a point P of a space of curvature $\leqslant K$ has a compact neighbourhood and each shortest curve PX can be extended to a shortest curve PX' of length r (where r is a fixed positive number), then a shortest curve goes out in each direction at P. The space of directions at P is compact.*

Remark. The example of a closed disc on a Euclidean plane that is a domain R_0 with P on its bounding circle shows that the condition of extendability of the shortest curves PX is essential.

8.1.1. Plan of the Proof of Proposition 8.2. Suppose a curve has a definite direction at a point P. From the existence of a direction (for the definition see 3.1) it follows that there is a sequence of shortest curves with starting point P such that the angle between them tends to zero. Local compactness and extendability of shortest curves enables us to go over to a sequence of shortest curves that converge to a shortest curve different from the point P. Obviously the original curve and the limiting shortest curve have the same direction.

8.1.2. The Case when the Point P Has a Neighbourhood Homeomorphic to a Ball in a Finite-dimensional Euclidean Space. As the proposition stated below shows, the conditions of Proposition 8.2 are satisfied for topological manifolds with metric curvature $\leqslant K$.

Proposition 8.3 (Aleksandrov (1957b)). *If the point P is contained in a neighbourhood $\mathcal{U}$ that is a domain R_K and is homeomorphic to a ball in E_n, and r_0 is the distance from P to the boundary of $\mathcal{U}$, then any shortest curve starting from P can be extended to have length r_0.*

8.2. Intrinsic Metric in Ω_P.

8.2.1. The Angle of the Cone of Directions. A curve in Ω_P is usually called a cone of directions, and its length is called the angle of the cone of directions.

8.2.2. The Cone of Directions of a Surface Triangle. Let $T = ABC$ be an arbitrary triangle in R_K. Let D_t $(0 \leqslant t \leqslant 1, D_t \in BC)$ be a parametrization of BC. Consider all possible shortest curves AD_t, $0 \leqslant t \leqslant 1$. The resulting set of shortest curves forms a *surface triangle*, which we shall also denote by the letter T. We denote by $C_T(t)$ the direction specified by the shortest curve AD_t. The resulting cone of directions C_T is called the *cone of directions* of the surface triangle T.

The surface triangle T, with respect to the intrinsic metric induced from R_K, is a space of curvature $\leqslant K$ (see 9.3). We denote its angle at the vertex A by α. We can prove that the angle of the cone C_T is exactly equal to α. Hence it follows (see Theorem 5.1) that $\alpha \leqslant \alpha_K$. It was proved in Aleksandrov (1957b) that $\alpha = \alpha_K$ if and only if the surface triangle T and the corresponding surface triangle T^K from S_K are isometric.

8.2.3. Shortest Curves in Ω_P. Let $\xi, \zeta \in \Omega_P$ be two arbitrary directions, the angle α between which is less than π. Consider two shortest curves AB and AC, starting from A and specifying the directions ξ and ζ respectively. Let $X \in AB$ and $Y \in AC$ be arbitrary points. By virtue of 8.2.2 the angle of the cone C_T, where $T = \Delta AXY$ is arbitrarily small, is different from α. We make X and Y tend to A. We obtain a sequence of cones whose angles are arbitrarily close to the angle α of the triangle ABC at the vertex A. In the case when the conditions of Proposition 8.2 (or 8.3) are satisfied, it is easy to prove that there is a limit cone, which is obviously a shortest curve joining the directions ξ and ζ. From Theorem 8.1 stated below it follows that there are no other shortest curves joining ξ and ζ.

8.2.4. The Curvature of Ω_P.

Theorem 8.1 (Nikolaev (1978)). *If a point P of a space of curvature $\leqslant K$ has a neighbourhood homeomorphic to a ball in E_n, $n > 1$, then the space of directions $\Omega_P\mathfrak{M}$ at this point is a space of curvature $\leqslant 1$ (more precisely, any domain in $\Omega_P\mathfrak{M}$ of diameter less than π is a domain P_1).*

Remark. In Theorem 8.1 the case $n = 1$ is excluded. In this case the space of directions consists of two points.

8.3. Tangent Space. An immediate consequence of Theorem 8.1 is the following theorem.

Theorem 8.2 (Aleksandrov, Berestovskij and Nikolaev (1986)). *If some neighbourhood of a point P of a metric space $\mathfrak{M}$ of curvature $\leqslant K$ is a manifold of finite dimension or the conditions of Proposition* 8.2 *are satisfied, then the tangent space $\mathfrak{M}_P$ to $\mathfrak{M}$ at the point P is a space of curvature $\leqslant 0$ (more precisely, $\mathfrak{M}_P$ is a domain R_0).*

§ 9. Surfaces and their Areas

A space of curvature $\leqslant K$ can have a "rather poor" structure in comparison with a Riemannian manifold. Moreover, the situation is possible in which this space is not even a topological manifold (see 6.1.2). Nevertheless, direct synthetic methods enable us to solve interesting geometrical problems even in spaces of curvature $\leqslant K$. In this section we touch on the following ones: the area of a surface, Plateau's problem, and the "extrinsic" geometry of "ruled" surfaces in R_K.

9.1. The Definition of the Area of a Surface (Aleksandrov (1957b), Nikolaev (1979)). We denote by B a closed unit disc on a Euclidean plane. We understand a *parametrized surface* f in the metric space $\mathfrak{M}$ as any continuous map of the disc B into $\mathfrak{M}$.

By a *non-parametrized surface* we understand the image of a parametrized surface (Aleksandrov (1957b)).

A *triangulation of the disc* B is by definition a triangulation of an arbitrary polygon inscribed in B. Suppose we are given a triangulation of B, that is, a splitting into triangles of a polygon inscribed in B; we denote their vertices by A_i, $i = 1, 2, \ldots, l$. Let f be a surface in P_K. With each point A_i we associate a point $\tilde{A}_i \in M$, $i = 1, 2, \ldots, l$, where the points $\tilde{A}_i$ and $\tilde{A}_j$ coincide if and only if the points $f(A_i)$ and $f(A_j)$ coincide. We now join the points $\tilde{A}_i$ by shortest curves in the same order as the vertices of triangulation of the disc are joined. We thus obtain a set of triangles in P_K, which we call a *complex of the surface* f. We call the points $\tilde{A}_i$ the *vertices of the complex*. We say that a sequence $\{\Phi_m\}$ of complexes belongs to $\Phi(f)$ or approximates the surface f if the maximum of the distances between $\tilde{A}_i$ and $f(A_i)$ and the largest side of the triangles tends to zero with the growth of m.

If Φ is a complex, then as the area of Φ we take the number $\sigma(\Phi)$, which is equal to the sum of the areas of Euclidean triangles with the same lengths of sides as for the triangles of the complex Φ. Now the *area* $\sigma(f)$ of the surface f is defined like the area of a surface in Euclidean space:

$$\sigma(f) = \inf\{\underline{\lim}\, \sigma(\Phi_m)\}, \quad \{\Phi_m\} \in \Phi(f).$$

This definition was given in Nikolaev (1979) and modifies the definition of the area in Aleksandrov (1957b). In the case of Euclidean space the definition we have given turns into the definition of area according to Lebesgue (see Cesari (1956)).

If F is a non-parametrized surface, then by the definition in Aleksandrov (1957b)

$$\sigma(F) = \inf\{\sigma(f)\}, \quad f(B) = F.$$

9.2. Properties of Area. As in the case of Lebesgue area, for surfaces in Euclidean space we can prove the following properties of area:

1) There is a sequence of complexes that approximate a given surface in P_K, the limit of whose areas is the area of the given surface.

2) (*Semicontinuity*). If the sequence of surfaces $\{f_m\}$ in P_K converges uniformly to the surface f, then

$$\sigma(f) \leqslant \underline{\lim}\, \sigma(f_m).$$

3) (*Kolmogorov's principle*). If p is a non-expanding map from one domain P_K to another, and f is a surface in the first domain, then

$$\sigma(p \circ f) \leqslant \sigma(f).$$

9.3. Ruled Surfaces in R_K. The general definition was given in Aleksandrov (1957a). Here we consider a special case.

Let Γ be a closed curve in R_K (whose length when $K > 0$ is less than $2\pi/\sqrt{K}$), C the bounding circle of the disc B, and $f\colon C \to R_K$ a parametrization of the curve Γ. Let O be an arbitrary point on C. Consider the surface whose parametrization f is specified as follows: for $X \in B$, lying on the interval OY, $Y \in C$, $f(X)$ lies on the shortest curve $O'Y'$ ($O' = f(O)$, $Y' = f(Y)$), and

$$O'f(X) : O'Y' = OX : OY.$$

Because of the condition on the length of the curve Γ, $O'Y'$ depends continuously on $Y' \in \Gamma$ and f is a parametrized surface. In this case we shall say that a ruled surface with vertex at O' is spanned on f. In the general case a ruled surface is a surface formed by shortest curves. We note that in the case when Γ is a triangle, the surface spanned on Γ at a vertex of Γ is a surface triangle (see 8.2.2). The next theorem extends to spaces of curvature $\leqslant K$ the known "extrinsic" geometrical properties of ruled surfaces in Riemannian manifolds.

Theorem 9.1 (Aleksandrov (1957a)). *A ruled surface in a space of curvature $\leqslant K$ is itself a space of curvature $\leqslant K$ in the sense of its intrinsic metric induced from the space* (*so long as it has an intrinsic metric at all*; *see the corresponding example in* 4.1.1).

In other words, the "extrinsic curvature" of a ruled surface is non-positive. Theorem 9.1 was proved in Aleksandrov (1967a) by the method of approximating the ruled surface by finite sequences of its generators and constructing the geometry of finite sequences of intervals.

Here we mention the following application of the area of a surface triangle.

Proposition 9.1 (Aleksandrov (1957b)). *The area of a surface triangle T in R_K is not greater than the area of the corresponding triangle T^K on S_K and is equal to it if and only if T and T^K are isometric.*

9.4. Isoperimetric Inequality. From Proposition 9.1 we can derive the following theorem.

Theorem 9.2 (Aleksandrov (1957b)). *The area of a ruled surface spanned on a closed rectifiable curve Γ in R_K (the length of Γ in the case $K > 0$ is less than $2\pi/\sqrt{K}$) with vertex at a point on Γ is not greater than the area of a disc on S_K with length of circle equal to the length of Γ. Equality holds only when the disc and the ruled surface are isometric.*

9.5. Plateau's Problem. Let Γ be a closed curve in P_K. We denote by $S(\Gamma)$ the set of those parametrized surfaces in P_K whose restriction to the boundary of the disc B is a closed curve equivalent in the sense of Fréchet to the curve Γ. We introduce the notation

$$a(\Gamma) = \inf\{\sigma(f)\}, \quad f \in S(\Gamma).$$

Plateau's problem is posed as follows: in the class $S(\Gamma)$ is there a surface whose area is equal to $a(\Gamma)$? (A minimal surface.)

Theorem 9.3 (Nikolaev (1979)). *Suppose that P_K is finitely compact (that is, in P_K each bounded closed set is compact), its diameter is less than $\pi/2\sqrt{K}$ when $K > 0$, and Γ is a closed Jordan curve in P_K. Then there is an $f \in S(\Gamma)$ such that $\sigma(f) = a(\Gamma)$.*

Remark. By Theorem 9.2, $a(\Gamma) < +\infty$.

9.5.1. Plan of the Proof of Theorem 9.3.
The proof relies on an analogue of the Busemann-Feller theorem (about projection onto a convex body in Euclidean space) and the lemma based on it about the cutting of "crusts". The latter consists in the fact that from each surface of a "minimizing" sequence of surfaces by means of "projection" onto a ball in P_K we "cut off" those "crusts" that interfere with the equicontinuity of the surfaces. The area of each surface is not increased, by virtue of the Kolmogrov principle and the generalized Busemann-Feller theorem. Applying Arzelà's theorem, we obtain the required surface.

§ 10. Spaces of Curvature both $\leqslant K$ and $\geqslant K'$

Here we consider spaces of curvature $\leqslant K$ for which we assume in addition that their curvature $\geqslant K'$, where $K' \leqslant K$. For such surfaces, of couse, all the properties of spaces of curvature $\leqslant K$ are satisfied, and the lower bound on the curvature implies the existence of properties that are in a certain sense opposite to those that hold by virtue of the upper boundedness of the curvature. For example, together with the property of K-concavity (5.2.2) there is the property of K'-convexity (10.2.1) and so on. Section 10 is devoted to the presentation of these properties. We should mention that the results presented here are a basis for obtaining significantly stronger assertions for spaces of curvature both $\leqslant K$ and $\geqslant K'$ in what follows. Namely, it turns out that these spaces "almost" coincide with ordinary Riemannian manifolds.

In conclusion we point out that it is possible to define and investigate spaces whose curvature is only bounded below. We shall touch on this question in § 11 (11.3).

10.1. Definition of a Space of Curvature both $\leqslant K$ and $\geqslant K'$. Let K and K' be real numbers with $K' \leqslant K$. A metric space is called a space of curvature both $\leqslant K$ and $\geqslant K'$ if each point of it has a neighbourhood G for which the following conditions are satisfied:

a) any two points of G are joined by a shortest curve, and this shortest curve lies in G;

b) for any triangle T in G its K- and K'-excesses have the bounds

$$\bar{\delta}_K(T) \leqslant 0, \quad \bar{\delta}_{K'}(T) \geqslant 0;$$

c) when $K > 0$ the perimeter of each triangle of G is less than $2\pi/\sqrt{K}$.

Any neighbourhood of a metric space for which conditions a), b), c) are satisfied is called a domain $R_{K',K}$.

Remark. In b) the lower boundedness of the curvature should have been defined in terms of strong lower angles, and not upper angles (see 11.3). However, in the case that we consider this is not important, because from the upper boundedness of the curvature there follows the existence of an angle in the strong sense (see Proposition 5.2) and so the upper angle coincides with the strong lower angle.

10.2. Basic Properties of a Domain $R_{K',K}$. A domain $R_{K',K}$ is, in particular, a domain R_K, and the results of § 5 are satisfied for it. In particular, in $R_{K',K}$ between any shortest curves there is a strong angle. The angle comparison theorem in $R_{K',K}$ follows immediately from Theorem 2.2.

Theorem 10.1 (Aleksandrov (1957b)). *The angle α of any triangle T in $R_{K',K}$ has the bounds*

$$\alpha_{K'} \leqslant \alpha \leqslant \alpha_K,$$

where $\alpha_{K'}$ and α_K are the corresponding angles of the triangles $T^{K'}$ and T^K on $S_{K'}$ and S_K respectively.

Using Theorem 10.1, we can prove the following lemma:

Lemma 10.1 (Berestovskij (1975)). *For any three shortest curves in $R_{K',K}$ with common starting point the sum of the angles between pairs of them does not exceed 2π.*

It follows from Lemma 10.1 that in $R_{K',K}$ the sum of adjacent angles is not greater than π. Taking account of Proposition 2.2, we obtain the following lemma.

Lemma 10.2. *In $R_{K',K}$ the sum of adjacent angles is equal to π.*

10.2.1. *K*′-Convexity. The angle comparison theorem in $R_{K',K}$ and Lemma 10.2 enable us to prove the following theorem.

Theorem 10.2 (Aleksandrov (1957b)). *For any two shortest curves L and M in $R_{K',K}$, starting from a common point O, the angle $\gamma_{LM}^{K'}(x, y)$ (see 1.4) is a non-increasing function of x and y (that is, $\gamma_{LM}^{K'}(x, y) \geqslant \gamma_{LM}^{K'}(x_0, y_0)$ when $x \leqslant x_0$, $y \leqslant y_0$).*

The proof of Theorem 10.2 is completely analogous to the proof of Theorem 5.2. We only need to mention the following:

We keep the notation of Theorem 5.2. Then the angles α' and β' of the triangles $T_1^{K'}$ and $T_2^{K'}$ at the vertex Y_1' are not greater than the angles α and β in T_1 and T_2 at the vertex Y_1 (Theorem 10.1). By Lemma 10.2, $\pi = \alpha + \beta \geqslant \alpha' + \beta'$ and the quadrangle Q formed from $T_1^{K'}$ and $T_2^{K'}$ has angle at Y_1' not greater than π (see Fig. 9). Straightening the polygonal line $O'Y_1'Y'$, we obtain the required bound.

10.2.2. Consequences of *K*′-Convexity. We shall give just two direct consequences of K'-convexity.

Proposition 10.1. *Let X and Y be points on the sides AB and AC of the triangle $T = ABC$ in $R_{K',K}$ and X' and Y' the corresponding points on the sides of $T^{K'} = A'B'C'$ (that is, $A'X' = AX$, $A'Y' = AY$). Then $XY \geqslant X'Y'$.*

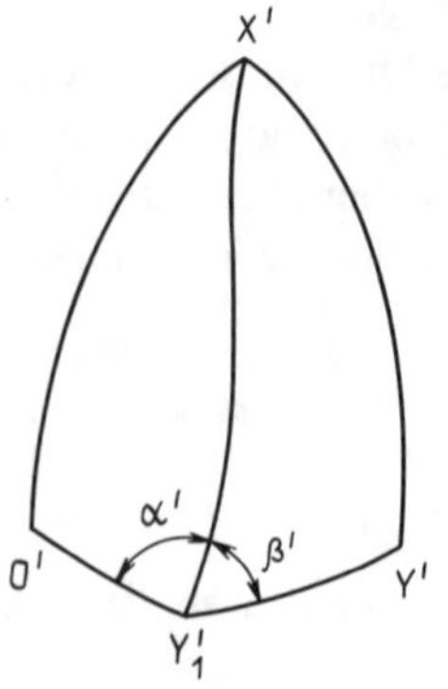

Fig. 9

From Proposition 10.1 it follows that in $R_{K',K}$ the *non-overlapping condition for shortest curves* is satisfied:

Proposition 10.2. If $AC \supset AB$ and $AC_1 \supset AB$ *are satisfied for shortest curves in* $R_{K',K}$, *then either* $AC_1 \subseteq AC$ or $AC \subseteq AC_1$.

10.3. Equivalent Definitions of Boundedness of Curvature. We give variants of restrictions on the curvature, similar to those in § 7, for spaces of curvature both $\leqslant K$ and $\geqslant K'$.

Theorem 10.3. *Let* $\mathfrak{M}$ *be a metric space in which points are locally joined by shortest curves. Then the following assertions are equivalent*:

1) $\mathfrak{M}$ *is a space of curvature both* $\leqslant K$ *and* $\geqslant K'$;

2) *for each point* $P \in \mathfrak{M}$ *and each sequence of triangles* T_m *of* $\mathfrak{M}$ *that contract in an arbitrary way to* P *we have*

$$K' \leqslant \varliminf_{T_m \to P} \bar{\delta}(T_m)/\sigma(T_m) \leqslant \varlimsup_{T_m \to P} \bar{\delta}(T_m)/\sigma(T_m) \leqslant K$$

(*in the case when* $\sigma(T_m) = 0$ *it is required that* $\bar{\delta}(T_m) = 0$);

3) *each point of* $\mathfrak{M}$ *has a neighbourhood, any quadruple of points of which is isometrically embedded in* $S^3_{K_0}$ *with* $K' \leqslant K_0 \leqslant K$, *depending on the chosen quadruple*;

4) *in* $\mathfrak{M}$ *the condition of* K*-concavity and* K'*-convexity is locally satisfied.*

Remark. The proof of the equivalence of 1) and 3) is contained in Berestovskij (1986).

10.3.1. Comparison of the Angles of Triangles in $R_{K',K}$ with the Angles of the Corresponding Euclidean Triangles (Aleksandrov, Berestovskij and Nikolaev (1986)).

Proposition 10.3. *For angles of the triangle* T *in* $R_{K',K}$ *we have the bound*

$$|\alpha - \alpha_0| \leqslant \mu \cdot \sigma(T), \tag{10.1}$$

where μ *is a positive constant that depends on* K' *and* K.

§ 11. Remarks, Examples

11.1. Spaces of Curvature $\leqslant K$ as a Generalization of Riemannian Spaces. By Corollary 7.1, any Riemannian manifold whose sectional curvatures at all points in all two-dimensional directions have the bound $K_\sigma \leqslant K$ is a space of curvature $\leqslant K$ in the sense of Aleksandrov. At the same time, in the definition of a space of curvature $\leqslant K$ there are no requirements on the smoothness of the metric, moreover, the space may not be a manifold; see the example in 6.1.2 (or it may be a manifold, but infinite-dimensional). Despite this, spaces of curvature $\leqslant K$ have a number of important properties in common with Riemannian spaces; a local angle comparison theorem for triangles, the existence of angles between shortest curves, the local uniqueness of shortest curves, and so on. We observe that shortest curves in a space of curvature $\leqslant K$ may "ramify", that is, coincide on a certain part, and then diverge. For example, the shortest curves AOC and AOD in Fig. 5 behave in this way. This effect occurs also for shortest curves on a cone with total angle at the vertex $> 2\pi$, passing through the vertex of the cone. Such a cone is a space of curvature $\leqslant 0$.

11.1.1. Spaces of Curvature $\leqslant K$ and Finsler Spaces.

The following assertion was mentioned by Aleksandrov in Aleksandrov (1957a), p. 7:

A Finsler manifold of curvature $\leqslant K$ is a Riemannian manifold.

Thus, the essence of the restrictions on the curvature of a space of curvature $\leqslant K$ is not in the differential properties of the metric (the metric tensor of a Finsler space may even be infinitely differentiable), but in its infinitesimal properties.

11.1.2. The Domain R_K and the Radius of Injectivity of a Riemannian Manifold. We recall that the radius of injectivity $i_{\mathfrak{M}}(P)$ of a Riemannian manifold $\mathfrak{M}$ at a point $P \in \mathfrak{M}$ is the supremum of all δ such that the map $\exp_P$, contracted onto the ball in $\mathfrak{M}_P$ with centre at the origin and radius δ, is injective. The number $i(\mathfrak{M})$, which is equal to $\inf\{i_{\mathfrak{M}}(P)\}$, $P \in \mathfrak{M}$, is called the *radius of injectivity* of the manifold $\mathfrak{M}$.

Proposition 11.1. *Let $\mathfrak{M}$ be a complete Riemannian manifold for which*:

1) *the radius of injectivity $i(\mathfrak{M})$ is positive*;

2) *the sectional curvatures of $\mathfrak{M}$ at all points in all two-dimensional directions have the bound $K_\sigma \leqslant K$.*

Then an open ball in $\mathfrak{M}$ with centre at an arbitrary point $P \in \mathfrak{M}$ and radius r_0, equal to $i(\mathfrak{M})/4$ when $K \leqslant 0$ and $\min\{\pi/3\sqrt{K}, i(\mathfrak{M})/4\}$ when $K > 0$, is a domain R_K.

Remark. In the case when the curvature of $\mathfrak{M}$ is additionally subject to the restriction $K_\sigma \geqslant K'$, the ball mentioned in Proposition 11.1 is a domain $R_{K',K}$.

The proof of Proposition 11.1 consists in the following.

Obviously any two points of $B(P, 2r_0)$ are joined by a unique shortest curve. But then, in view of the completeness of $\mathfrak{M}$, the shortest curves with ends in $B(P, 2r_0)$ depend continuously on their ends. Obviously the perimeter of any triangle with vertices in $B(P, r_0)$ is less than $2\pi/\sqrt{K}$ when $K > 0$, and any shortest curve with ends in $B(P, r_0)$ lies in $B(P, 2r_0)$. By virtue of the conditions on the curvature of $\mathfrak{M}$ and Proposition 7.1, we deduce that for each triangle with vertices in $B(P, r_0)$ the angle comparison theorem holds for the triangle. Hence $B(P, r_0)$ is a domain R_K.

We mention that from some bounds on the curvature it is impossible to obtain a lower bound for the radius of injectivity, as, for example, on an arbitrarily "narrow" torus we can specify a planar metric ($K \equiv 0$). The requirements that $\mathfrak{M}$ is simply-connected are also insufficient. Namely, there is an example of a compact Riemannian manifold diffeomorphic to a three-dimensional sphere for which all sectional curvatures lie within the limits $0 < K_\sigma \leqslant 1$ and on which there is a closed geodesic of length $< 2\pi$, that is, in the manifold there is a triangle with perimeter less than 2π and with angles equal to π (Gromoll, Klingenberg and Meyer (1968), § 6.4, Remark 4) and the angle comparison theorem is not true in the large on $\mathfrak{M}$, that is, $\mathfrak{M}$ is not R_1.

However, if there are restrictions on the volume, the diameter and the curvature of a compact Riemannian manifold, then a positive bound on the radius of injectivity exists.

Proposition 11.2 (see in Peters (1986)). *For each compact Riemannian manifold $\mathfrak{M}$ for which*

1) *the diameter of $\mathfrak{M}$ is not greater than d;*

2) *for the sectional curvatures of $\mathfrak{M}$ at all points in all two-dimensional directions we have $|K_\sigma| \leqslant C_0$;*

3) *the volume of $\mathfrak{M}$ has the bound $\mathrm{Vol}(\mathfrak{M}) \geqslant V_0 > 0$, the radius of injectivity $i(\mathfrak{M})$ has the bound $i(\mathfrak{M}) \geqslant i_0$, where i_0 is a positive constant depending only on d, C_0, V_0 and* $\dim \mathfrak{M}$.

11.1.3. Spaces of Curvature $\leqslant K$ and Limits of Riemannian Metrics. In many cases metrics of curvature $\leqslant K$ in the sense of Aleksandrov arise as the limit of Riemannian metrics with curvatures $\leqslant K$, for example, such is the metric of a cone with total angle at the vertex $> 2\pi$. We give conditions under which the limit of Riemannian metrics of curvature $\leqslant K$ is a metric of curvature $\leqslant K$. Let us first recall the following concepts.

We say that a Riemannian metric ρ is specified on a set $\mathfrak{M}$ if the metric space $(\mathfrak{M}, \rho)$ is isometric to some Riemannian manifold $\mathfrak{M}_\rho$ with respect to its standard intrinsic metric; see 1.1.

A space is said to be *finitely compact* if each set in it that is bounded in the sense of the metric of this space has compact closure. For Riemannian manifolds completeness is equivalent to finite compactness.

Proposition 11.3. *We assume that on the set $\mathfrak{M}$ there are specified Riemannian matrics ρ_m defined in $\mathfrak{M}$ by "one and the same topology", and that the following conditions are satisfied:*

1) *the metric space $(\mathfrak{M}, \rho_m)$ is finitely compact;*

2) *the radius of injectivity $i_m(\mathfrak{M})$ of $\mathfrak{M}$ with respect to the Riemannian metric ρ_m has the bound*

$$i_m(\mathfrak{M}) \geqslant i_0 > 0;$$

3) *the sectional curvatures of $\mathfrak{M}_{\rho_m}$ at all points in all two-dimensional directions have the bound $K_\sigma^m \leqslant K$;*

4) *the metrics ρ_m converge to the metric ρ uniformly on each compact set of $\mathfrak{M}$.*

Then the limit space $(\mathfrak{M}, \rho)$ is a space of curvature $\leqslant K$ in the sense of Aleksandrov.

Remark. If in Proposition 11.3 we replace 3) by the condition $K' \leqslant K_\sigma^m \leqslant K$, then in the limit we obtain a space of curvature both $\leqslant K$ and $\geqslant K'$.

The proof of Proposition 11.3 consists in the following.

By Lemma 7 of Aleksandrov and Zalgaller (1962) (§ 1 of Chapter 4) the metric ρ of the space $(\mathfrak{M}, \rho)$ is intrinsic and any two points in $(\mathfrak{M}, \rho)$ can be joined by a shortest curve. By Proposition 11.1 the ball $B_m(P, r_0) \subseteq (\mathfrak{M}, \rho_m)$, where $r_0 = i_0/4$ when $K \leqslant 0$ and $\min\{i_3/4, \pi/3\sqrt{K}\}$ when $K > 0$, is a domain R_K. Hence the condition of K-concavity is satisfied in $B_m(P, r_0)$ (see 5.2.2). By means of a limiting process we deduce that the property of K-concavity is satisfied for the ball $B(P, r_0)$ in $(\mathfrak{M}, \rho)$. But then $B(P, r_0)$ is a domain R_K in $(\mathfrak{M}, \rho)$ (see 7.3). Since the point P is arbitrary, we deduce that $(\mathfrak{M}, \rho)$ is a space of curvature $\leqslant K$.

In conclusion we mention the following: those spaces of curvature $\leqslant K$ that are two-dimensional topological manifolds are special cases of the well studied two-dimensional manifolds of bounded integral curvature (Aleksandrov and Zalgaller (1962)). From the results of that work (see also Part I of this volume) it follows that the metric of these spaces is in turn the limit of smooth Riemannian metrics of curvature $\leqslant K_m$, where $K_m \to K$ as $m \to \infty$. It is not known whether a similar assertion is true for the multidimensional case.

11.2. Polyhedral Metrics. We can say that a metric is polyhedral if it is obtained by "gluing" finitely many l-dimensional polyhedra in S_K^l along isometric $(l-1)$-dimensional faces.

For two-dimensional polyhedral metrics a simple criterion holds: a polyhedron formed from triangles S_K is a space of curvature $\leqslant K$ if and only if the sum of its angles at each vertex not lying on the boundary is $\geqslant 2\pi$. We give an example showing that for multidimensional polyhedral metrics the situation is not so simple.

11.2.1. One Example of a Three-dimensional Polyhedral Metric. We give a slightly modified example from Ionin (1972).

In E_3 we consider a set R' included between a tetrahedron $S' = A'B'C'D'$ and a tetrahedron $S_1' = A_1'B_1'C_1'D_1'$, whose base lies inside the triangle $A'B'C'$ and whose vertex D_1' is strictly inside S_1', with intrinsic metric induced from E_3. We now consider the set R obtained as a result of "doubling" the set R' with intrinsic metric defined in a similar way, that is, we consider a second copy R'' of the space R' and "glue together" R' and R'' along the congruent tetrahedra $D_1'A_1'B_1'C_1'$ and $D_1''A_1''B_1''C_1''$ (see Fig. 10).

We denote the glued vertices A_1', B_1', C_1', D_1' and A_1'', B_1'', C_1'', D_1'' by A_1, B_1, C_1, D_1 respectively. Obviously R has the following properties:

1) R can be obtained by gluing finitely many tetrahedra in E_3, that is, the intrinsic metric of R is a three-dimensional polyhedral metric;

2) with respect to the intrinsic metric arising as a result of the gluing, R is homeomorphic to a closed ball in E_3;

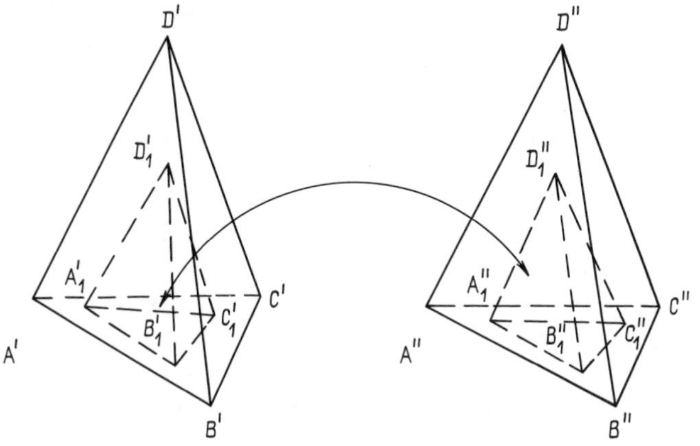

Fig. 10

3) the sum of the dihedral angles at each of the edges A_1D_1, B_1D_1, C_1D_1 is at least 2π, and the sum of the solid angles at each of the vertices D_1, A_1, B_1, C_1 is greater than the area of a unit sphere, that is, 4π.

However, R is not a space of curvature $\leqslant 0$ (and is not a space of curvature $\leqslant K$ for any K) in the sense of Aleksandrov, since in an arbitrarily small neighbourhood of the vertex D_1' there are pairs of points that are joined by two shortest curves.

11.3. Spaces of Curvature $\geqslant K'$. We can define a space whose curvature is bounded below: a space of curvature $\geqslant K'$. In order that the angle comparison theorem should be fulfilled, the lower bound on the curvature must be imposed in terms of strong lower angles (see Example 4.3): each point of $\mathfrak{M}$ has a neighbourhood G such that the strong lower excess of an arbitrary triangle of G is non-negative: $\underline{\alpha}_s + \underline{\beta}_s + \underline{\gamma}_s - (\alpha_{K'} + \beta_{K'} + \gamma_{K'}) \geqslant 0$ (Aleksandrov and Berestovskij (1984), Aleksandrov, Berestovskij and Nikolaev (1986)).

As in Theorem 5 on p. 28 of Aleksandrov and Zalgaller (1962), we can prove the inequalities $\underline{\alpha} \geqslant \alpha_{K'}$, $\underline{\beta} \geqslant \beta_{K'}$, $\underline{\gamma} \geqslant \gamma_{K'}$ for the lower angles $\underline{\alpha}$, $\underline{\beta}$, $\underline{\gamma}$ and the property of non-overlapping of shortest curves (see Proposition 10.2). In Aleksandrov and Berestovskij (1984) and Aleksandrov, Berestovskij and Nikolaev (1986) (§ 8) it was erroneously asserted that there is an angle in the strong sense between shortest curves lying in G (see Example 4.2, and also Aleksandrov (1948), pp. 55–56). Under certain additional conditions (for example, assuming that in G the sum of lower adjacent angles does not exceed π) we can prove that the metric of G is K'-convex (for the definition see 10.2.1), so it folows that there is an ordinary angle between arbitrary shortest curves lying in G. Condition (A_2^+) in Aleksandrov, Berestovskij and Nikolaev (1986), § 8, is actually equivalent to the K'-convexity of the metric. It is easy to prove that conditions (A^+) and (A_1^+) are equivalent if we add to what we said above for G the condition that shortest curves lying in G depend continuously on their ends. The authors do not know whether the conditions imposed on G in addition to the conditions on the excesses of triangles in G are essential.

11.3.1. Global Lower Bound on the Angles of Triangles in a Space of Curvature $\geqslant K'$. As we saw above, a bound for the angles of a triangle $\alpha \geqslant \alpha_K$ and so on is fulfilled in the general case only for sufficiently small triangles. It is very important, however, to know conditions that guarantee the fulfilment of such a bound for all triangles of a space of curvature $\geqslant K'$.

For two-dimensional metrized manifolds the corresponding assertion was first obtained by Aleksandrov (Aleksandrov (1948)). In particular, by Aleksandrov's theorem on complete convex surfaces that are spaces of curvature $\geqslant 0$ the bound $\alpha \geqslant \alpha_0$ is satisfied for angles of arbitrary triangles.

In the multidimensional case a generalization of Aleksandrov's theorem holds only for smooth Riemannian manifolds. Namely, a theorem due to Toponogov (Toponogov (1959) asserts that the required bound on the angles of a triangle is satisfied in the large in complete simply-connected Riemannian manifolds whose sectional curvatures have the bound $K_\sigma \geqslant K'$ at all points for all plane elements σ in the tangent space at the corresponding point.

We observe that the proposition that Toponogov's theorem can be extended to the case of the spaces with bounded curvature considered in Chapter 3 is very likely (for this see Theorem 15.1).[1]

Chapter 3
Spaces with Bounded Curvature

In Chapter 3 we continue the study of properties of spaces of curvature both $\leqslant K$ and $\geqslant K'$. The main question of interest to us here is the comparison of these spaces with classical Riemannian spaces. We naturally restrict our consideration to that of metrics of curvature both $\leqslant K$ and $\geqslant K'$ specified on finite-dimensional topological manifolds. For this we add to the axioms of a space of curvature both $\leqslant K$ and $\geqslant K'$ the conditions of local compactness and local extendability of a shortest curve (for the definition see 12.1). As we shall show, these conditions are necessary and sufficient for a space of curvature both $\leqslant K$ and $\geqslant K'$ to be a topological manifold of finite dimension. Spaces of curvature both $\leqslant K$ and $\geqslant K'$ with the properties stated above we shall call spaces with bounded curvature.

The spaces with bounded curvature that we investigate turn out to be Riemannian, but in a certain generalized sense in comparison with the generally accepted point of view: in accordance with 1.1 their intrinsic metric is specified

[1] After our article had been published, the general Toponogov theorem for spaces of curvature bounded below appeared in a paper by Yu. Burago, M. Gromov and G. Perel'man (Russian Math. Survey 47: 2 (1992), 1–58).

by a metric tensor; however, the differential properties of this tensor, generally speaking, are worse than those usually required in Riemannian geometry (in the general case the metric tensor of a space with bounded curvature is not twice continuously differentiable). Our main results are concerned with determining the differential properties of the metric of spaces with bounded curvature. The results obtained in this direction enable us to assert that the metric of a space with bounded curvature is in a certain natural sense the limit of smooth Riemannian metrics (for example, of class C^∞) whose sectional curvatures at all points in all two-dimensional directions are bilaterally uniformly bounded.

§ 12. C^0-Riemannian Structure in Spaces with Bounded Curvature

In this section we show that, as in the case of classical Riemannian manifolds, the intrinsic metric of a space with bounded curvature is specified by means of a certain metric tensor. There will be no question of the differential properties of this tensor. Here we just mention that the metric tensor of a space with bounded curvature is continuous. Hence in a space with bounded curvature we specify the structure of a C^0-smooth Riemannian manifold.

12.1. Definition of a Space with Bounded Curvature. We shall define a space with bounded curvature as a metric space $\mathfrak{M}$ with intrinsic metric in which the following axioms are satisfied:

a) $\mathfrak{M}$ is a locally compact metric space;

b) the condition of *local extendability of a shortest curve* is satisfied in $\mathfrak{M}$: for each point $P \in \mathfrak{M}$ some open ball $B(P, r_P)$ has the property that any shortest curve XY with ends $X, Y \in B(P, r_P)$ can be extended to a shortest curve $X_1 Y_1$ in $\mathfrak{M}$ for which X and Y are internal points;

c) the curvature of $\mathfrak{M}$ is locally bounded in the sense of Aleksandrov: each point $P \in \mathfrak{M}$ is contained in some neighbourhood $\mathscr{U} \subseteq \mathfrak{M}$ that is a domain $R_{K',K}$ (see 10.1) for some K', K (where $K' \leqslant K$) that depend on $\mathscr{U}$ (or equivalently conditions 2) and 3) of Theorem 10.3 are satisfied).

Remark. From conditions a), b), c) it will follow (see Theorem 12.1) that a space with bounded curvature is a topological manifold of finite dimension. We note that conversely a space of curvature both $\leqslant K$ and $\geqslant K'$ that is a topological manifold of finite dimension satisfies conditions a), b), c), that is, it is a space with bounded curvature. Obviously in the verification we need only condition b), whose fulfilment follows from Proposition 8.3.

Example 12.1 (Peters (1986)). Part of a circular cylinder in E_3 closed by two hemispheres, considered with the intrinsic metric induced from E_3, is a space with bounded curvature (see Fig. 11). The components of the metric tensor of this space in some coordinate system are not twice continuously differentiable functions (otherwise the space under consideration would have constant curva-

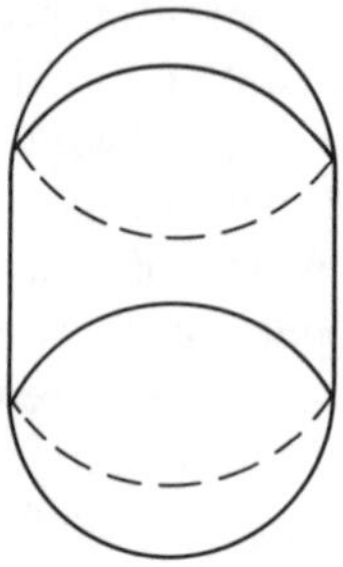

Fig. 11

ture). Moreover, it is easy to construct a coordinate system in which the components of the metric tensor are differentiable and their first derivatives satisfy a Lipschitz condition. Thus, the space with bounded curvature under consideration is a Riemannian manifold of class $C^{1,1}$ but not C^2.

12.2. The Tangent Space at a Point of a Space with Bounded Curvature. In 8.3 we mentioned that the curvature of the tangent space $\mathfrak{M}_P$ at a point P of a space of curvature $\leqslant K$ that is a topological manifold is not greater than zero. For spaces with bounded curvature, using in addition the lower boundedness of their curvature, we can deduce that the curvature of $\mathfrak{M}_P$ is not less than zero. But then $\mathfrak{M}_P$ is a space of zero curvature. Thus we have the following result.

Proposition 12.1 (Berestovskij (1975)). *The tangent space at an arbitrary point P of a space with bounded curvature is isometric to a finite-dimensional Euclidean space.*

12.2.1. Exponential Map. The results of 12.2 enable us to define an exponential map in spaces with bounded curvature.

Consider the ball $B(P, r)$, $P \in \mathfrak{M}$, where r $(0 < r < r_P)$ is so small that the closed ball $\bar{B}(P, r)$ is compact. Then we can define the map

$$\exp_P\colon B(0, r) \subset \mathfrak{M}_P \to B(P, r) \subseteq \mathfrak{M}$$

(here $B(0, r)$ is the ball in the Euclidean space $\mathfrak{M}_P$ of radius r with centre at the origin), which associates with each element $[D, t]$ of $B(0, r)$, where $D \in \Omega_P\mathfrak{M}$, $0 \leqslant t \leqslant r$ (for the definition of a tangent element see 3.3) the end of the unique shortest curve of length t that specifies the direction D (see Proposition 8.2). The uniqueness of the shortest curve is guaranteed by the non-overlapping of shortest curves (Proposition 10.2) and Theorem 10.1 (comparison of angles). By part b) of the definition of a space with bounded curvature, any shortest curve with origin at P can be extended at least to length $r_P > t$. The map $\exp_P$ is obviously a bijective map of the ball $B(0, r)$ onto $B(P, r)$. From the results of § 8 we can deduce that $\exp_P$ is a continuous map, and so by what we said above it is a homeomorphism. We thus arrive at the following theorem.

Theorem 12.1 (Berestovskij (1975)). *In a space $\mathfrak{M}$ with bounded curvature all the tangent spaces $\mathfrak{M}_P$, $P \in \mathfrak{M}$, are Euclidean spaces of the same finite dimension. $\mathfrak{M}$ is a topological manifold of the same dimension.*

Remark. If in the definition of a space with bounded curvature we reject local compactness, then we can assert that $\mathfrak{M}$ is an infinite-dimensional manifold (Berestovskij (1981)).

12.3. Introduction of C^0-Smooth Riemannian Structure.
12.3.1. Statement of the Main Result.

Theorem 12.2 (Berestovskij (1975)). *A space $\mathfrak{M}$ with bounded curvature is a C^0-smooth Riemannian manifold of finite dimension.*

More precisely, Theorem 12.2 asserts the following.

1) In a neighbourhood of each point of $\mathfrak{M}$ we can construct a coordinate system of special form – a distance coordinate system (12.3.2) – and the distance coordinates specify on $\mathfrak{M}$ the structure of a C^1-smooth differentiable manifold.

2) On $\mathfrak{M}$ we can specify a symmetric positively defined tensor field g of type (2, 0) of class C^0. In the distance coordinate system g is written as a matrix $(g_{ij}(x))_{i,j=1,\dots,n}$, whose coefficients are continuous. For the length of an arbitrary C^1-smooth curve γ we have the standard formula

$$l(\gamma) = \int_\alpha^\beta \left[\sum_{i,j=1}^n g_{ij}(\gamma(t)) \cdot \dot{\gamma}^i(t) \cdot \dot{\gamma}^j(t) \right]^{1/2} dt, \tag{12.1}$$

where $\gamma(t)$, $\alpha \leqslant t \leqslant \beta$, is a parametrization of the curve γ, and $\dot{\gamma}$ is the field of vectors tangent to γ with respect to the parametrization $\gamma(t)$.

3) The distance between points P, $Q \in \mathfrak{M}$ in the original metric coincides with the greatest lower bound of the lengths of all C^1-smooth curves joining P and Q, calculated from (12.1), that is, it is equal to the infimum of the integrals in (12.1).

Remark 1. In fact the distance coordinates specify on $\mathfrak{M}$ the structure of a $C^{1,1}$-smooth differentiable manifold, and the components of the metric tensor in the distance coordinate system $g_{ij}(x)$ satisfy a Lipschitz condition: $g_{ij} \in C^{0,1}$, see Theorem 13.2.

12.3.2. Distance Coordinates and the Metric Tensor. We denote by n the dimension of a space $\mathfrak{M}$ with bounded curvature. We shall say that points $P_0, P_1, \dots, P_n \in R_{K',K} \subset \mathfrak{M}$ are in general position if the vectors are linearly independent.

Suppose Q is an arbitrary point and the points P_i, $i = 0, 1, \dots, n$, are in general position. We denote $\rho(Q, P_i)$ by $u^i(Q)$, $i = 1, 2, \dots, n$. We can prove that the map $u = (u^1, u^2, \dots, u^n)$ maps a small neighbourhood of P_0 homeomorphically onto a domain of E_n, that is, it is a local coordinate system in $\mathfrak{M}$ in a neighbourhood of P_0. We call a coordinate system of this kind a *distance coordinate system*.

We denote by $A(Q)$ the matrix

$$A(Q) = (\cos \measuredangle (X_{P_i}(Q), X_{P_j}(Q)))_{i,j=1,2,\ldots,n},$$

where $\measuredangle$ denotes the angle between corresponding vectors, and $Q = P_0$. We can verify directly (Berestovskij (1975)) that the distance coordinates are connected to each other by transition functions of class C^1. Direct calculation shows that in distance coordinates the components of the metric tensor, calculated for the standard formulae, have the form

$$G(u^1, u^2, \ldots, u^n) = [A(u^{-1}(u^1, \ldots, u^n))]^{-1}. \tag{12.2}$$

Thus the components of the metric tensor in distance coordinates are continuous functions. Relying on (10.1), by means of (12.2) we can prove that formula (12.1) and 3) hold.

§ 13. Parallel Translation in Spaces with Bounded Curvature

Parallel translation is concerned with a number of fundamental concepts of differential geometry. By specifying on a manifold the operation of parallel translation of vectors, we obtain the corresponding geometry on the manifold. In Riemannian geometry a metric is an original concept, and a parallel translation consistent with the Riemannian structure is restored uniquely from the metric (the Levi-Cività theorem).

In this section, by means of a purely geometrical construction we construct an operation that subsequently turns out to be a parallel translation uniquely corresponding to the metric.

In the case of a classical Riemannian manifold our construction gives the parallel translation of Levi-Cività.

13.1. Construction of a Parallel Translation (Nikolaev (1980), (1983)). We shall carry out all the constructions in a sufficiently small convex domain $\mathscr{U} \subseteq \mathfrak{M}$, for which the bounds on the dimensions will be stated in the course of the presentation.

At the basis of the construction of a parallel translation lies the symmetry map.

Symmetry with respect to a point $O \in \mathfrak{M}$ is the map $\mathscr{S}_0 \colon \mathscr{U} \to \mathfrak{M}$ that sets up a correspondence between a point $P \in \mathscr{U}$ and the point $P' \in \mathfrak{M}$ lying on the extension of the shortest curve PO beyond O so that $P'O = PO$. The points P and P' are said to be symmetrical with respect to the point O. By virtue of the axioms of a space with bounded curvature, the symmetry map is well defined in a small domain $\mathscr{U} \subseteq \mathfrak{M}$.

Consider a shortest curve $AA' \subset \mathscr{U}$ and the tangent spaces to $\mathfrak{M}$ at A and A'. We construct a map $\Pi_{AA'}$ of the ball $B(O, \varepsilon) \subset \mathfrak{M}_A$ of radius $\varepsilon > 0$ with centre at the origin (in $\mathfrak{M}_{A'}$). Let $\xi \in B(O, \varepsilon)$. From A we draw the shortest curve AH in

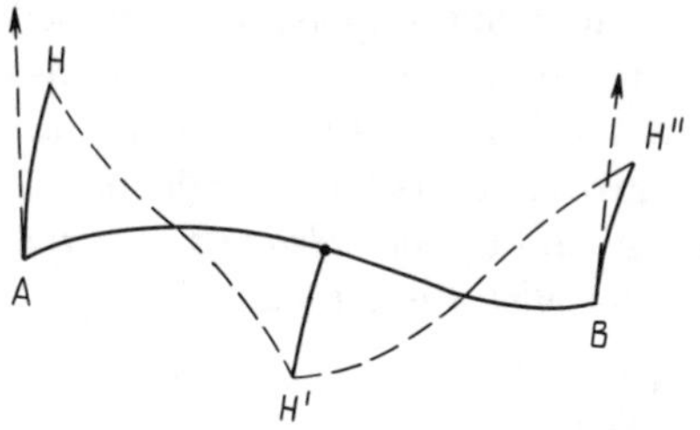

Fig. 12

the direction of the vector ξ and of length $|\xi|$ ($H = \exp_A \xi$). Let H' be the point symmetrical to H about the midpoint of the shortest curve AA'. Then $\Pi_{AA'}(\xi) = \exp_{A'}^{-1} H'$. If the domain $\mathscr{U}$ is sufficiently small, and ε is close to zero, then all our constructions are obviously possible.

Now let AB be an arbitrary shortest curve lying in $\mathscr{U}$. We split it into 2^m equal intervals by points $A_0 = A, A_1, \ldots, A_{2^m} = B$. We denote the map $\Pi_{A_l A_{l+1}}$ constructed above by $\Pi_{l,l+1}$, and put $h_m = AB/2^m$.

Let us define a map Π_m: $\mathfrak{M}_A \to \mathfrak{M}_B$. Let $\xi \in \mathfrak{M}_A$ be an arbitrary vector; we put $\xi' = h_m \xi/|\xi|$. Consider the map

$$\Pi'_m(\xi) = \Pi_{2^m-1,2^m} \circ \cdots \circ \Pi_{1,2} \circ \Pi_{0,1}(\xi')$$

(for $m = 1$ see Fig. 12). When $m \geqslant 1$ we take the vector $\Pi_m(\xi)$ equal to a vector of length $|\xi|$ directed like the vector $\Pi'_m(\xi)$ (when $m = 0$ the direction of $\Pi'_m(\xi)$ is changed to the opposite). By means of (10.1) we can show that for each $\xi \in \mathfrak{M}_A$ the sequence $\Pi_m(\xi)$ has a limit, which we denote by $\Pi(\xi)$ (Nikolaev (1983a)). We call the resulting map Π: $\mathfrak{M}_A \to \mathfrak{M}_B$ a *parallel translation* along the shortest curve AB. After this a parallel translation is defined along any rectifiable curve in the space with bounded curvature.

Remark. E. Cartan considered symmetry in smooth Riemannian manifolds, and by means of Cartan symmetry he interpreted parallel translation in Riemannian manifolds (Cartan (1928), pp. 205, 206)). Our construction is based on this interpretation of Cartan.

13.2. Statement of the Main Results. The geometrically introduced parallel translation Π preserves the scalar product of vectors, that is, it is Riemannian. Namely, we have the following result.

Proposition 13.1 (Nikolaev (1983a)). *In a space with bounded curvature a parallel translation Π along an arbitrary rectifiable curve is an isometric map of the corresponding tangent spaces.*

It is known that in the smooth case the curvature tensor is expressed in terms of the ratio of the increment in the vector, on going round a closed contour lying on a surface touching a given two-dimensional direction, and the area of this surface. Therefore, in the non-smooth case the boundedness of curvature of a space with a parallel translation specified in it (in terms of the parallel translation) is naturally understood in the sense that all such ratios are bounded. Theorem 13.1 stated below connects the boundedness of curvature of a space according to Aleksandrov and the *boundedness of curvature* in the sense mentioned above *in terms of parallel translation.*

Theorem 13.1 (Nikolaev (1983b)). *Any point P in a space with bounded curvature $\mathfrak{M}$ has a neighbourhood $\mathscr{V} \subset \mathfrak{M}$ such that for an arbitrary closed rectifiable curve γ in $\mathscr{V}$ ($P \in \gamma$) the increment $\Delta\xi$ of a vector $\xi \in \mathfrak{M}_P$ under a parallel translation Π along γ has the bound*

$$|\Delta\xi| \leqslant \mu(K', K)\cdot\sigma(P, \gamma)\cdot|\xi|,$$

where $\mu(K', K)$ is a positive constant depending on K' and K, and $\sigma(P, \gamma)$ is the area of the ruled surface spanned on γ (see 9.3).

By means of the construction of a parallel translation we can prove the following theorem, which is necessary for further examination of the parallel translation Π that we have introduced.

Theorem 13.2 (Nikolaev (1983b)). *The components of the metric tensor (g_{ij}), $i, j = 1, 2, \ldots, n$, in an arbitrary distance coordinate sytem in a space with bounded curvature satisfy a Lipschitz condition, that is,*

$$|g_{ij}(x) - g_{ij}(y)| \leqslant L|x - y|,$$

where x and y belong to the range of distance coordinates, and L is a positive constant.

Let $G \subseteq \mathbb{R}^n$ be the range of distance coordinates. From Theorem 13.2 it follows that at almost all points (in the sense of Lebesgue measure) of G there is a total differential of the components of the metric tensor g_{ij} in the distance coordinate system. This gives us the possibility of defining analytically the parallel translation Π' along "almost all" differentiable curves in $\mathfrak{M}$. We say that a curve $\gamma \subset \mathfrak{M}$ is differentiable if it is specified in the distance coordinates by functions of class C^1. We assume that the differentiable curve γ is such that at almost all points of γ the g_{ij} are differentiable (by Theorem 13.2 this is satisfied for "almost all" differentiable curves). Then along γ in the distance coordinate system we can define the Christoffel symbols

$$\Gamma_{ij}^{l} = \frac{1}{2} g^{kl}\left(\frac{\partial g_{jk}}{\partial x^i} + \frac{\partial g_{ki}}{\partial x^j} - \frac{\partial g_{ij}}{\partial x^k}\right). \tag{13.1}$$

We now define a parallel translation along γ, $\Pi' : \mathfrak{M}_{\gamma(t_0)} \to \mathfrak{M}_{\gamma(t)}$, by the formula

$$\xi^r(t) = \xi_0^r - \int_{t_0}^{t} \Gamma_{il}^{r}(\gamma(s))\xi^i(s)\cdot\dot{\gamma}^l(s)\,ds, \tag{13.2}$$

where $\xi_0^1, \ldots, \xi_0^n$ are the coordinates of the parallel displaceable vector $\xi_0 \in \mathfrak{M}_{\gamma(t_0)}$ in the distance coordinate system, $\dot{\gamma}^1(s), \ldots, \dot{\gamma}^n(s)$ are the coordinates of the vector tangent to the curve γ for the value s of the parameter, and $(\xi^1(t), \ldots, \xi^n(t))$ is the resulting field of parallel vectors.

From the Lipschitz condition for $g_{ij}(x)$ it follows easily that the integral equation (13.2) has a unique solution $\xi^r(s)$ on condition that $\xi^r(0) = \xi_0^r$.

The next theorem connects the geometrically defined parallel translation Π and the parallel translation Π' introduced analytically.

Theorem 13.3 (Nikolaev (1983b)). *Along any differentiable curve in* $\mathfrak{M}$, *at almost all points of which the components of the metric tensor in the distance coordinate system are differentiable, the geometrically defined parallel translation* Π *and the analytically defined parallel translation* Π' *coincide.*

Remark. From Theorem 14.1 stated in § 14 it follows that by using "harmonic" coordinates we can define a parallel translation analytically (that is, by means of (13.2)) for any differentiable curve in $\mathfrak{M}$ and it also coincides with the geometrically defined parallel translation Π.

13.3. Plan of the Proof of the Main Results of § 13.

13.3.1. Isometry of the map Π. First of all we mention the properties we need for the symmetry map that lies at the basis of the construction of the map Π.

Lemma 13.1. *We consider arbitrary points* $P, Q \in \mathfrak{M}$ *and points* $P', Q' \in \mathcal{U}$ *symmetrical to them with respect to some point* $O \in \mathcal{U}$. *Then*

$$|PQ^2 - P'Q'^2| \leqslant \mu'[\max\{PO, QO\}]^4,$$

where μ' *is a positive constant depending on* K' *and* K.

Lemma 13.2. *We retain the notation of Lemma* 13.1. *We also consider a point* $A \in \mathcal{U}$ *and the point* A' *symmetrical to it with respect to* O. *Then*

$$|\cos \measuredangle PAQ - \cos \measuredangle P'A'Q'| \leqslant \mu'' \cdot M^2(1 + M^2/m^2 + M^4/m^4),$$

where $M = \max\{AO, AP, AQ\}$, $m = \min\{AP, AQ, A'P', A'Q'\}$, *and* μ'' *is a constant depending on* K' *and* K.

Lemmas 13.1 and 13.2 were proved in Aleksandrov, Berestovskij and Nikolaev (1986). They follow easily from (10.1) and the Euclidean cosine theorem.

From Lemmas 13.1 and 13.2 we obtain the following preliminary bound for $\Pi_{AA'}$.

Lemma 13.3 (Nikolaev (1983a)). *Let* $A'A$ *be a shortest curve of length* $h > 0$, L, *and* L_2 $(0 < L_1 < L_2)$ *certain fixed constants, and* $\Pi_{AA'}\colon \mathfrak{M}_A \to \mathfrak{M}_{A'}$ *the map constructed in* 13.1. *Then for arbitrary vectors* $\xi, \zeta \in \mathfrak{M}_A$ *whose lengths satisfy the inequalities*

$$h \cdot L_1 \leqslant |\xi|, \quad |\zeta| \leqslant h \cdot L_2$$

we have the bound

$$|\measuredangle(\xi, \zeta) - \measuredangle(\Pi_{AA'}(\xi), \Pi_{AA'}(\zeta))| \leqslant \mu''' \cdot h^2,$$

where μ''' is a constant depending on L_1, L_2, K' and K.

The proof of Lemma 13.3 is obtained from Lemma 13.2 by considering in $\mathfrak{M}_A$ the orthonormal basis X_i, $i = 1, 2, \ldots, n$, and in $\mathfrak{M}_{A'}$ the basis $X_i' = \Pi_{AA'}(X_i)$ (orthonormal to within $O(h^2)$).

We now consider an arbitrary shortest curve $AB \subset \mathscr{U}$ and vectors $\xi, \zeta \in \mathfrak{M}_A$. Then, by Lemma 13, 3, for the maps Π_m' constructed in 13.1 we have the following bounds (the fulfilment of the bound required in Lemma 13.3 with constants L_1 and L_2 depending on K' and K is guaranteed by Lemma 13.1):

$$|\measuredangle(\Pi_{2^m-1,2^m} \circ \cdots \circ \Pi_{0,1}(\xi), \Pi_{2^m-1,2^m} \circ \cdots \circ \Pi_{0,1}(\zeta))$$
$$- \measuredangle(\Pi_{2^m-2,2^m-1} \circ \cdots \circ \Pi_{0,1}(\xi), \Pi_{2^m-2,2^m-1} \circ \cdots \circ \Pi_{0,1}(\zeta))| \leqslant \mu'''/4^m,$$
$$|\measuredangle(\Pi_{2^m-2,2^m-1} \circ \cdots \circ \Pi_{0,1}(\xi), \Pi_{2^m-2,2^m-1} \circ \cdots \circ \Pi_{0,1}(\zeta))$$
$$- \measuredangle(\Pi_{2^m-3,2^m-2} \circ \cdots \circ \Pi_{0,1}(\xi), \Pi_{2^m-3,2^m-2} \circ \cdots \circ \Pi_{0,1}(\zeta))| \leqslant \mu'''/4^m, \ldots$$

from which it follows that

$$|\measuredangle(\xi, \zeta) - \measuredangle(\Pi_m'(\xi), \Pi_m'(\zeta))| \leqslant \mu'''/2^m.$$

Making m in the last inequality tend to infinity, we obtain the assertion of Proposition 13.1.

13.3.2. Boundedness of the Curvature in Terms of Parallel Translation. We observe that it is sufficient to prove Theorem 13.1 for the case when γ is a triangle. The general case reduces to consideration of a parallel translation along a K-fan (see 6.1.3) inscribed in γ, which in turn reduces to consideration of a parallel translation along the triangles that form the K-fan.

We also observe that there is no need to consider triangles that can be arbitrarily close to degenerate ones. We can always "split" such "poor" triangles into "proper" triangles whose angles differ from zero and π by some constant, and represent a parallel translation along a "poor" triangle as a sum of parallel translations along "proper" triangles.

In considering a parallel translation along a "proper" triangle $T = OBC$ it is sufficient to develop bounds for the maps Π_{OB}, Π_{BC}, Π_{CO} constructed in 13.1. These bounds can be obtained synthetically by using (10.1).

13.3.3. The Lipschitz Condition for the Components of the Metric Tensor in a Distance Coordinate System. Let us state a lemma from which Theorem 13.2 follows directly. As a preliminary we introduce the following concept: let $B \in \mathscr{U}$ be a fixed point. When $Q \in \mathscr{U} \setminus \{B\}$ we denote by $X_B(Q)$ a vector of $\mathfrak{M}_Q$ such that $B = \exp_Q X_B(Q)$.

Lemma 13.4 (Nikolaev (1983b)). *Let $\gamma\colon I \to \mathscr{U}$ be a differentible curve, and let B be a fixed point of $\mathscr{U}$ such that $\rho_{\mathfrak{M}}(\gamma(t), B) \geqslant \delta > 0$ for any $t \in I$ and some $\delta > 0$*

(here $\rho_{\mathfrak{M}}$ denotes the intrinsic metric of $\mathfrak{M}$). Then for any t, $t_0 \in I$ we have the bound

$$|(X_B)_t^{t_0} - X_B(t_0)| \leqslant \mu(\delta) \cdot l(\gamma; t, t_0).$$

Here, as usual, $(X_B)_t^{t_0}$ denotes the vector of $\mathfrak{M}_{\gamma(t_0)}$ obtained as a result of parallel translation of the vector $X_B(t)$ along the curve γ from the point $\gamma(t)$ to the point $\gamma(t_0)$, $l(\gamma; t, t_0)$ denotes the length of the arc of γ corresponding to values of the parameter lying between t_0 and t, and $\mu(\delta)$ denotes a positive constant depending only on δ, K' and K.

The proof of this lemma is based on the following geometrical assertion, which can be obtained purely synthetically by using (10.1).

Consider the vector field $X_B(Q)$ for Q belonging to the shortest curve AC ($B \notin AC$). Let L be an arbitrary point for which $AL = AC$, and L' the point symmetrical to L with respect to the midpoint O of the shortest curve AC. We have the bound

$$|\cos \measuredangle LAB + \cos \measuredangle BCL'| \leqslant \mu(\delta) \cdot AC.$$

A direct consequence of Lemma 13.4 is the following lemma.

Lemma 13.5 (Nikolaev (1983b)). *Consider a convex domain $\mathscr{V} \subseteq \mathscr{U}$ and points $A, B \in \mathscr{U} \backslash \mathscr{V}$ for which $\rho_{\mathfrak{M}}(A, \mathscr{V}), \rho_{\mathfrak{M}}(B, \mathscr{V}) \geqslant \delta > 0$ for some δ. Then for the function $\varphi(P) = \langle X_A(P), X_B(P) \rangle$ we have the bound*

$$|\varphi(P) - \varphi(Q)| \leqslant \mu(\delta) \cdot PQ$$

for arbitrary $P, Q \in \mathscr{V}$. Here $\mu(\delta)$ is a constant depending on δ, K' and K.

For the points A and B we consider the points P_i, $i = 1, 2, \ldots, n$, that take part in the construction of the distance coordinate system (see 12.3.2). Then, taking account of (12.2), we obtain the assertion of Theorem 13.2 from Lemma 13.5.

13.3.4. Coincidence of Geometrically and Analytically Defined Parallel Translations. It is known that to a parallel translation there corresponds the concept of covariant differentiation of the vector field along a differentiable curve. Namely, let $\gamma: I \to \mathfrak{M}$ be a differentible curve, and X the vector field along γ. We denote by $(X)_t^{t_0}$ the result of parallel translation of the vector $X(t)$ along γ from the point $\gamma(t)$ to the point $\gamma(t_0)$. We call the limit

$$\lim_{t \to t_0} [(X)_t^{t_0} - X(t_0)]/(t - t_0),$$

if it exists, the *covariant derivative of the vector field X* at the point t_0 along γ and denote it by $\nabla_\gamma X|t_0$.

Obviously in the case of a parallel translation defined analytically Π' specifies a Levi-Cività connection on $\mathfrak{M}$. It is known that on $(\mathfrak{M}, \rho)$ there is only one Levi-Cività connection (Gromoll, Klingenberg and Meyer (1968)). Therefore, to prove that the parallel translations Π and Π' coincide we need to prove that Π specifies a connection ∇ on $\mathfrak{M}$ that is a Levi-Cività connection. This means that apart from the usual properties of a connection we need to establish that ∇ is a Riemannian connection (that is, the following rule for differentiating the scalar product along γ is satisfied:

$$\langle X, Y\rangle' = \langle \nabla_\gamma X, Y\rangle + \langle X, \nabla_\gamma Y\rangle$$

and the connection ∇ has zero torsion: $\nabla_X Y = \nabla_Y X + [X, Y]$).

It is sufficient to carry out the verification of the necessary properties of ∇ for vector fields of the form X_B for arbitrary $B \in \mathfrak{M}$. The geometrical character of the definition of the fields X_B (see 13.3.3) enables us to do this on the whole synthetically, relying on the boundedness of the curvature of $\mathfrak{M}$ in the form (10.1).

§ 14. Smoothness of the Metric of Spaces with Bounded Curvature

In this section we describe the differential properties of the metric tensor of a space with bounded curvature.

Example 12.1 shows that we cannot expect the metric tensor of a space with bounded curvature to be twice continuously differentiable. Nevertheless we can assert that the metric tensor of a space with bounded curvature has second derivatives in some generalized sense. This turns out to be very important, since it gives the possibility of writing the components of the curvature tensor of the metric formally by means of the components of the metric tensor and its first and second deriatives by well-known formulae, and in many cases to carry over the apparatus of "smooth Riemannian geometry" almost automatically to the case of a space with bounded curvature. In the next chapter we give results showing that the formally written curvature coincides with the curvature defined in the natural geometrical way.

The differential properties of the metric of spaces with bounded curvature are described in terms of the Sobolev function spaces W_p^l. This enables us to use the apparatus of the theory of functions with generalized derivatives.

14.1. Statement of the Main Result.

14.1.1. Some Function Spaces. Let Ω be a domain of $\mathbb{R}^n$; we denote by $L_p(\Omega)$, $p \geqslant 1$, the normed space of all functions on Ω that are Lebesgue integrable in degree p with norm

$$\|f\|_{L_p(\Omega)} = \left[\int_\Omega |f|^p \, dx\right]^{1/p}.$$

We define the space of functions $W_p^l(\Omega)$, p, $l \geqslant 1$, as the subspace of those functions of $L_p(\Omega)$ that have in Ω all generalized derivatives with respect to x^i, $i = 1, 2, \ldots, n$, up to order l inclusive that are integrable in degree p on Ω. For

$f \in W_p^l(\Omega)$ we introduce the norm

$$\|f\|_{W_p^l(\Omega)} = \left[\int_\Omega \left(|f|^p + \sum_{k=1}^{l} \sum_{|s|=k} |D^{(s)}f|^p \right) dx \right]^{1/p},$$

where $s = (s_1, s_2, \ldots, s_n)$, $s_i \geqslant 0$, are integers.

In the case of a domain Ω with smooth boundary, $W_p^l(\Omega)$ is the closure, in the norm indicated above, of all smooth functions on Ω.

We also need the class of functions $\mathring{W}_p^1(\Omega)$. We denote by $\mathring{W}_p^1(\Omega)$ the closure in the norm W_p^1 of the set of all smooth functions that are compactly supported in Ω.

As usual, we understand by $C^{r,\alpha}(\Omega)$, $r = 0, 1, 2, \ldots$, $0 < \alpha < 1$, the space of r times continuously differentiable functions on Ω, all r-th derivatives of which satisfy a Hölder condition which exponent α: $|D^{(r)}f(x) - D^{(r)}f(x_0)| \leqslant C|x - x_0|^\alpha$.

For the function spaces under consideration we refer the reader to the monographs Sobolev (1950), Gol'dshtein and Reshetnyak (1983), and Ladyzhenskaya and Ural'tseva (1964).

14.1.2. What Do We Understand by the Smoothness of the Metric of a Riemannian Manifold? The metric of a Riemannian manifold belongs to some function space H if for each point of $\mathfrak{M}$ there is a local coordinate system of differential structure given on $\mathfrak{M}$ such that the components of the metric tensor, written in this coordinate system, are functions of the class H.

14.1.3. Harmonic Coordinates. We recall that a coordinate system $\xi^1, \xi^2, \ldots, \xi^n$, specified in a domain of an n-dimensional Riemannian manifold, is called *harmonic* if

$$\Delta_2 \xi^l = 0, \quad l = 1, 2, \ldots, n,$$

where Δ_2 is the Laplace operator on $\mathfrak{M}$ (see 14.2).

In the investigation of the question of the smoothness of the metric of a Riemannian manifold, not all coordinate systems are in an equivalent position. Thus, for example, in normal Riemannian coordinates there is a loss of two derivatives of the components of the metric tensor. In many cases harmonic coordinates are optimal in the sense of smoothness of the metric (for example, in the case $H = C^{r,\alpha}$). Sometimes harmonic coordinates give the worst smoothness of the metric in comparison with what it really is (Sabitov and Shefel' (1976), p. 924).

Harmonic coordinate systems were first used by Einstein (Einstein (1916)). In Riemannian geometry harmonic coordinate systems were first applied to questions connected with the smoothness of the metric in Sabitov and Shefel' (1976); these results were proved again later in DeTurck and Kazdan (1981).

14.1.4. Statement of the Main Result.

Theorem 14.1 (Nikolaev (1983b)). *Let $\mathfrak{M}$ be a space with bounded curvature. Then in a neighbourhood of each point of it we can introduce a harmonic coordi-*

nate system. The components of the metric tensor in any harmonic coordinate system in $\mathfrak{M}$ are continuous functions of class W_p^2, where for p we can take any number not less than one. Harmonic coordinate systems specify on $\mathfrak{M}$ an atlas of class $C^{3,\alpha}$ for any $0 < \alpha < 1$.

Remark 1. From Sobolev's embedding theorem (see Ladyzhenskaya and Ural'tseva (1964), Theorem 2.1 on p. 64) it follows that in harmonic coordinates the components of the metric tensor belong to the class $C^{1,\alpha}$ for any $0 < \alpha < 1$.

Remark 2. The components of the metric tensor in a harmonic coordinate system have an ordinary second differential almost everywhere (in the sense of n-dimensional Lebesgue measure) (this follows from the fact that $g_{ij} \in W_p^2$ when $p > n$ and Theorem 5.2 in Chapter 2 of Gol'dshtein and Reshetnyak (1983)).

14.2. Plan of the Proof of Theorem 14.1.

14.2.1. Harmonic Coordinates in Spaces with Bounded Curvature. The *Laplace operator* Δ_2 on $\mathfrak{M}$ has the form

$$\Delta_2 u = g^{ij} \frac{\partial^2 u}{\partial x^i \partial x^j} - g^{ij} \Gamma_{ij}^p \frac{\partial u}{\partial x^p}.$$

By Theorem 13.2 the coefficients in Δ_2 for the higher derivatives belong to the class $C^{0,1}$, and the coefficients for the lower derivatives are bounded with respect to the distance coordinate system. In this connection, the equation $\Delta_2 \xi = 0$ has solutions $\xi^1, \xi^2, \ldots, \xi^n$ that satisfy at the initial point the condition

$$\frac{\partial \xi^l}{\partial x^i}(x_0) = \delta_i^l, \quad i, l = 1, 2, \ldots, n,$$

(see Theorem 2 in Chapter 5 of Bers, John and Schechter (1964); the assertion of this theorem remains true when we require not continuity but only boundedness of the lowest coefficients of the equation).

Now let g_{ij}, Γ_{ij}^k and so on be the components of the metric tensor and the Christoffel symbols in a harmonic coordinate system. From the "harmonicity" of the coordinates it follows that

$$g^{pl} \Gamma_{pl,j} = 0. \tag{14.1}$$

From (14.1), by analogy with what was done in Sabitov and Shefel' (1976), we can derive an equation of elliptic type on the metric. Since $\xi^l(x)$ (the transition functions from the distance coordinate system to the harmonic) belong to the class W_p^2 for any $p \geqslant 1$ (Ladyzhenskaya and Ural'tseva (1964), Theorem 15.1 of Chapter III), we have

$$g_{ij} \in W_p^1, \quad \Gamma_{ij}^l, \Gamma_{pr,j} \in L_p. \tag{14.2}$$

The conditions (14.2) enable us to define the generalized function

$$(R_{ij}, \eta) = -\frac{1}{2} \int_\Omega g^{pl} (D_j g_{pl} \cdot D\eta - D_p g_{il} \cdot D_j \eta$$
$$+ D_l g_{ij} \cdot D_p \eta - D_l g_{pj} \cdot D_i \eta) \, d\xi, \quad i, j = 1, 2, \ldots, n,$$

that is, a functional on $\mathring{W}_p^1$. Here $D_l = \partial/\partial\xi^l$, and Ω is the range of the harmonic coordinates. We observe that up to the first derivatives of g_{ij}, (R_{ij}, η) in the smooth case is the integral with respect to Ω of the product of η and the corresponding Ricci curvature. Differentiating (14.1), we can obtain the equality

$$\int_\Omega g^{pl} D_l g_{ij} \cdot D_p \eta \, d\xi = -(R_{ij} + R_{ji}, \eta).$$

If the generalized function (R_{ij}, η) is represented in the form

$$(R_{ij}, \eta) = \int_\Omega R_{ij}(\xi) \cdot \eta(\xi) \, d\xi, \tag{14.3}$$

then g_{ij} is the generalized solution of the equation

$$\frac{\partial}{\partial \xi^p}\left(g^{pl} \frac{\partial g_{ij}}{\partial \xi^l}\right) = 2R_{ij}. \tag{14.4}$$

Equation (14.4) is the required elliptic equation on the metric.

14.2.2. The Bound in L_p of the Curvature of the Averaged Metric. In order to prove that the generalized function (R_{ij}, η) that we have introduced can be represented in the form (14.3), we need to prove that the norm of the functional R_{ij} is bounded. To this end we consider the *Sobolev averaging of the metric* g_{ij} and prove that the curvature of the averaged metric is bounded.

Let Ω_0 be a convex domain, lying strictly inside the domain Ω, where the harmonic coordinates are specified. We take the Sobolev average of the components of the metric tensor g_{ij} (Gol'dshtein and Reshetnyak (1983)). Then for the resulting metric g_{ij}^h, $i, j = 1, 2, \ldots, n$, we denote the Christoffel symbols by Γ_{ij}^{hk}, $k = 1, 2, \ldots, n$.

Proposition 14.1 (Nikolaev (1983b)). *For each $q \geqslant 1$ we have the bound*

$$\left\| \frac{\partial \Gamma_{jl,k}^h}{\partial \xi^i} - \frac{\partial \Gamma_{il,k}^h}{\partial \xi^j} \right\|_{L_q(\Omega_0)} \leqslant \mu(q), \tag{14.5}$$

where $\mu(q)$ is a constant depending on q, K', K (and the chosen harmonic coordinate system).

The plan of the proof of Proposition 14.1 is as follows.

Consider a square in Ω whose sides are parallel to the i-th and j-th coordinate curves in Ω and have length $\varepsilon > 0$. Let x denote the vertex of the square with the least i-th and j-th coordinates. Let $K_{\varepsilon,x,i,j}$ denote the curve in $\mathfrak{M}$ whose harmonic coordinates form the square in Ω mentioned above. We now specify the function $\Delta_{\varepsilon,ijl}^r(x)$ for $x \in \Omega_0$: we carry out a parallel translation of the l-th coordinate vector $\Xi_l(x)$ of the harmonic coordinates along $K_{\varepsilon,x,i,j}$. Then $\Delta_{\varepsilon,ijl}^r$ is the r-th coordinate of the increment of $\Xi_l(x)$. By Theorem 13.1, $\Delta_{\varepsilon,ijl}^r(x) = O(\varepsilon^2)$, on the other hand, $\Delta_{\varepsilon,ijl}^r(x)$ can be expressed in terms of the components of the metric tensor g_{ij} by means of (13.2), from which we deduce (14.5).

14.2.3. Smoothness of the Metric. From Proposition 14.1 we deduce that the generalized function (R_{ij}, η) can be represented in the form (14.3) and that $R_{ij}(\xi) \in L_p$ for any $p \geqslant 1$. Then the smoothness of the metric is obtained from Theorem 15.1 in Ladyzhenskaya and Ural'tseva (1964), p. 236. The existence of a $C^{3,\alpha}$-smooth atlas follows from Sabitov and Shefel' (1976).

§ 15. Spaces with Bounded Curvature and Limits of Smooth Riemannian Metrics

From Theorem 14.1 on the smoothness of the metric of a space with bounded curvature we can deduce an important corollary that characterizes the metric of such a space as the limit of smooth Riemannian metrics with uniformly bilaterally bounded sectional curvatures at all points in all two-dimensional directions. As we mentioned in 11.1.3, if certain conditions are satisfied, the *converse assertion* is also true: the uniform limit of such smooth Riemannian metrics is a metric with bounded curvature.

It is convenient to illustrate the main results of this section on the *class of Riemannian manifolds* $\mathfrak{M}(n, d, \Lambda, V)$ introduced below (see 15.2). Roughly speaking, $\mathfrak{M}(n, d, \Lambda, V)$ is a class of n-dimensional compact Riemannian manifolds whose curvatures have the bound $|K_\sigma| \leqslant \Lambda$ and for which certain conditions of "normalizing" character are satisfied.

In $\mathfrak{M}(n, d, \Lambda, V)$ we introduce a metric in a natural way (see 15.2.1). With respect to this metric $\mathfrak{M}(n, d, \Lambda, V)$ is not a complete metric space. Its completion consists precisely of the corresponding spaces with bounded curvature.

In this section we also discuss Gromov's compactness theorem for the class of Riemannian manifolds $\mathfrak{M}(n, d, \Lambda, V)$ (Gromov (1981)).

15.1. Approximation of the Metric of a Space with Bounded Curvature by Smooth Riemannian Metrics.

15.1.1. A Local Variant of Approximation. Smooth metrics that converge to the original metric in a small domain of a space with bounded curvature are constructed by means of the operation of Sobolev averaging (Gol'dshtein and Reshetnyak (1983)).

Let $\Omega \subset \mathbb{R}^n$ be a domain in which harmonic coordinates $\xi^1, \xi^2, \ldots, \xi^n$ are specified, and Ω_h the subdomain of Ω at a distance h from the boundary of Ω. Then in Ω_h we define the averaged metric g_{ij}^h:

$$g_{ij}^h(\xi) = \frac{1}{h^n} \int_\Omega g_{ij}(u)\omega\left(\frac{\xi - u}{h}\right) du,$$

where $\omega: \mathbb{R}^n \to \mathbb{R}$ is the averaging kernel, that is, a function for which the following conditions are satisfied:

1) the support of ω is contained in the unit ball $B(O, 1) \subset \mathbb{R}^n$;
2) the function ω is infinitely differentiable;
3) $\int_{\mathbb{R}^n} \omega(x)\, dx = 1$.

The resulting metric g_{ij}^h is infinitely differentiable. From Theorem 17.1, and also from the standard properties of the *Sobolev averaging*, it follows that the curvatures K_σ^h of the metric g_{ij}^h have the bounds

$$K' - \varepsilon'(h) \leqslant K_\sigma^h \leqslant K + \varepsilon(h),$$

where K' and K are such that $\mathfrak{M}$ in the neighbourhood under consideration is a space of curvature both $\leqslant K$ and $\geqslant K'$, and $\varepsilon(h)$ and $\varepsilon'(h)$ tend to zero as $h \to 0$.

On the other hand, obviously $\|g_{ij} - g_{ij}^h\|_{W_p^2} \to 0$ as $h \to 0$, and from Sobolev's embedding theorems (Ladyzhenskaya and Ural'tseva (1964)) it follows that g_{ij}^h converge uniformly to g_{ij} in an arbitrary subdomain Ω_0 of the domain Ω.

Putting $h_m = 1/m$, we obtain the required local approximation of the metric g_{ij} by smooth metrics $g_{ij}^m = g_{ij}^{hm}$ with sectional curvatures that are bilaterally bounded in aggregate.

15.1.2. Global Approximation. The global approximation of the metric of a space with bounded curvature can be constructed by means of the *de Rham averaging operator* (de Rham (1955)), which has as its basis the Sobolev averaging operator. Necessary bounds for the de Rham averaging operator can be obtained by starting from its form. Some of these bounds were obtained in Gol'dshtein, Kuz'minov and Shvedov (1984).

Theorem 15.1. *Let M be a space with bounded curvature for which the bounds of the curvatures are K_0 and K (that is, $\mathfrak{M}$ is a space of curvature both $\leqslant K$ and $\geqslant K_0$). Then for every $K' > K$ and $K_0' < K$ on $\mathfrak{M}$ there is a sequence $\{g_m\}$ of infinitely differentable Riemannian metrics for which*

1) *the sectional curvatures of the Riemannian manifold $(\mathfrak{M}, g_m)$ at all points in all two-dimensional directions have the bounds*

$$K_0' \leqslant K_\sigma^m \leqslant K';$$

2) *the sequence $\{\rho_m\}$ of intrinsic metrics of the Riemannian manifolds $(\mathfrak{M}, g_m)$ converges uniformly to the intrinsic metric ρ of the space $\mathfrak{M}$ on each compact subset $\mathscr{A} \subset \mathfrak{M}$: $\rho_m(P, Q) \rightrightarrows \rho(P, Q)$, $P, Q \in \mathscr{A}$.*

Remark 1. In fact we can assert that the metrics g_m obtained from the metric g by de Rham averaging converge to g in the norm W_p^2, $p \geqslant 1$.

Remark 2. As we have already mentioned, in a space $\mathfrak{M}$ with bounded curvature there is a $C^{3,\alpha}$-smooth differential structure. By a well-known theorem of Whitney, this structure contains a C^∞-smooth atlas. We have in mind that the metric tensors g_m belong to the class C^∞ with respect to the charts of this atlas.

Remark 3. If the space $(\mathfrak{M}, \rho)$ is complete, then, beginning with a certain number, the spaces $(\mathfrak{M}, \rho_m)$ are also complete.

Remark 4. The authors do not know whether it is possible to approximate the metric g by smooth Riemannian metrics g_m with the same bounds for the sectional curvatures (that is, whether one can take $K_0 = K_0'$ and $K = K'$ in Theorem 15.1).

15.2. A Space of Riemannian Manifolds with Sectional Curvatures Bounded in Aggregate. We restrict ourselves to the consideration of compact Riemannian manifolds whose sectional curvatures have the bound

$$|K_\sigma| \leqslant \Lambda. \tag{15.1}$$

We observe that by multiplying the metric of an arbitrary compact manifold by a suitable constant we can arrange that (15.1) is satisfied for a Riemannian manifold conformally equivalent to the given one. To avoid this, we introduce a restriction on the diameter of the manifold

$$d(\mathfrak{M}) \leqslant d. \tag{15.2}$$

We also restrict ourselves to the consideration of manifolds of constant dimension n:

$$\dim(\mathfrak{M}) = n. \tag{15.3}$$

In the class of Riemannian manifolds that satisfy conditions (15.1)–(15.3) we can find a sequence of manifolds that "converge" to a manifold of lower dimension as shown, for example, in Fig. 13. To remove the possibility of this situation, we introduce a condition on the volume $V(\mathfrak{M})$:

$$V(\mathfrak{M}) \geqslant V > 0 \tag{15.4}$$

or on the radius of injectivity $i(\mathfrak{M})$ of $\mathfrak{M}$:

$$i(\mathfrak{M}) \geqslant i_0 > 0. \tag{15.4$'$}$$

We denote by $\mathfrak{M}(n, d, \Lambda, V)$ the class of compact Riemannian manifolds that satisfy conditions (15.1)–(15.4). Correspondingly, $\mathfrak{M}(n, d, \Lambda, i_0)$ is the class of compact Riemannian manifolds that satisfy conditions (15.1)–(15.4)$'$.

Remark. From Proposition 11.2 it follows that in the class $\mathfrak{M}(n, d, \Lambda, V)$ for all $\mathfrak{M} \in \mathfrak{M}(n, d, \Lambda, V)$ there is a uniform bound of type (15.4)$'$ on the radius of injectivity $i(\mathfrak{M})$.

We have taken the above heuristic arguments for the definition of the classes $\mathfrak{M}(n, d, \Lambda, V)$ and $\mathfrak{M}(n, d, \Lambda, i_0)$ from Peters (1987).

We now introduce a distance in the space $\mathfrak{M}(n, d, \Lambda, V)$ that specifies a "natural" convergence of Riemannian manifolds.

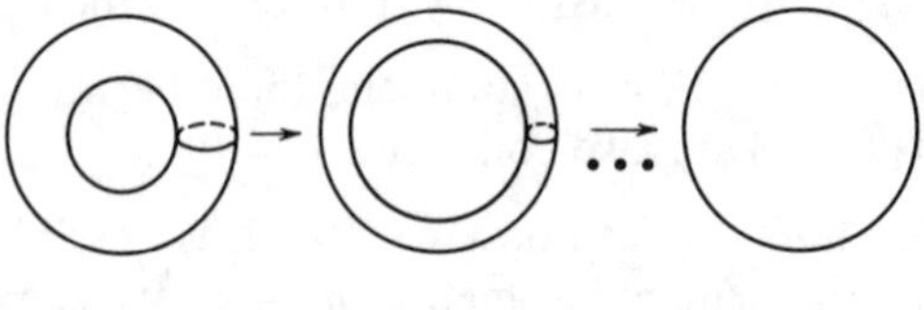

Fig. 13

15.2.1. Lipschitz Distance. *Let* $f: X \to Y$ be the Lipschitz map of the metric spaces X and Y. Then the *dilatation* of f is the quantity

$$\operatorname{dil} f = \sup_{x \neq x'} \rho(f(x), f(x'))/\rho(x, x').$$

The Lipschitz distance between the metric spaces X and Y is taken to be

$$d_L(X, Y) = \inf\{|\ln \operatorname{dil} f| + |\ln \operatorname{dil} f^{-1}|\},$$

where the infimum is considered over all bi-Lipschitz maps $f: X \to Y$. If there is no bi-Lipschitz map $f: X \to Y$, we take the distance to be $+\infty$.

15.2.2. Hausdorff Distance. The classical Hausdorff distance between subspaces of a given metric space is defined as follows.

Let Z be a metric space, and X and Y subspaces of Z. We denote the set $\{z \in Z | \rho(z, X) < \varepsilon\}$ by $U_\varepsilon^Z(X)$. Then $d_H^Z(X, Y) = \inf\{\varepsilon | U_\varepsilon^Z(X) \supseteq Y, U_\varepsilon^Z(Y) \supseteq X\}$.

Now the *Hausdorff distance* between arbitrary metric spaces X and Y is taken to be $\inf\{d_H^Z(f(X), g(Y))\}$, where the infimum is considered over all metric spaces Z and all isometric embeddings $f: X \to Z$ and $g: Y \to Z$. The notation is $d_H(X, Y)$.

15.2.3 Connection between the Topologies Specified by the Lipschitz Distance and the Hausdorff Distance. Obviously, in the general case these topologies do not coincide (for an example, see Peters (1986)), namely, the topology specified by the Hausdorff distance is coarser than the topology specified by the Lipschitz distance. However, for the class of Riemannian manifolds $\mathfrak{M}(n, d, \Lambda, V)$ the following assertion is true.

Proposition 15.1 (Gromov (1981)). *In* $\mathfrak{M}(n, d, \Lambda, V)$ *the topologies specified by the Lipschitz and Hausdorff distances coincide.*

In other words, convergence of a sequence of Riemannian manifolds of $\mathfrak{M}(n, d, \Lambda, V)$ with respect to the Hausdorff distance is equivalent to the convergence of the same sequence with respect to the Lipschitz distance.

15.2.4. Gromov's Compactness Theorem.

Theorem 15.2 (Gromov (1981)). *The class of Riemannian manifolds* $\mathfrak{M}(n, d, \Lambda, V)$ *with respect to the topology defined by the Lipschitz distance (or the Hausdorff distance) is a relatively compact set in a wider class of* $C^{1,1}$*-smooth* n*-dimensional manifolds with* C^0*-smooth metric.*

Thus, Gromov's compactness theorem asserts that from any sequence of Riemannian manifolds in the class $\mathfrak{M}(n, d, \Lambda, V)$ we can choose a sequence that converges with respect to Lipschitz (or Hausdorff) distance to a Riemannian manifold with continuous metric tensor.

15.2.5. Limiting metrics for $\mathfrak{M}(n, d, \Lambda, V)$. As we have seen, Gromov's theorem asserts only the continuity of the limiting metric tensor. In fact, the properties of the limiting metric are considerably better. This is caused by the fact that the limiting manifold is a space with bounded curvature (see the remark to Proposition 11.3).

Theorem 15.3 (Peters (1986)). *Suppose that* $0 < \alpha < 1$. *Then an arbitrary sequence of* $\mathfrak{M}(n, d, \Lambda, V)$ *contains a subsequence that converges to an n-dimensional differentiable manifold* $\mathfrak{M}$ *with metric g of Hölder class* $C^{1,\alpha}$.

Since the limiting manifold is a space with bounded curvature, we can strengthen Theorem 15.3 by applying Theorem 14.1:

Theorem 15.4 (Peters (1987)). *The limiting Riemannian manifold in the compactness theorem* 15.2 *is a space with bounded curvature, so the components of the metric tensor of the limiting metric in a harmonic coordinate system belong to the class* W_p^2 *for any* $p \geqslant 1$.

By Sobolev's embedding theorems, Theorem 15.3 obviously follows from Theorem 15.4.

15.2.6. Why is it Important to Know the Smoothness of the Metric of the Limiting Riemannian Manifold? To answer this question we give one of the characteristic examples of the application of Gromov's compactness theorem.

First example. The rigidity theorem and δ-pinched manifolds with $\delta < 1/4$. The rigidity theorem says that if $\mathfrak{M}$ is a compact simply-connected Riemannian manifold of even dimension n with δ-pinched curvature for $\delta = 1/4$ (that is, at all points $P \in \mathfrak{M}$ for all plane elements $\sigma \subseteq \mathfrak{M}_P$ the sectional curvatures $K_\sigma(P)$ satisfy the bounds $1/4 \leqslant K_\sigma(P) \leqslant 1$), then $\mathfrak{M}$ is either homeomorphic to an n-dimensional sphere or is isometric to one of the symmetric spaces of rank 1: the complex projective space $\mathbb{C}P^n$, the quaternion projective space HP^n, or the Cayley projective space CaP^2 (see Berger (1983)).

In the case $\delta < 1/4$ there is the conjecture that there are no other differentiable manifolds apart from the symmetric spaces of rank 1 mentioned above. In this connection Berger proved the following theorem.

Theorem 15.5 (Berger (1983)). *For each even n there is a number* $\varepsilon(n) < 1/4$ *such that all simply-connected compact Riemannian manifolds that have an* $\varepsilon(n)$-*pinched metric are either homeomorphic to* S^n *or diffeomorphic to one of the symmetric spaces of rank* 1.

There is great interest in the method of proof of Theorem 15.5. The proof can be carried out by contradiction.

Suppose there is a sequence of $\varepsilon_m(n)$-pinched Riemannian manifolds $(\mathfrak{M}_m, g_m)$, $m = 1, 2, \ldots$, where $\varepsilon_m(n) \to 1/4$ as $m \to \infty$, for which the assertion of Theorem 15.5 does not hold.

By means of Cheeger's finiteness theorem (see Peters (1986)) we can arrange matters so that all the manifolds $\mathfrak{M}_m$ are diffeomorphic to some manifold $\mathfrak{M}$. The theorems of Meyer and Klingenberg give the necessary bounds on the diameter and the radius of injectivity of the Riemannian manifolds $(\mathfrak{M}, g_m)$. Consequently, we can apply Gromov's compactness theorem to the sequence $(\mathfrak{M}, g_m)$. If the metric of the limiting manifold $(\mathfrak{M}, g_0)$ is sufficiently smooth, then by applying the rigidity theorem to $(\mathfrak{M}, g_0)$ we obtain the necessary contradiction. The main difficulty in the proof of Theorem 15.5 just consists in establishing that the metric g_0 is sufficiently smooth (in Berger (1983) it was proved that g_0 belongs to the class C^∞).

Another example of this kind of the application of Gromov's theorem is given by Brittain's theorem:

Theorem (Peters (1986)). *There is a positive number ε, depending only on n, $\max|K_{\mathfrak{M}}|$ and a positive number V_0, such that if $\operatorname{Ric}(\mathfrak{M}) \geqslant n-1$, $\operatorname{Vol}(\mathfrak{M}) \geqslant V_0$ and $d(\mathfrak{M}) \geqslant \pi - \varepsilon$ (where $d(\mathfrak{M})$ is the diameter of $\mathfrak{M}$), then $\mathfrak{M}$ is diffeomorphic to S^n.*

For applications of Gromov's compactness theorem we refer the reader to the articles of Peters (Peters (1986), (1987)).

Chapter 4
Existence of the Curvature of a Metric Space at a Point and the Axioms of Riemannian Geometry

In this chapter we state the geometrical conditions that distinguish Riemannian manifolds among general metric spaces. The most essential part of these conditions is that "continuous curvature exists" for a given metric space.

Thus, an important feature is a suitable "Riemannian" definition of the curvature at a point of the space. The curvature defined here, as in the case of multidimensional Riemannian manifolds, depends not only on the point, but also on a pair of directions at the point. We call it *non-isotropic Riemannian curvature.*

The value of the curvature is calculated by means of the limit of the ratios of the excesses of triangles and their "area", namely $\bar{\delta}(T)/\sigma(T)$, that contract to a given point, so that the directions of fixed pairs of sides of these triangles "converge" to a given "admissible" pair of directions at the point. Continuity of the non-isotropic Riemannian curvature is defined by means of the distance d, which is introduced between the directions at different points of the space (see § 16).

On the other hand, at points of two-fold differentiability of a Riemannian metric, from the metric tensor we can purely formally calculate the sectional curvature $K_\sigma(P)$ of the space $\mathfrak{M}$. Our main result involving the curvature that we have introduced is that "almost everywhere" $K_\sigma(P)$ can be calculated by means of the limit of the ratios $\bar{\delta}(T)/\sigma(T)$.

We also introduce the *isotropic Riemannian curvature*, that is, the curvature that depends only on the point, and not on the directions at this point.

At the end of the chapter we state a theorem asserting that *isotropic metric spaces* (that is, metric spaces at each point of which the isotropic Riemannian curvature exists) of dimension greater than two (the dimension can be understood, for example, in the sense of Menger and Uryson, see Hurewicz and Wallman (1941)) are spaces of constant curvature.

§ 16. The Space of Directions of an Arbitrary Metric Space

In this section we define the function of the distance d between directions specified at different points of a metric space. We shall call the set of all directions of a metric space, considered together with the distance function d, the *space of directions* of the given metric space.

In contrast to the space of directions at a point (§ 3), the space of directions is not a metric space. The distance function d that we have defined is semimetric (the triangle inequality is not satisfied for d). We give an example showing that this is due to the essence of the question.

We need the concept of the space of directions to define the continuity of the non-isotropic Riemannian curvature of a metric space.

16.1. Distance between Directions (Nikolaev (1987)). The following requirements are imposed on the functions of distance between arbitrary directions: this function makes sense for directions in an arbitrary metric space; the distance we have introduced "induces" in the case of a Riemannian manifold the standard metric in the spherical bundle to the manifold (the Sasaki metric, Sasaki (1958), (1962)). We recall that in the case of a Riemannian manifold the set of directions coincides exactly with the spherical bundle of the manifold.

16.1.1. Heuristic Arguments. Let us recall how to define the *Sasaki metric* in the spherical bundle. Consider two unit tangent vectors v and $v + dv$ that touch a Riemannian manifold $\mathfrak{M}$ at points P and dP. Then the square of the distance between these vectors is taken to be

$$d\alpha^2 = ds^2 + d\theta^2,$$

where ds denotes the distance between the points P and $P + dP$, and $d\theta$ denotes the angle between the vector v and the vector obtained as a result of parallel translation of the vector $v + dv$ along the shortest curve joining the points $P + dP$ and P (see Fig. 14).

An expression for the components of the metric tensor of the Sasaki metric in terms of the metric of the Riemannian manifold was given in Sasaki (1962). Here we just give a formula for the calculation of the lengths of differentiable curves

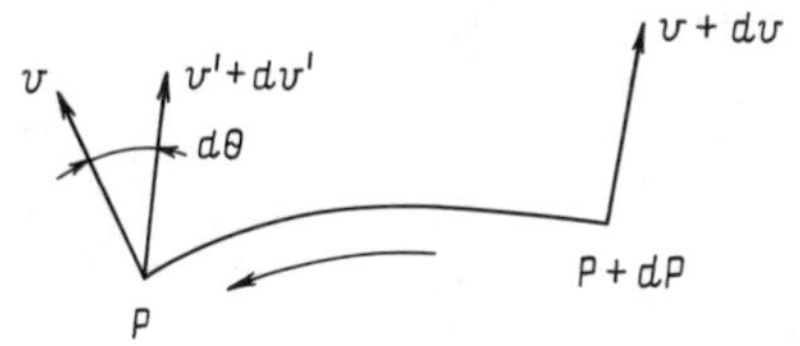

Fig. 14

in the spherical bundle with respect to the Sasaki metric:

$$l(\Xi) = \int_a^b (\langle \dot{c}, \dot{c}\rangle + \langle \nabla_{\dot{c}}\xi, \nabla_{\dot{c}}\xi\rangle)^{1/2}\, dt. \tag{16.1}$$

Here $\Xi(t) = (c(t), \xi(t))$, $a \leqslant t \leqslant b$, is the field of unit vectors defined along the differentiable curve $c(t)$ in $\mathfrak{M}$, that is, a differentiable curve in $\Omega(\mathfrak{M})$, $\dot{c} = c_*(d/dt)$ the field of tangent vectors to $c(t)$, $\langle\ ,\ \rangle$ the scalar product in the tangent space to $\mathfrak{M}$, and ∇ the Levi-Cività connection (Gromoll, Klingenberg and Meyer (1968)).

From what we have said it follows that the basis for the calculation of the lengths of curves in $\Omega(\mathfrak{M})$ must be the distance, defined between close vectors as

$$d'(\xi, \zeta) = (PQ^2 + \theta^2(\xi, \zeta))^{1/2}, \quad \xi \in \Omega_P\mathfrak{M}, \quad \zeta \in \Omega_Q\mathfrak{M}.$$

As before, $\theta(\xi, \zeta)$ here denotes the angle between the unit vector ξ and the vector obtained as a result of parallel translation of the vector ζ along the unique shortest curve joining P and Q.

Namely, in order to calculate the length of the curve $\Xi(t)$ in $\Omega(\mathfrak{M})$ we need to split it by arbitrary points $t_0 = a < t_1 < t_2 < \cdots < t_m = b$ and consider the limit over all such splittings of the interval $[a, b]$:

$$l_{d'}(\Xi) = \lim_{\substack{m\to\infty \\ \max|t_i - t_j|\to 0}} \sum_{i=0}^{m-1} d'(\Xi(t_i), \Xi(t_{i+1})), \tag{16.2}$$

which coincides exactly with the quantity defined in (16.1). In this sense the distance d' induces the Sasaki metric in $\Omega(\mathfrak{M})$.

We give an example showing that the distance d' is not a metric.

Example 16.1. On S_K we consider a non-degenerate triangle $T = ABC$. Let $K = -k^2 < 0$. Consider the unit vectors l_1, l_2, l_3 touching the sides AB, AC, CB, respectively, of the triangle T (see Fig. 15). Then obviously

$$\theta(l_1, l_2) = \alpha, \quad \theta(l_1, l_3) = \beta, \quad \theta(l_2, l_3) = \pi - \gamma.$$

But $\theta(l_2, l_3) - \theta(l_1, l_2) - \theta(l_1, l_3) = -\delta(T) > 0$, that is, the triangle inequality for θ is not satisfied.

We now suppose that the triangle T is equilateral: $AB = BC = AC = c_0$, $\alpha = \beta = \gamma = \alpha_0$. We observe that α_0 is close to zero for very large k. Now taking k sufficiently large, and c_0 sufficiently small, we deduce that $\theta(l_2, l_3) - \theta(l_1, l_2) - \theta(l_1, l_3)$ is close to π, that is, the triangle inequality is violated for d'.

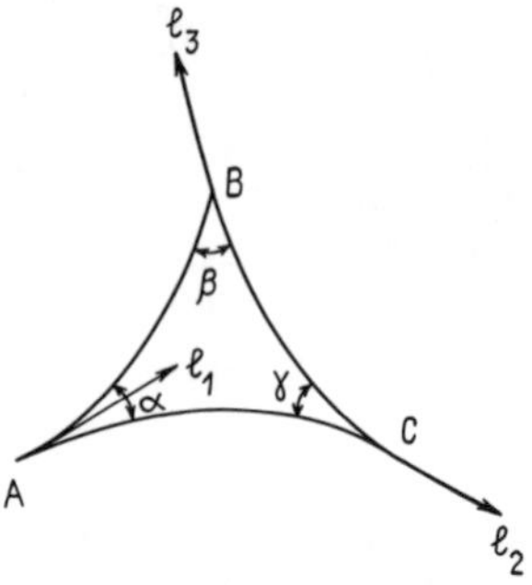

Fig. 15

To conclude this subsection we make the following important remark.

We assume that in $\Omega(\mathfrak{M})$ we have defined the distance d, connected with d' as follows:

$$|d(\xi, \zeta) - d'(\xi, \zeta)| \leqslant \mu \cdot PQ^q, \quad \xi \in \Omega_P\mathfrak{M}, \quad \zeta \in \Omega_Q\mathfrak{M},$$

where μ is a positive constant and $q > 1$.

Then obviously by calculating the length of the curve $\Xi(t)$ in $\Omega(\mathfrak{M})$ from (16.2), but with respect to the distance d, we obtain the same value. Thus, the distances d and d' induce the same metric in $\Omega(\mathfrak{M})$.

In an arbitrary metric space there is no concept of parallel translation. Therefore, in order to introduce on the set of directions of an arbitrary metric space the distance induced in the case of a smooth Riemannian manifold by the Sasaki metric, we consider another distance d, which differs from d' by $O(PQ^2)$.

At the basis of the definition of the distance d is the formula for the cosine of the angle between rays in Euclidean space.

Suppose that a ray l with starting point P passes through a point X, and that a ray l' with starting point Q passes through a point Y. Then for the cosine of the angle α between the rays l and l' we have

$$\cos \alpha = (XQ^2 + YP^2 - PQ^2 - XY^2)/(2PX \cdot QY). \tag{16.3}$$

16.1.2. Definition of the Distance d. We denote the set of directions of a metric space $\mathfrak{M}$ by $\Omega(\mathfrak{M})$, that is,

$$\Omega(\mathfrak{M}) = \bigcup_{P \in \mathfrak{M}} \Omega_P\mathfrak{M}.$$

Suppose that $\xi, \zeta \in \Omega(\mathfrak{M})$, and that $\xi \in \Omega_P\mathfrak{M}$, $\zeta \in \Omega_Q\mathfrak{M}$, where P and Q are points of $\mathfrak{M}$ and $P \neq Q$. We consider curves L and M, starting at P and Q respectively, that specify the directions ξ and ζ (see 3.1).

Suppose that $X \in L$, $X \neq P$, and $Y \in M$, $Y \neq Q$, are arbitrary points. In accordance with (16.3) we introduce into consideration the function

$$h_{LM}(X, Y) = (XQ^2 + YP^2 - PQ^2 - XY^2)/(2PX \cdot QY).$$

To obtain a local characterization, we introduce the quantity

$$h_{LM} = \lim_{\substack{a\to 0+\\ b\to+\infty}} \varlimsup_{\substack{X\to P\\ Y\to Q}} {}_{a\leqslant PX/QY\leqslant b}\, h_{LM}(X, Y).$$

We observe that h_{LM}, when $P = Q$, is equal to the cosine of the upper angle between these curves (see Aleksandrov and Zalgaller (1962), Theorem 1 on p. 250).

Finally, to define the function $h(\xi, \zeta)$ depending only on the directions ξ and ζ, and not on each of the curves that specify them, we put $h(\xi, \zeta) = \inf\{h_{LM}\}$, where the infimum is considered over all curves L and M such that L specifies the direction ξ and M specifies the direction ζ.

We introduce into consideration the function

$$\varphi(\xi, \zeta) = \left\{ \sup_{\eta\in\Omega_P\mathfrak{M}} [h(\xi, \eta) - h(\zeta, \eta)] + \sup_{\eta'\in\Omega_Q\mathfrak{M}} [h(\xi, \eta') - h(\zeta, \eta')] \right\}/2,$$

which coincides with $|\xi - \zeta|$ in the case when ξ and ζ are unit vectors in E_n. Finally we define the *distance between the directions* $\xi \in \Omega_P\mathfrak{M}$ and $\zeta \in \Omega_Q\mathfrak{M}$ in the case of an arbitrary metric space as

$$d(\xi, \zeta) = [PQ^2 + \varphi^2(\xi, \zeta)]^{1/2}.$$

16.1.3. Remarks on the Definition of the Function h_{LM}. As we have seen, a key role in the definition of the distance d is played by the function h_{LM}. We mentioned above that in Euclidean space h_{LM} coincides with the cosine of the angle between L and M. In the general case h_{LM} can exceed one and even be equal to infinity. Using the formulae for the first and second variation of the length of a geodesic, we can prove, however, that in the case of a Riemannian manifold h_{LM} differs slightly, to the cosine of the angle between the vector ξ and the vector ζ translated in a parallel way to the same point.

Proposition 16.1. *Let G be a domain $R_{K',K}$ of a C^2-smooth Riemannian manifold $\mathfrak{M}$ whose diameter has the bound*

$$\operatorname{diam} G \leqslant \min\{1, (6nC)^{-1}\},$$

where $C = \max\{|K|, |K'|\}$, $n = \dim \mathfrak{M}$. *Let* $\xi \in \Omega_P\mathfrak{M}$, $\zeta \in \Omega_Q\mathfrak{M}$ *be arbitrary directions, where* $P, Q \in G$. *Then*

$$|h(\xi, \zeta) - \cos\theta(\xi, \zeta)| \leqslant C(K', K)\cdot PQ^2,$$

where the constant $C(K', K)$ *depends only on* K, K' *and* n.

From Proposition 16.1 it follows that we have the bound

$$|d(\xi, \zeta) - d'(\xi, \zeta)| \leqslant \theta^3(\xi, \zeta)/24 + C'(K', K)\cdot PQ^2,$$

where the constant $C'(K', K)$ depends on K, K' and n. Using the remark made at the end of 16.1.1 and the last inequality, we can deduce that the semimetric d also induces the Sasaki metric in $\Omega(\mathfrak{M})$.

We now discuss the behaviour of h_{LM} in "non-Riemannian" metric spaces. Namely, we give the following examples.

Example 16.2. Consider the normed space $(\mathbb{R}^2, \| \ \|_1)$ (see Example 4.1) and in it the rays $l_1 = (0, t)$, $l_2 = (1, 1 + t)$, $t \geqslant 0$. Then $h_{LM} = +\infty$.

Example 16.3. On the set $\mathbb{R}^2 = \{(x, y) | x, y \in \mathbb{R}\}$ we specify the norm $\|(x, y)\|_\infty = \max\{|x|, |y|\}$. In the resulting normed space we consider the rays $l_1 = (0, -t)$, $l_2 = (1, 1 + t)$, $t \geqslant 0$. Then $h_{LM} = -1$.

Example 16.4. In the normed space $(\mathbb{R}^2, \| \ \|_\infty)$ we consider the rays $l_1 = (0, t)$, $l_2 = (1, 1 - t)$, $t \geqslant 0$. Then $h_{LM} = 0$.

Examples 16.2–16.4 show that the distance we have introduced is slightly connected with parallelism in the "non-Riemannian" case.

16.2. Space of Directions. Let us recall the concept of a semimetric space. A set $\mathfrak{M}$ with a non-negative function $d_{\mathfrak{M}}$ defined on a Cartesian square of $\mathfrak{M}$ is called a *semimetric space* if $d_{\mathfrak{M}}$ satisfies the following conditions:

1) $d_{\mathfrak{M}}(P, Q)$ is equal to zero if and only if $P = Q$;
2) $d_{\mathfrak{M}}(P, Q) = d_{\mathfrak{M}}(Q, P)$;

see Blumenthal (1970), § 5. In contrast to Blumenthal (1970) we admit the case $d_{\mathfrak{M}}(P, Q) = +\infty$. The function $d_{\mathfrak{M}}$ is called a *semimetric*. The notation for a semimetric space $\mathfrak{M}$ with semimetric $d_{\mathfrak{M}}$ is $(\mathfrak{M}, d_{\mathfrak{M}})$. For the topology of semimetric spaces and a discussion of possible pathologies we refer the reader to Blumenthal (1970).

Directly from the definition it follows that the function d introduced in 16.1.2 is a semimetric.

The semimetric space $(\Omega(\mathfrak{M}), d)$ will be called the *space of directions* of the metric space $\mathfrak{M}$.

§ 17. Curvature of a Metric Space

Here we discuss questions connected with the determination of the existence of the curvature at a point of a metric space, we discuss the concept of continuity of the curvature in a metric space, and we state a theorem which asserts that "almost everywhere" the formally written sectional curvature can be calculated by means of the ratios $\bar{\delta}(T_m)/\sigma(T_m)$.

17.1. Definition of Non-isotropic Riemannian Curvature (Nikolaev (1987)).

17.1.1. Convergence of Triangles to a Point in a Given Pair of Directions. Let $(\mathfrak{M}, \rho)$ be a metric space, $P \in \mathfrak{M}$, and $\{T_m\}$ a sequence of triangles in $\mathfrak{M}$, one of whose vertices coincides with P. Consider a pair of directions $\xi, \zeta \in \Omega_P \mathfrak{M}$ satisfying the condition

$$0 < \bar{\alpha}(\xi, \zeta) < \pi. \tag{17.1}$$

A sequence of triangles $\{T_m = PB_mC_m\}$ *converges to the point P in a pair of directions* (ξ, ζ) (notation $T_m \overset{\xi,\zeta}{\to} P$) if

a) the vertices of the triangles T_m converge (in the sense of the metric ρ) to the point P as $m \to \infty$;

b) the directions $\xi_m, \zeta_m \in \Omega_P\mathfrak{M}$ specified by the sides a_m and b_m of the triangle T_m that start from P converge (in the sense of the metric $\bar{\alpha}$ in 3.1) to the directions ξ and ζ respectively.

17.1.2. Non-isotropic Riemannian Curvature. Let P be a point of the metric space $(\mathfrak{M}, \rho)$, and $\xi, \zeta \in \Omega_P\mathfrak{M}$ a pair of directions satisfying (17.1). We say that at the point P of the metric space $(\mathfrak{M}, \rho)$ in the direction of the pair (ξ, ζ) there is a *non-isotropic Riemannian curvature* $K(P; \xi, \zeta)$ if the following conditions are satisfied:

a) there is a sequence $\{T_m = PB_mC_m\}$ of non-degenerate triangles in $(\mathfrak{M}, \rho)$ (that is, $\sigma(T_m) \neq 0$) that converge to P in the pair of directions (ξ, ζ) for which a limit exists:

$$K(P; \xi, \zeta) = \lim_{m\to\infty} \bar{\delta}(T_m)/\sigma(T_m).$$

b) if for a sequence of non-degenerate triangles $\{T'_m = PB'_mC'_m\}$ that converge to P in the pair of directions (ξ, ζ) there exists a limit of the quantities $\bar{\delta}(T'_m)/\sigma(T'_m)$, then it coincides with $K(P; \xi, \zeta)$.

Remark 1. In part b) of the definition above we can take into consideration degenerate triangles ($\sigma(T_m) = 0$ for them). In this case the existence of the limit of $\bar{\delta}(T_m)/\sigma(T_m)$ is understood in the sense that from $\sigma(T_m) = 0$ for sufficiently large m it follows that $\bar{\delta}(T_m) = 0$.

Remark 2. It may turn out that there is no sequence of non-degenerate triangles that converge to P in the pair of directions (ξ, ζ). In this case we assume that $K(P; \xi, \zeta)$ does not exist.

Remark 3. In the case of classical Riemannian manifolds our definition of non-isotropic Riemannian curvature obviously gives the sectional curvature of a Riemannian manifold at a point in the direction of the plane element specified by the directions ξ and ζ.

17.2. Existence of Curvature at a Point. The example of a convex cone in E_3, considered in the intrinsic metric induced from E_3, shows that for the metric space to be Riemannian it is insufficient to require the existence of non-isotropic Riemannian curvature for all pairs $\xi, \zeta \in \Omega_P\mathfrak{M}$ satisfying (17.1) (the "specific curvature" of the cone at its vertex is unbounded).

In this connection, below we make more precise how we should understand the existence of the curvature of the metric space at a point of it.

Before giving a definition of the existence of curvature we mention the upper and lower curvatures of a metric space, which are necessary for this definition.

17.2.1. Upper (Lower) Curvature at a Point (Aleksandrov (1957a)). Let T be a triangle in the metric space $(\mathfrak{M}, \rho)$. We first define the *"averaged" curvature* $\overline{K}(T)$ of the triangle T (Aleksandrov (1951), p. 6):

$$\overline{K}(T) = \overline{\delta}(T)/\sigma(T) \quad \text{when } \sigma(T) \neq 0,$$

$$\overline{K}(T) = +\infty \quad \text{when } \sigma(T) = 0 \text{ and } \overline{\delta}(T) > 0.$$

$$\overline{K}(T) = -\infty \quad \text{when } \sigma(T) = 0 \text{ and } \overline{\delta}(T) \leqslant 0.$$

Here $\overline{\delta}(T)$ is the excess of the triangle T, calculated from the upper angles.

Similarly we can define the "averaged" curvature $\underline{K}(T)$: $\underline{K}(T) = \underline{\delta}_s(T)/\sigma(T)$ when $\sigma(T) \neq 0$; $\underline{K}(T) = +\infty$ when $\sigma(T) = 0$ and $\underline{\delta}_s(T) \geqslant 0$; $\underline{K}(T) = -\infty$ when $\sigma(T) = 0$ and $\underline{\delta}_s(T) < 0$.

Here $\underline{\delta}_s(T)$ is the excess of the triangle T, calculated from the strong lower angles.

We define the *upper* (*lower*) *curvature* of a metric space at a point P as

$$\overline{K}_{\mathfrak{M}}(P) = \varlimsup_{T \to P} \overline{K}(T) \quad (\underline{K}_{\mathfrak{M}}(P) = \varliminf_{T \to P} \underline{K}(T)),$$

where the limits are considered over all triangles that contract to P (in the case when there is no sequence of triangles that contract to P we put $\overline{K}_{\mathfrak{M}}(P) = +\infty$, $\underline{K}_{\mathfrak{M}}(P) = -\infty$.

17.2.2. Existence of the Curvature of a Metric Space at a Point. We say that the curvature exists for a metric space $(\mathfrak{M}, \rho)$ at a point $P \in \mathfrak{M}$ if the following conditions are satisfied:

a) at the point P of the metric space $(\mathfrak{M}, \rho)$ the upper and lower curvatures satisfy the inequalities

$$-\infty < \underline{K}_{\mathfrak{M}}(P), \quad \overline{K}_{\mathfrak{M}}(P) < +\infty.$$

b) for arbitrary pairs of directions $\xi, \zeta \in \Omega_P\mathfrak{M}$ satisfying (17.1) there is a non-isotropic Riemannian curvature $K(P; \xi, \zeta)$.

Remark. In the case when there is no pair of directions $\xi, \zeta \in \Omega_P\mathfrak{M}$ satisfying (17.1) we assume that the metric space $(\mathfrak{M}, \rho)$ does not have curvature at P.

17.3. Geometrical Meaning of Sectional Curvature. As we have already mentioned, Theorem 14.1 enables us to calculate the sectional curvature $K_\sigma(P)$ "at almost all points" of a space with bounded curvature from the standard formulae of "smooth Riemannian geometry". However, in view of the fact that up to now the restrictions on the curvature and its existence have been defined by means of the excesses of triangles, it is important to establish that the formally introduced sectional curvature has the same geometrical meaing: it admits calculation by means of the limits of the ratios $\overline{\delta}(T)/\sigma(T)$.

In classical Riemannian geometry this fact is well known and, as we have already mentioned, it lies at the basis of the construction of the generalized Riemannian spaces considered in this article.

The geometrical meaning for sectional curvature mentioned above is perserved for spaces with bounded curvature. However, in view of the fact that the metric of these spaces has substantially "worse" differential properties, in the proof we have to overcome substantial difficulties.

Thus our main result is the following theorem.

Theorem 17.1. *Let $\mathfrak{M}$ be a space with bounded curvature. Then there is a set $\mathcal{O}_2 \subset \mathfrak{M}$ of zero n-dimensional Hausdorff measure* ($n = \dim \mathfrak{M}$) *that includes the set $\mathcal{O}_1$ of all points in $\mathfrak{M}$ at which the metric tensor does not have second derivatives. At each point $P \in \mathfrak{M} \setminus \mathcal{O}_2$ the following condition is satisfied.*

For arbitrary pairs of directions $\xi, \zeta \in \Omega_P \mathfrak{M}$ satisfying (17.1) *there is a sequence $\{T_m = PB_m C_m)$ of non-degenerate triangles that contract to P in the pair of directions (ξ, ζ) (see* 17.1.1) *such that the limit of the ratios $\bar{\delta}(T_m)/\sigma(T_m)$ exists and*

$$K_\sigma(P) = \lim_{m \to \infty} \frac{\bar{\delta}(T_m)}{\sigma(T_m)}.$$

Here $K_\sigma(P)$ denotes the sectional curvature of $\mathfrak{M}$, calculated at P in the direction of the two-dimensional element of area $\sigma \subset \mathfrak{M}_P$ given by the bivector $\xi \wedge \zeta$.

Remark. By Remark 2 to Theorem 14.1, in a harmonic coordinate system the components of the metric tensor have a second differential almost everywhere. However, this is not sufficient to prove Theorem 17.1. The set of points at which the sectional curvature can be calculated as the limit of the ratios $\bar{\delta}(T_m)/\sigma(T_m)$ may differ from the set of points of two-fold differentiability of the metric tensor by a set of zero n-dimensional Hausdorff measure.

In the two-dimensional case the assertion of Theorem 17.1 follows from the theorem on differentability of measures. The multidimensional case differs substantially.

We also note that the assertion of Theorem 17.1 can be strengthened (see Nikolaev (1987)). However, the statement given here is sufficient for us.

17.3.1. Plan of the Proof of Theorem 17.1. Let us introduce the following notation: we denote the angles of the triangle $T_m = PB_m C_m$ at the vertices P, B_m, C_m by α_m, β_m, γ_m respectively, the corresponding angles in a planar triangle with the same lengths of sides as T_m by α_{0m}, β_{0m}, γ_{0m}, and we denote $\sigma(T_m)$ by σ_m.

For the proof of Theorem 17.1 it is sufficient to establish the existence of the "almost everywhere" limits

$$(\alpha_m - \alpha_{0m})/\sigma_m, \quad (\beta_m - \beta_{0m})/\sigma_m,$$
$$(\gamma_m - \gamma_{0m})/\sigma_m$$

and the fact that they are equal to $K_\sigma(P)/3$.

In Cartan (1928) (see Chapter IX) it was shown that in order to prove that the last of the above limits is equal to $K_\sigma(P)/3$ it is sufficient to verify that

$$\lim_{m\to\infty} \frac{z_m^2 - z_{0m}^2}{\sigma_m \cdot x_m \cdot y_m \cdot \sin\gamma_m} = -\frac{2}{3}K_\sigma(P),$$

where $z_m = PB_m$, $z_{0m} = x_m^2 + y_m^2 - 2x_m \cdot y_m \cdot \cos\alpha_m$ ($x_m = B_mC_m$, $y_m = PC_m$).

In the case of "smooth Riemannian spaces" the investigation of the above limit can be carried out by using Taylor's formula and the formula for second variation of length (*Synge's formula*, see Gromoll, Klingenberg and Meyer (1968)). For the triangle $T = PBC$ we need to consider the variation V, where $V(s, t)$, $s, t \in [0, 1)$, is a point X lying on the shortest curve joining P and B_t at a distance $s \cdot PB_t$ from P, and the point B_t in turn lies on the shortest curve BC at a distance $t \cdot BC$ from B.

We observe that after transformations in the remainder term of the Taylor expansion for $z^2(t) = PB_t$, taken in integral form, under the integral there will be differences of the components of the tensor curvature taken at the points P and B_t respectively.

In view of the fact that in the case of a space with bounded curvature these differences, generally speaking, do not tend to zero when $P - B_t \to 0$, we use the following construction.

We consider a pair of vectors $e, e' \in \mathfrak{M}_O$ for which $e \wedge e' \neq 0$. Here O is a fixed point of $\mathfrak{M}$.

Let B_m and C_m denote the points given by

$$B_m = \exp_O(m^{-1} \cdot e), \quad C_m = \exp_O(m^{-1} \cdot e').$$

For sufficiently large m these points are well defined.

Let $\mathscr{E}(P)$ and $\mathscr{E}'(P)$ denote the vectors in $\mathfrak{M}_p$ obtained from the vectors e and e' as a result of parallel translation of these vectors along the shortest curve joining O and P. In a small neighbourhood of O such a shortest curve is unique.

Finally, let $T_m(P)$ denote the triangle with vertices

$$P, \quad B_m(P) = \exp_P(m^{-1} \cdot \mathscr{E}(P)), \quad C_m(P) = \exp_P(m^{-1} \cdot \mathscr{E}'(P)).$$

If P is sufficiently close to O, then such a triangle is well defined.

Let us introduce the function

$$h_m(P) = \left| \frac{z_m^2(P) - z_{0m}^2(P)}{\sigma_m(T_m(P)) \cdot x_m(P) \cdot y_m(P) \cdot \sin\gamma_m(P)} + \frac{2}{3}K_\sigma(P) \right|,$$

where the notations $z_m(P)$ and so on have the same sense for the triangle $T_m(P)$ as for the triangle T_m.

Relying on the integral bounds that follow from Theorems 13.2 and 14.1, we can prove that

$$\lim_{m\to\infty} |h_m|_{L_p} = 0, \quad p \geqslant 1,$$

from which it follows that there is a subsequence $\{h_{m_k}\}$ of the sequence $\{h_m\}$ for which $\lim h_{m_k}(P) = 0$ as $k \to \infty$ for almost all $P \in \mathfrak{M}$; from this we obtain the assertion of Theorem 17.1.

In the proof that $|h_m|_{L_p} \to 0$ as $m \to \infty$ we needed to use Synge's formula for the variation V, considered in a small domain of a space with bounded curvature. This formula is deduced by means of Theorems 13.2 and 14.1. Its specific character in the case of a space with bounded curvature is that it can be applied only to "almost all" triangles.

17.4. Isotropic Riemannian Curvature. Let us recall that a neighbourhood of a point of a metric space is said to be *linear* if it can be isometrically embedded in a straight line.

We shall say that a metric space $(\mathfrak{M}, \rho)$ at a point $P \in \mathfrak{M}$ has *isotropic Riemannian curvature* if

a) no neighbourhood of P is linear;

b) for each sequence of triangles $\{T_m\}$ that contract to P in an arbitrary way (that is, the vertices of T_m converge to P in the sense of the metric ρ) there exists the limit

$$K(P) = \lim_{m\to\infty} \bar{\delta}(T_m)/\sigma(T_m),$$

which does not depend on the choice of the sequence of triangles that converge to P.

In the case when there is no such sequence of non-degenerate triangles $\{T_m\}$ we suppose that there is no isotropic Riemannian curvature at P.

We also observe that as in Remark 1 in 17.1.2 we need to mention the case of degenerate triangles separately.

The *curvature* $R(P)$ was introduced in Kirk (1964):

A metric space $\mathfrak{M}$ with intrinsic metric has curvature $\check{R}(P)$ at an accumulation point P if it satisfies condition a) of the previous definition and

b′) for each $\varepsilon > 0$ there is a number $\rho > 0$ such that the ball with centre at P and radius ρ is a domain $R_{K-\varepsilon, K+\varepsilon}$ (see 10.1), where $K = R(P)$.

Reyling on the connection between the boundedness of curvature in terms of the ratios $\bar{\delta}(T)/\sigma(T)$ and the domains $R_{K',K}$ (for the case when the curvature is only bounded above, see Theorem 7.1) we can assert that in metric spaces with intrinsic metric and condition a) at an accumulation point we have $K(P) = R(P)$ on condition that one of these curvatures exists.

17.5. Wald Curvature and its Connection with Isotropic Riemannian Curvature (Wald (1935)). Let us recall that a quadruple of points of a metric space has *curvature of immersion equal to* K if it is isometric to some quadruple in S_K.

A metric space $\mathfrak{M}$ has *Wald curvature* $K_W(P)$ at an accumulation point P if $\mathfrak{M}$ satisfies condition a) of 17.4 and

b″) for each $\varepsilon > 0$ there is a number $\rho > 0$ such that each quadruple of points Q of the ball in M with centre P and radius ρ has curvature of immersion $K(Q)$ and

$$|K(Q) - K_W(P)| < \varepsilon.$$

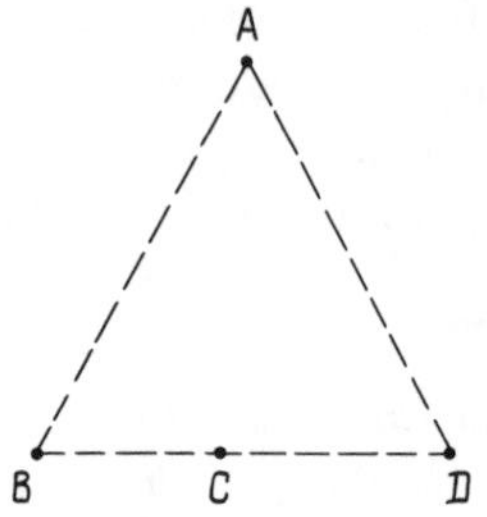

Fig. 16

In Wald (1935) it was shown that a non-linear quadruple of distinct points has two unequal curvatures of immersion, while a quadruple of distinct points that contains a linear triple of points (that is, a triple of points such that the distances between them "form a degenerate triangle", see Fig. 16) has no more than one curvature of immersion.

Wald weakened the original definition of the curvature $K_W(P)$ by admitting only quadruples of points that contain a linear triple of points. The resulting curvature is also called Wald curvature and denoted by $K'_W(P)$.

In Kirk (1964) it was shown that $K'_W(P) = R(P)$. Thus for metric spaces with intrinsic metric and with condition a) of 7.4 the existence of isotropic Riemannian curvature at an accumulation point is equivalent to the existence of Wald curvature $K'_W(P)$ at this point, and we have

$$K(P) = K'_W(P).$$

17.6. Continuity of Curvature. Let $(\mathfrak{M}, \rho)$ be a metric space. We say that the curvature of $(\mathfrak{M}, \rho)$ is continuous at a point $P \in \mathfrak{M}$ if

a) there is a number $r_P > 0$ such that for all points $Q \in B(P, r_P) \subset \mathfrak{M}$ of $(\mathfrak{M}, \rho)$ the curvature exists at Q (see 17.2.2);

b) for an arbitrary $\varepsilon > 0$ we can find a number δ, $0 < \delta < r_P$, such that for all points Q belonging to $B(P, \delta)$ and all pairs of directions $\xi, \zeta \in \Omega_P\mathfrak{M}$, $\xi', \zeta' \in \Omega_Q\mathfrak{M}$ satisfying the inequality

$$\max|\{d(\xi, \xi'), d(\zeta, \zeta')\} < \delta$$

and condition (17.1) we have

$$|K(P; \xi, \zeta) - K(Q; \xi', \zeta')| < \varepsilon.$$

The continuity of the curvature of $(\mathfrak{M}, \rho)$ on a subset $\mathscr{A}$ of $\mathfrak{M}$ implies its continuity at each point $P \in \mathscr{A}$.

The curvature of $(\mathfrak{M}, \rho)$ satisfies a *Hölder condition with exponent* α $(0 < \alpha < 1)$ and constant L on a subset $\mathscr{A} \subseteq \mathfrak{M}$ if it is continuous at each point $P \in \mathscr{A}$ and if the non-isotropic Riemannian curvature has the bound

$$|K(P; \xi, \zeta) - K(Q; \xi', \zeta')| \leqslant L \cdot [d^2(\xi, \xi') + d^2(\zeta, \zeta')]^{\alpha/2}$$

for arbitrary P, $Q \in \mathscr{A}$ and all pairs of directions ξ, $\zeta \in \Omega_P\mathfrak{M}$, ξ', $\zeta' \in \Omega_Q\mathfrak{M}$ satisfying the condition (17.1).

§ 18. Axioms of Riemannian Geometry

More precisely speaking, it is a question of the axioms of classical Riemannian geometry. We give a list of axioms of synthetic character that distinguish among general metric spaces $C^{m,\alpha}$- and C^∞-smooth Riemannian manifolds ($m = 2, 3, \ldots$, $\alpha \in (0, 1)$).

18.1. Synthetic Description of $C^{2,\alpha}$-Smooth Riemannian Manifolds.

Theorem 18.1. *Let* $(\mathfrak{M}, \rho)$ *be a metric space for which the following conditions are satisfied*:

1) ρ *is the intrinsic metric*;

2) $(\mathfrak{M}, \rho)$ *is a locally compact metric space*;

3) *in* $(\mathfrak{M}, \rho)$ *the condition of local extendability of shortest curves is satisfied* (*see* 12.1);

4) *the curvature exists at each point of* $(\mathfrak{M}, \rho)$ (*see* 17.2.2);

5) *the curvature of* $(\mathfrak{M}, \rho)$ *satisfies a Hölder condition with exponent* $\alpha \in (0, 1)$ *in a small neighbourhood of an arbitrary point of* $\mathfrak{M}$ (*see* 17.6).

Then we can specify the structure of a $C^{4,\alpha}$*-smooth differentiable manifold on* $\mathfrak{M}$, *and we can specify the metric* ρ *with respect to the charts of this structure by means of a* $C^{2,\alpha}$*-smooth metric tensor.*

Remark 1. The above conditions are obviously necessary in order that the given metric space should be a $C^{2,\alpha}$-smooth Riemannian manifold.

Remark 2. The $C^{4,\alpha}$-smooth differentiable structure on $\mathfrak{M}$ that occurs in Theorem 18.1 contains an atlas formed by harmonic coordinate systems on $(\mathfrak{M}, \rho)$. The metric tensor that specifies ρ has maximal possible smoothness.

Remark 3. In Peters (1987) (see p. 14) an example is constructed from which it follows that, when $\alpha = 0$, in the harmonic coordinate system in $(\mathfrak{M}, \rho)$ the components of the metric tensor that specifies ρ may have discontinuous second derivatives.

In this case, however, we can assert that in harmonic coordinates the metric tensor of $(\mathfrak{M}, \rho)$ belongs to the Sobolev class W_p^2 for each $1 \leqslant p < +\infty$, and the definition of the curvature tensor of $(\mathfrak{M}, \rho)$ is extended to a tensor specified at each point of $\mathfrak{M}$ whose components with respect to the harmonic coordinates are continuous functions. Thus when $\alpha = 0$ we can assert that $(\mathfrak{M}, \rho)$ is a Riemannian manifold with continuous curvature tensor.

18.1.1. Plan of the Proof of Theorem 18.1. From the conditions of Theorem 18.1 it follows that $(\mathfrak{M}, \rho)$ is a space with bounded curvature. By Theorem 17.1

the non-isotropic Riemannian curvature "at almost all points" of $\mathfrak{M}$ coincides with the sectional curvature, calculated formally from the metric tensor. In particular, the sectional curvature also has a Hölder bound.

Now, reverting to equation (15.4), we can make use of a theorem on the smoothness of solutions of elliptic equations (Ladyzhenskaya and Ural'tseva (1964)). The existence of the required differential structure follows from the results of Sabitov and Shefel' (1976).

18.2. Synthetic Description of $C^{m,\alpha}$-Smooth Riemannian Manifolds ($m = 3, 4, \ldots$). As a preliminary we introduce the concepts of the tangent space $T(\mathfrak{M})$ for an arbitrary metric space $(\mathfrak{M}, \rho)$ and the i-th tangent space $T^{(i)}(\mathfrak{M})$, which play an important role in the synthetic description of $C^{m,\alpha}$-smooth Riemannian manifolds.

18.2.1. The Tangent Space $T(\mathfrak{M})$ and the Spaces $T^{(i)}(\mathfrak{M})$.
The set

$$T(\mathfrak{M}) = \bigcup \mathfrak{M}_P, \quad P \in \mathfrak{M}$$

is called the *set of tangent elements* of the metric space $(\mathfrak{M}, \rho)$ (see 3.3). As in 16.1.2, for $[\xi, s] \in \mathfrak{M}_P$, $[\zeta, t] \in \mathfrak{M}_Q$ we introduce the function

$$\Phi([\xi, s], [\zeta, t]) = \left\{ \sup_{\eta \in \Omega_P \mathfrak{M}} [s \cdot h(\xi, \eta) - t \cdot h(\zeta, \eta)] \right. \\ \left. + \sup_{\eta' \in \Omega_Q \mathfrak{M}} [s \cdot h(\xi, \eta') - t \cdot h(\zeta, \eta')] \right\} \Big/ 2.$$

The semimetric in $T(\mathfrak{M})$ between the tangent elements $[\xi, s] \in \mathfrak{M}_P$, $[\zeta, t] \in \mathfrak{M}_Q$ is introduced by means of the constant c:

$$c([\xi, s], [\zeta, t]) = [PQ^2 + \Phi^2([\xi, s], [\zeta, t])]^{1/2}. \tag{18.1}$$

We call the semimetric space $(T(\mathfrak{M}), c)$ the *tangent space of the metric space* $(\mathfrak{M}, \rho)$.

We now go over to the definition of the spaces $T^{(i)}(\mathfrak{M})$.

We denote $\mathfrak{M}$ by $T^{(0)}(\mathfrak{M})$ and ρ by $c_0 = \mathscr{C}_0$. We denote the semimetric space $(T(\mathfrak{M}), c)$ by $(T^{(1)}(\mathfrak{M}), c_1)$. We assume that the semimetric c_1 induces the metric $\mathscr{C}_1$ in $T^{(1)}(\mathfrak{M})$ (see 16.1.1), that is, for an arbitrary pair of points of $T^{(1)}(\mathfrak{M})$ the infimum of the lengths of all curves (measured in the semimetric c_1) joining these points is defined and is a finite quantity, and it specifies a metric in $T^{(1)}(\mathfrak{M})$. Then for the metric space $(T^{(1)}(\mathfrak{M}), \mathscr{C}_1)$ we can consider the semimetric space $(T^{(2)}(\mathfrak{M}), c_2)$, where $T^{(2)}(\mathfrak{M}) = T(T^{(1)}(\mathfrak{M}))$, and the semimetric c_2 is determined from the metric $\mathscr{C}_1$ by means of formula (18.1).

Suppose we have constructed the spaces $(T^{(i)}(\mathfrak{M}), c_i)$, $i = 0, 1, 2, \ldots, k-1$, and the semimetrics c_i induce the metrics $\mathscr{C}_i$, $i = 1, 2, \ldots, k-2$. If c_{k-1} induces the metric $\mathscr{C}_{k-1}$, then for the metric space $(T^{(k-1)}(\mathfrak{M}), \mathscr{C}_{k-1})$ we can again consider the semimetric space $(T^{(k)}(\mathfrak{M}), c_k)$, where $T^{(k)}(\mathfrak{M}) = T(T^{(k-1)}(\mathfrak{M}))$, and the semimetric c_k is determined from the metric $\mathscr{C}_{k-1}$ in accordance with (18.1).

18.2.2. Statement of the Main Result. As a preliminary we introduce the following concepts.

Let $(\mathfrak{M}, \rho)$ be a metric space. We shall say that $(\mathfrak{M}, \rho)$ has *Hölder continuous curvature* with exponent $\alpha \in (0, 1)$ if $(\mathfrak{M}, \rho)$ satisfies conditions 4) and 5) of Theorem 18.1.

If $(\mathfrak{M}, \rho)$ satisfies all the conditions of Theorem 18.1, then from this theorem, and also from the bounds of Proposition 16.1, we can deduce that the semimetric c_1 induces the Sasaki metric $\mathscr{C}_1$ in $T^{(1)}(\mathfrak{M})$ (see 16.1). If in turn the metric space $(T^{(1)}(\mathfrak{M}), \mathscr{C}_1)$ has Hölder continuous curvature, then similarly we deduce that the semimetric c_2 induces in $T^{(2)}(\mathfrak{M})$ the Riemannian metric $\mathscr{C}_2$, which coincides with the Sasaki metric for $T(T^{(1)}(\mathfrak{M}))$, and so on.

Thus for a metric space $(\mathfrak{M}, \rho)$ that satisfies the conditons of Theorem 18.1 we can give the following definitions.

$\{T^{(i)}(\mathfrak{M})\}$, $i = 1, 2, \ldots, m$, is a sequence of spaces with Hölder continuous curvature with exponent α $(0 < \alpha < 1)$ if there is a sequence of numbers α_i $(0 < \alpha_i < 1)$, $i = 1, 2, \ldots, m$, $\alpha_m = \alpha$, such that the metric spaces $(T^{(i)}(\mathfrak{M}), \mathscr{C}_i)$, $i = 1, 2, \ldots, m$, have Hölder continuous curvarture with exponent α_i.

In a similar way we define what is meant by an infinite sequence of spaces $\{T^{(i)}(\mathfrak{M})\}$, $i = 1, 2, \ldots$, being a sequence of spaces with Hölder continuous curvature.

Theorem 18.2. *Let $(\mathfrak{M}, \rho)$ be a metric space for which conditions* 1)–5) *of Theorem* 18.1 *are satisfied, and suppose that in addition the following condition is satisfied*:

6) $\{T^{(i)}(\mathfrak{M})\}$, $i = 1, 2, \ldots, m$, *is a sequence of spaces with Hölder continuous curvature with exponent* $\alpha \in (0, 1)$.

Then on $\mathfrak{M}$ we can specify the structure of a $C^{m+4,\alpha}$-smooth differentiable manifold, and we can specify a metric ρ with respect to the charts of this structure by means of a $C^{m+2,\alpha}$-smooth metric tensor, $m = 1, 2, 3$.

Corollary 18.1. *If we replace condition* 6) *in Theorem* 18.2 *by the condition*

6′) $\{T^{(i)}(\mathfrak{M})\}$, $i = 1, 2, \ldots$, *is an infinite sequence of spaces with Hölder continuous curvature, then we can specify on $\mathfrak{M}$ the structure of a C^∞-smooth differentiable manifold, and we can specify a metric ρ with respect to the charts of this structure by means of a C^∞-smooth metric tensor.*

Remark 1. The conditions in Theorem 18.2 and Corollary 18.1 are also necessary in order that a given metric space should be a $C^{m+2,\alpha}$-smooth or respectively C^∞-smooth Riemannian manifold.

Remark 2. The differential structure on $\mathfrak{M}$ that occurs in Theorem 18.2 and Corollary 18.1 contains an atlas formed by harmonic coordinate systems in $\mathfrak{M}$.

Remark 3. The proof of Theorem 18.2 is obtained as a consequence of Theorem 18.1 and the results of Sasaki (1958), (1962).

18.3. Isotropic Metric Spaces. For isotropic metric spaces, that is, for metric spaces at each point of which there is isotropic Riemannian curvature (see 17.4) we have the following theorem, which extends the well-known *theorem of Schur* to the case of metric spaces.

Theorem 18.3 (Nikolaev (1989)). *Let* $(\mathfrak{M}, \rho)$ *be a locally compact metric space with intrinsic metric, whose Menger-Uryson dimension is greater than two. We assume that the condition of local extendability of shortest curves is satisfied in* $\mathfrak{M}$ *and that* $\mathfrak{M}$ *is isotropic at all its points. Then* $(\mathfrak{M}, \rho)$ *is isotropic to a Riemannian manifold of constant curvature.*

18.3.1. The Consequence for "Multidimensional" Metric Spaces at each Point of which the Wald Curvature $K_W'(P)$ Exists. The next result follows directly from Theorem 18.3 and the results of 17.5.

Theorem 18.4. *Let* $(\mathfrak{M}, \rho)$ *be a locally compact metric space with intrinsic metric, whose Menger-Uryson dimension is greater than two. We assume that the condition of local extendability of shortest curves is satisfied in* $\mathfrak{M}$ *and that the Wald curvature* $K_W'(P)$ *exists at each point* $P \in \mathfrak{M}$. *Then* $(\mathfrak{M}, \rho)$ *is isometric to a Riemannian manifold of constant curvature.*

Remark. Theorem 18.4 gives the answer to a question in Kirk (1964).

18.3.2. Plan of the Proof of Theorem 18.3. The standard proof of Schur's theorem is based on the use of the Bianchi identity. Hence we assume that the components of the metric tensor of a Riemannian manifold are at least thrice differentiable. From the fact that the metric space is isotropic (and the remaining conditions of the theorem) we can deduce that $(\mathfrak{M}, \rho)$ is a space with bounded curvature. Therefore by Theorem 14.1 the components of the metric tensor of $\mathfrak{M}$ in harmonic coordinates belong to the Sobolev class W_p^2 $(1 \leqslant p < +\infty)$. Thus we need to modify the proof of Schur's theorem in order to bring into consideration the derivatives of a metric tensor of order not greater than two.

This is achieved as follows. As we know (Cartan (1928)) Bianchi's identity implies that the absolute extrinsic derivative of the curvature form of a manifold is equal to zero. As for the usual extrinsic differentation, for the absolute differential we can introduce the concept of generalized differentiation, and this in turn enables us to rewrite Bianchi's identity in a generalized sense, not using the third derivatives of the metric tensor. The truth of Bianchi's identity on $(\mathfrak{M}, \rho)$ is established by means of Theorem 14.1. Using the isotropy of the metric, Bianchi's identity on $(\mathfrak{M}, \rho)$ reduces to the vanishing of the derivative of the distribution on $\mathfrak{M}$ defined by K (Hörmander (1983)):

$$\int_{\Omega} K(x) \frac{\partial \phi}{\partial x_k}(x)\, dx^1\, dx^2 \ldots dx^n = 0, \quad k = 1, 2, \ldots, n, \tag{18.2}$$

where $x\colon \mathscr{U} \subset \mathfrak{M} \to \mathbb{R}^n$ $(x(\mathscr{U}) = \Omega)$ is an arbitrary chart on $\mathfrak{M}$, and φ is an arbitrary smooth function compactly supported in Ω.

By virtue of the continuity of the function $K(x)$ (this follows from the definition of isotropic curvature as in the two-dimensional case in Aleksandrov (1948)) and Theorem 3.1.4′ in Hörmander (1983), in view of the arbitrariness of the chart x we deduce that the function $K(x)$ is identically constant on $\mathfrak{M}$.

From a theorem of Aleksandrov (see Aleksandrov (1957b)) it then follows that $(\mathfrak{M}, \rho)$ is isometric to a Riemannian manifold of constant curvature, as required.

References*

Aleksandrov, A.D. [1948]: Intrinsic Geometry of Convex Surfaces. Gostekhizdat, Moscow-Leningrad, Zbl.38,352. German transl.: Akademie Verlag, Berlin 1955, Zbl.65,151

Aleksandrov, A.D. [1951]: A theorem on triangles in a metric space and some applications of it. Tr. Mat. Inst. Steklova *38*, 5–23 [Russian], Zbl.49,395

Aleksandrov, A.D. [1957a]: Ruled surfaces in metric spaces. Vestn. Leningr. Univ. *12*, No. 1, 5–26 [Russian], Zbl.96,166

Aleksandrov, A.D. [1957b]: Über eine Verallgemeinerung der Riemannschen Geometrie. Schr. Forschungsinst. Math. *1*, 33–84, Zbl.77,357

Aleksandrov, A.D. and Berestovskij, V.N. [1984]: Generalized Riemannian space. Mathematical Encyclopaedia, vol. 4, 1022–1026. English transl.: Klūwer Acad. Publ., Dordrecht.

Aleksandrov, A.D., Berestovskij, V.N. and Nikolaev, I.G. [1986]: Generalized Riemannian spaces. Usp. Mat. Nauk *41*, No. 3, 3–44. English transl.: Russ. Math. Surv. *41*, No. 3, 1–54 (1986), Zbl.625.53059

Aleksandrov, A.D. and Zalgaller, V.A. [1962]: Two-dimensional manifolds of bounded curvature. Tr. Mat. Inst. Steklova *63*. English transl.: Intrinsic geometry of surfaces. Transl. Math. Monographs *15*, Am. Math. Soc. (1967), Zbl.122,170

Berestovskij V.N. [1975]: Introduction of a Riemannian structure in certain metric spaces. Sib. Mat. Zh. *16*, 651–662. English transl.: Sib. Math. J. *16*, 499–507 (1976), Zbl.325.53059

Berestovskij, V.N. [1981]: Spaces with bounded curvature. Dokl. Akad. Nauk SSSR *258*, 269–271. English transl.: Sov. Math., Dokl. *23*, 491–493 (1981), Zbl.511.53074

Berestovskij, V.N. [1986]: Spaces with bounded curvature and distance geometry. Sib. Mat. Zh. *27*, No. 1, 11–25. English transl.: Sib. Math. J. *27*, 8–19 (1986), Zbl.596.53029

Berger, M. [1983]: Sur les variétés Riemanniennes pincées juste au dessous de 1/4. Ann. Inst. Fourier *33*, 135–150, Zbl.497.53044

Bers, L., John, F. and Schechter, M. [1964]: Partial Differential Equations. Interscience, New York, Zbl.128,93 and Zbl.128,94

Blumenthal, L.M. [1970]: Theory and Applications of Distance Geometry. 2nd. ed., Chelsea Publ. Co., New York, Zbl.50,385

Busemann, H. [1955]: The Geometry of Geodesics. Academic Press, New York-London, Zbl.112,370

Cartan, E. [1928]: Leçons sur la géométrie des espaces de Riemann. 2nd ed., Gauthier-Villars, Paris, Jbuch 54,755

Cesari, L. [1956]: Surface Area. Princeton Univ. Press, Princeton, N.J., Zbl.73,41

*For the convenience of the reader, references to reviews in Zentralblatt für Mathematik (Zbl.), compiled using the MATH database, and Jahrbuch über die Fortschritte der Mathematik (Jbuch) have, as far as possible, been included in this bibliography.

DeTurck, D.M. and Kazdan, J.L. [1981]: Some regularity theorems in Riemannian geometry. Ann. Sci. Éc. Norm. Super., IV. Ser. *14*, 243–260, Zbl.486.53014

Einstein, A. [1916]: Näherungsweise Integration der Feldgleichungen der Gravitation. Sitzungsber. Preussische Akad. Wiss. *1*, 688–696, Jbuch 46,1293

Gol'dshtein, V.M., Kuz'minov, V.I. and Shvedov, I.A. [1984]: A property of de Rham regularization operators. Sib. Mat. Zh. *25*, No. 2, 104–111. English transl.: Sib. Math. J. *25*, 251–257 (1984), Zbl.546.58002

Gol'dshtein, V.M. and Reshetnyak, Yu.G. [1983]: Introduction to the Theory of Functions with Generalized Derivatives and Quasiconformal Mappings. Nauka, Moscow. English transl.: Quasiconformal Mappings and Sobolev's spaces. Math. and its Appl., Sov. Ser. *54*. Kluwer Acad. Publ., Dordrecht 1990, Zbl.591.46021

Gromoll, D., Klingenberg, W. and Meyer, W. [1968]: Riemannsche Geometrie im Grossen. Lect. Not. Math. *55*, Springer, Berlin-Heidelberg-New York, Zbl.155,307

Gromov, M. [1981]: Structures métriques pour les variétés Riemanniennes. Cedic, Paris, Zbl.509.53034

Hörmander, L. [1983]: The Analysis of Linear Partial Differential Operators. I. Distribution Theory and Fourier Analysis. Springer, Berlin-Heidelberg-New York, Zbl.521.35001

Hurewicz, W. and Wallman, H. [1941]: Dimension Theory. Princeton Univ. Press, Princeton, N.J., Zbl.60,398

Ionin, V.K. [1972]: Isoperimetric inequalities in simply-connected Riemannian spaces of nonpositive curvature. Dokl. Akad. Nauk SSSR *203*, 282–284. English transl.: Sov. Math., Dokl. *13*, 378–381 (1972), Zbl.258.52011

Kirk, W.A. [1964]: On the curvature of a metric space at a point. Pac. J. Math. *14*, 195–198, Zbl.168,434

Ladyzhenskaya, O.A. and Ural'tseva, N.N. [1964]: Linear and Quasilinear Elliptic Equations. Nauka, Moscow. English transl.: Academic Press, New-York-London 1968, Zbl.143,336

Nikolaev, I.G. [1978]: Space of directions at a point in a space of curvature not greater than *K*. Sib. Mat. Zh. *19*, 1341–1348. English transl.: Sib. Math. J. *19*, 944–949 (1979), Zbl.405.53044

Nikolaev, I.G. [1979]: Solution of the Plateau problem in spaces of curvature not greater than *K*. Sib. Mat. Zh. *20*, 345–353. English transl.: Sib. Math. J. *20*, 246–252 (1979), Zbl.406.53044

Nikolaev, I.G. [1980]: Parallel translation and smoothness of the metric of spaces with bounded curvature. Dokl. Akad. Nauk SSSR *250*, 1056–1058. English transl.: Sov. Math., Dokl. *21*, 263–265 (1980), Zbl.505.53015

Nilolaev, I.G. [1983a]: Parallel translation of vectors in spaces with curvature that is bilaterally bounded in the sense of A.D. Aleksandrov. Sib. Mat. Zh. *24*, No. 1, 130–145. English transl.: Sib. Math. J. *24*, 106–119 (1983), Zbl.539.53030

Nikolaev, I.G. [1983b]: Smoothness of the metric of spaces with curvature that is bilaterally bounded in the sense of A.D. Aleksandrov. Sib. Mat. Zh. *24*, No. 2, 114–132. English transl.: Sib. Math. J. *24*, 247–263, Zbl.547.53011

Nikolaev, I.G. [1987]: The curvature of a metric space at a point. All-Union Conf. on Geometry "in the large". Novoskbirsk [Russian]

Nikolaev, I.G. [1989]: Isotropic metric spaces. Dokl. Akad. Nauk SSSR *305*, 1314–1317. English transl.: Sov. Math., Dokl. *39*, 408–410, Zbl.714.54031

Peters, S. [1986]: Konvergenz Riemannscher Mannigfaltigkeiten. Bonner Math. Schr. *169*, Zbl.648.53024

Peters, S. [1987]: Convergence of Riemannian manifolds. Compos. Math. *62*, 3–16, Zbl.618,53036

Reshetnyak, Yu.G. [1960a]: Isothermal coordinates in manifolds of bounded curvature. I, II. Sib. Mat. Zh. *1*, 88–116, 248–276 [Russian], Zbl.108,338

Reshetnyak, Yu.G. [1960b]: On the theory of spaces of curvature not greater than *K*. Mat. Sb., Nov. Ser. *52*, 789–798 [Russian], Zbl.101,402

Reshetnyak, Yu.G. [1968]: Non-expanding maps in spaces of curvature not greater than *K*. Sib. Mat. Zh. *9*, 918–927. English transl.: Sib. Math. J. *9*, 683–689, Zbl.167,508

Rham de, G. [1955]: Variétés différentiables: Formes courants, formes harmoniques. Herrmann, Paris, Zbl.65,324

Sabitov, I.Kh. and Shefel', S.Z. [1976]: Connections between the order of smoothness of a surface and that of its metric. Sib. Mat. Zh. *17*, 916–925. English transl.: Sib. Math. J. *17*, 687–694, Zbl.386.53014

Sasaki, S. [1958]: On the differential geometry of tangent bundles of Riemannian manifolds. I. Tôhoku Math. J., II. Ser. *10*, 338–354, Zbl.86,150

Sasaki, S. [1962]: On the differential geometry of tangent bundles of Riemannian manifolds. II. Tôhoku Math. J., II. Ser. *14*, 146–155, Zbl.109,405

Sobolev, S.L. [1950]: Some Applications of Functional Analysis to Mathematical Physics. Izdat. Leningrad. Univ., Leningrad [Russian]

Toponogov, V.A. [1959]: Riemannian spaces of curvature bounded below. Usp. Mat. Nauk *14*, No. 1, 87–130. English transl.: Transl., II. Ser., Am. Math. Soc. *37*, 291–336 (1964), Zbl.114,375

Wald, A. [1935]: Begründung einer koordinatenlosen Differentialgeometrie der Flächen. Ergeb. Math. Kolloq., Vol. 7, 24–46, Zbl.14,230

Author Index

Subject Index